U0038839

Anhui
安徽
卷·中卷
珍稀收藏版
Special Edition
编号
No. 0276

# 中国手工纸文库

Library of Chinese Handmade Paper

汤书昆
总主编

# 徽卷

·中卷

珍稀收藏版

# Anhui II

Special Edition

汤书昆　黄飞松

主 编

中国科学技术大学出版社

University of Science and Technology of China Press

**图书在版编目（CIP）数据**

中国手工纸文库.安徽卷.中卷：珍稀收藏版/汤书昆，黄飞松主编.—合肥：中国科学技术大学出版社，2021.5

国家出版基金项目

“十三五”国家重点出版物出版规划项目

ISBN 978-7-312-04970-5

Ⅰ.中…　Ⅱ.①汤…　②黄…　Ⅲ.手工纸—介绍—安徽　Ⅳ.TS766

中国版本图书馆CIP数据核字（2020）第090656号

中国手工纸文库

安徽卷·中卷

项目负责　伍传平　项赟飚

责任编辑　郭红建　高哲峰　张　灿

艺术指导　吕敬人

书籍设计　敬人书籍设计<br>吕　旻+黄晓飞

出版发行　中国科学技术大学出版社<br>地址　安徽省合肥市金寨路96号<br>邮编　230026

印　　刷　北京雅昌艺术印刷有限公司

经　　销　全国新华书店

开　　本　880 mm×1230 mm　1/16

印　　张　24.75

字　　数　696千

版　　次　2021年5月第1版

印　　次　2021年5月第1次印刷

定　　价　4980.00元

# 总 序



造纸技艺是人类文明的重要成就。正是在这一伟大发明的推动下，我们的社会才得以在一个相当长的历史阶段获得比人类使用口语的表达与交流更便于传承的介质。纸为这个世界创造了五彩缤纷的文化记录，使一代代的后来者能够通过纸介质上绘制的图画与符号、书写的文字与数字，了解历史，学习历代文明积累的知识，从而担负起由传承而创新的文化使命。

中国是手工造纸的发源地。不仅人类文明中最早的造纸技艺发源自中国，而且中华大地上遍布着手工造纸的作坊。中国是全世界手工纸制作技艺提炼精纯与丰富的文明体。可以说，在使用手工技艺完成植物纤维制浆成纸的历史中，中国一直是人类造纸技艺与文化的主要精神家园。下图是中国早期造纸技艺刚刚萌芽阶段实物样本的一件遗存——西汉放马滩古纸。



西汉放马滩古纸残片
纸上绘制的是地图
1986年出土于甘肃省天水市
现藏于甘肃省博物馆

Map drawn on paper from
Fangmatan Shoals
in the Western Han Dynasty
Unearthed in Tianshui City,
Gansu Province in 1986
Kept by Gansu Provincial Museum

# Preface

Papermaking technique illuminates human culture by endowing the human race with a more traceable medium than oral tradition. Thanks to cultural heritage preserved in the form of images, symbols, words and figures on paper, human beings have accumulated knowledge of history and culture, and then undertaken the mission of culture transmission and innovation.

Handmade paper originated in China, one of the largest cultural communities enjoying advanced handmade papermaking techniques in abundance. China witnessed the earliest papermaking efforts in human history and embraced papermaking mills all over the country. In the history of handmade paper involving vegetable fiber pulping skills, China has always been the dominant centre. The picture illustrates ancient paper from Fangmatan Shoals in the Western Han Dynasty, which is one of the paper samples in the early period of papermaking techniques unearthed in China.

## 一
## 本项目的缘起

从2002年开始，我有较多的机缘前往东邻日本，在文化与学术交流考察的同时，多次在东京的书店街——神田神保町的旧书店里，发现日本学术界整理出版的传统手工制作和纸（日本纸的简称）的研究典籍，先后购得近20种，内容包括日本全国的手工造纸调查研究，县（相当于中国的省）一级的调查分析，更小地域和造纸家族的案例实证研究，以及日、中、韩等东亚国家手工造纸的比较研究等。如：每日新闻社主持编撰的《手漉和纸大鉴》五大本，日本东京每日新闻社昭和四十九年（1974年）五月出版，共印1 000套；久米康生著的《手漉和纸精髓》，日本东京讲谈社昭和五十年（1975年）九月出版，共印1 500本；菅野新一编的《白石纸》，日本东京美术出版社昭和四十年（1965年）十一月出版等。这些出版物多出自几十年前的日本昭和年间（1926~1988年），不仅图文并茂，而且几乎都附有系列的实物纸样，有些还有较为规范的手工纸性能、应用效果对比等技术分析数据。我阅后耳目一新，觉得这种出版物形态既有非常直观的阅读效果，又散发出很强的艺术气息。

## 1. Origin of the Study

Since 2002, I have been invited to Japan several times for cultural and academic communication. I have taken those opportunities to hunt for books on traditional Japanese handmade paper studies, mainly from old bookstores in Kanda Jinbo-cho, Tokyo. The books I bought cover about 20 different categories, typified by surveys on handmade paper at the national, provincial, or even lower levels, case studies of the papermaking families, as well as comparative studies of East Asian countries like Japan, Korea and China. The books include five volumes of *Tesukiwashi Taikan* (*A Collection of Traditional Handmade Japanese Papers*) compiled and published by Mainichi Shimbun in Tokyo in May 1974, which released 1 000 sets, *The Essence of Japanese Paper* by Kume Yasuo, which published 1 500 copies in September 1975 by Kodansha in Tokyo, Japan, *Shiraishi Paper* by Kanno Shinichi, published by Fine Arts Publishing House in Tokyo in November 1965. The books which were mostly published between 1926 and 1988 among the Showa reigning years, are delicately illustrated with pictures and series of paper samples, some even with data analysis on performance comparison. I was extremely impressed by the intuitive and aesthetic nature of the books.

我几乎立刻想起在中国看到的手工造纸技艺及相关的研究成果，在我们这个世界手工造纸的发源国，似乎尚未看到这种表达丰富且叙述格局如此完整出色的研究成果。对中国辽阔地域上的手工造纸技艺与文化遗存现状，研究界尚较少给予关注。除了若干名纸业态，如安徽省的泾县宣纸、四川省的夹江竹纸、浙江省的富阳竹纸与温州皮纸、云南省的香格里拉东巴纸和河北省的迁安桑皮纸等之外，大多数中国手工造纸的当代研究与传播基本上处于寂寂无闻的状态。

此后，我不断与国内一些从事非物质文化遗产及传统工艺研究的同仁交流，他们一致认为在当代中国工业化、城镇化大规模推进的背景下，如果不能在我们这一代人手中进行手工造纸技艺与文化的整体性记录、整理与传播，传统手工造纸这一中国文明的结晶很可能会在未来的时空中失去系统记忆，那真是一种令人难安的结局。但是，这种愿景宏大的文化工程又该如何着手？我们一时觉得难觅头绪。

《手漉和纸精髓》
附实物纸样的内文页
A page from *The Essence of Japanese Paper* with a sample

《白石纸》
随书的宣传夹页
A folder page from *Shiraishi Paper*

The books reminded me of handmade papermaking techniques and related researches in China, and I felt a great sadness that as the country of origin for handmade paper, China has failed to present such distinguished studies excelling both in presentation and research design, owing to the indifference to both papermaking technique and our cultural heritage. Most handmade papermaking mills remain unknown to academia and the media, but there are some famous paper brands, including Xuan paper in Jingxian County of Anhui Province, bamboo paper in Jiajiang County of Sichuan Province, bamboo paper in Fuyang District and bast paper in Wenzhou City of Zhejiang Province, Dongba paper in Shangri-la County of Yunnan Province, and mulberry paper in Qian'an City of Hebei Province.

Constant discussion with fellow colleagues in the field of intangible cultural heritage and traditional craft studies lead to a consensus that if we fail to record, clarify, and transmit handmade papermaking techniques in this age featured by a prevailing trend of industrialization and urbanization in China, regret at the loss will be irreparable. However, a workable research plan on such a grand cultural project eluded us.

2004年，中国科学技术大学人文与社会科学学院获准建设国家“985工程”的“科技史与科技文明哲学社会科学创新基地”，经基地学术委员会讨论，“中国手工纸研究与性能分析”作为一项建设性工作由基地立项支持，并成立了手工纸分析测试实验室和手工纸研究所。这一特别的机缘促成了我们对中国手工纸研究的正式启动。

2007年，中华人民共和国新闻出版总署的“十一五”国家重点图书出版规划项目开始申报。中国科学技术大学出版社时任社长郝诗仙此前知晓我们正在从事中国手工纸研究工作，于是建议正式形成出版中国手工纸研究系列成果的计划。在这一年中，我们经过国际国内的预调研及内部研讨设计，完成了《中国手工纸文库》的撰写框架设计，以及对中国手工造纸现存业态进行全国范围调查记录的田野工作计划，并将其作为国家“十一五”规划重点图书上报，获立项批准。于是，仿佛在不经意间，一项日后令我们常有难履使命之忧的工程便正式展开了。

2008年1月，《中国手工纸文库》项目组经过精心的准备，派出第一个田野调查组（一行7人）前往云南省的滇西北地区进行田野调查，这是计划中全中国手工造纸田野考察的第一站。按照项目设计，将会有很多批次的调查组走向全中国手工造纸现场，采集能获



In 2004, the Philosophy and Social Sciences Innovation Platform of History of Science and S&T Civilization of USTC was approved and supported by the National 985 Project. The academic committee members of the Platform all agreed to support a new project, "Studies and Performance Analysis of Chinese Handmade Paper". Thus, the Handmade Paper Analyzing and Testing Laboratory, and the Handmade Paper Institute were set up. Hence, the journey of Chinese handmade paper studies officially set off.

In 2007, the General Administration of Press and Publication of the People's Republic of China initiated the program of key books that will be funded by the National 11th Five-Year Plan. The former President of USTC Press, Mr. Hao Shixian, advocated that our handmade paper studies could take the opportunity to work on research designs. We immediately constructed a framework for a series of books, *Library of Chinese Handmade Paper*, and drew up the fieldwork plans aiming to study the current status of handmade paper all over China, through arduous pre-research and discussion. Our project was successfully approved and listed in the 11th Five-Year Plan for National Key Books, and then our promising yet difficult journey began.

The seven members of the *Library of Chinese Handmade Paper* Project embarked on our initial, well-prepared fieldwork journey to the northwest area of Yunnan

取的中国手工造纸的完整技艺与文化信息及实物标本。

2009年，国家出版基金首次评审重点支持的出版项目时，将《中国手工纸文库》列入首批国家重要出版物的资助计划，于是我们的中国手工纸研究设计方案与工作规划发育成为国家层面传统技艺与文化研究所关注及期待的对象。

此后，田野调查、技术分析与撰稿工作坚持不懈地推进，中国科学技术大学出版社新一届领导班子全面调动和组织社内骨干编辑，使《中国手工纸文库》的出版工程得以顺利进行。2017年，《中国手工纸文库》被列为“十三五”国家重点出版物出版规划项目。



## 二
## 对项目架构设计的说明

作为纸质媒介出版物的《中国手工纸文库》，将汇集文字记

调查组成员在香格里拉县白地村调查
2008年1月

Researchers visiting Baidi Village of Shangri-la County
January 2008

Province in January 2008. After that, based on our research design, many investigation groups would visit various handmade papermaking mills all over China, aiming to record and collect every possible papermaking technique, cultural information and sample.

In 2009, the National Publishing Fund announced the funded book list gaining its key support. Luckily, *Library of Chinese Handmade Paper* was included. Therefore, the Chinese handmade paper research plan we proposed was promoted to the national level, invariably attracting attention and expectation from the field of traditional crafts and culture studies.

Since then, field investigation, technical analysis and writing of the book have been unremittingly promoted, and the new leadership team of USTC Press has fully mobilized and organized the key editors of the press to guarantee the successful publishing of *Library of Chinese Handmade Paper*. In 2017, the book was listed in the 13th Five-Year Plan for the Publication of National Key Publications.

### 2. Description of Project Structure

*Library of Chinese Handmade Paper* compiles with many forms of ideography language: detailed descriptions and records, photographs, illustrations of paper fiber structure and transmittance images, data analysis, distribution of the papermaking sites, guide map

录与描述、摄影图片记录、样纸纤维形态及透光成像采集、实验分析数据表达、造纸地分布与到达图导引、实物纸样随文印证等多种表意语言形式，希望通过这种高度复合的叙述形态，多角度地描述中国手工造纸的技艺与文化活态。在中国手工造纸这一经典非物质文化遗产样式上，《中国手工纸文库》的这种表达方式尚属稀见。如果所有设想最终能够实现，其表达技艺与文化活态的语言方式或许会为中国非物质文化遗产研究界和保护界开辟一条新的途径。

项目无疑是围绕纸质媒介出版物《中国手工纸文库》这一中心目标展开的，但承担这一工作的项目团队已经意识到，由于采用复合度很强且极丰富的记录与刻画形态，当项目工程顺利完成后，必然会形成非常有价值的中国手工纸研究与保护的其他重要后续工作空间，以及相应的资源平台。我们预期，中国（计划覆盖34个省、市、自治区与特别行政区）当代整体的手工造纸业态按照上述记录与表述方式完成后，会留下与《中国手工纸文库》伴生的中国手工纸图像库、中国手工纸技术分析数据库、中国手工纸实物纸样库，以及中国手工纸的影像资源汇集等。基于这些伴生的集成资源的丰富性，并且这些资源集成均为首次，其后续的价值延展空间也不容小视。中国手工造纸传承与发展的创新拓展或许会给有志于继续关注中国手工造纸技艺与文化的同仁提供

to the papermaking sites, and paper samples, etc. Through such complicated and diverse presentation forms, we intend to display the technique and culture of handmade paper in China thoroughly and vividly. In the field of intangible cultural heritage, our way of presenting Chinese handmade paper was rather rare. If we could eventually achieve our goal, this new form of presentation may open up a brand-new perspective to research and preservation of Chinese intangible cultural heritage.

Undoubtedly, the *Library of Chinese Handmade Paper* Project developed with a focus on paper-based media. However, the team members realized that due to complicated and diverse ways of recording and displaying, there will be valuable follow-up work for further research and preservation of Chinese handmade paper and other related resource platforms after the completion of the project. We expect that when contemporary handmade papermaking industry in China, consisting of 34 provinces, cities, autonomous regions and special administrative regions as planned, is recorded and displayed in the above mentioned way, a Chinese handmade paper image library, a Chinese handmade paper technical data library, a Chinese handmade paper sample library, and a Chinese handmade paper video information collection will come into being, aside from the *Library of Chinese Handmade Paper*. Because of the richness of these byproducts, we should not overlook these possible follow-up

更多元的机遇。

毫无疑问，《中国手工纸文库》工作团队整体上都非常认同这一工作的历史价值与现实意义。这种认同给了我们持续的动力与激情，但在实际的推进中，确实有若干挑战使大家深感困惑。

## 三
## 我们的困惑和愿景

困惑一：

中国当代手工造纸的范围与边界在国家层面完全不清晰，因此无法在项目的田野工作完成前了解到中国到底有多少当代手工造纸地点，有多少种手工纸产品；同时也基本无法获知大多数省级区域手工造纸分布地点的情况与存活、存续状况。从调查组2008~2016年集中进行的中国南方地区（云南、贵州、广西、四川、广东、海南、浙江、安徽等）的田野与文献工作来看，能够提供上述信息支持的现状令人失望。这导致了项目组的田野工作规划处于“摸着石头过河”的境地，也带来了《中国手工纸文库》整体设计及分卷方案等工作的不确定性。

developments. Moving forward, the innovation and development of Chinese handmade paper may offer more opportunities to researchers who are interested in the techniques and culture of Chinese handmade papermaking.

Unquestionably, the whole team acknowledges the value and significance of the project, which has continuously supplied the team with motivation and passion. However, the presence of some problems have challenged us in implementing the project.

## 3. Our Confusions and Expectations

Problem One:

From the nationwide point of view, the scope of Chinese contemporary handmade papermaking sites is so obscure that it was impossible to know the extent of manufacturing sites and product types of present handmade paper before the fieldwork plan of the project was drawn up. At the same time, it is difficult to get information on the locations of handmade papermaking sites and their survival and subsisting situation at the provincial level. Based on the field work and literature of South China, including Yunnan, Guizhou, Guangxi, Sichuan, Guangdong, Hainan, Zhejiang and Anhui etc., carried out between 2008 and 2016, the ability to provide the information mentioned above is rather difficult. Accordingly, it placed the planning of the project's fieldwork into an obscure unplanned route,

困惑二：

中国正高速工业化与城镇化，手工造纸作为一种传统的手工技艺，面临着经济效益、环境保护、集成运营、技术进步、消费转移等重要产业与社会变迁的压力。调查组在已展开了九年的田野调查工作中发现，除了泾县、夹江、富阳等为数不多的手工造纸业态聚集地，多数乡土性手工造纸业态都处于生存的“孤岛”困境中。令人深感无奈的现状包括：大批造纸点在调查组到达时已经停止生产多年，有些在调查组到达时刚刚停止生产，有些在调查组补充回访时停止生产，仅一位老人或一对老纸工夫妇在造纸而无传承人……中国手工造纸的业态正陷于剧烈的演化阶段。这使得项目组的田野调查与实物采样工作处于非常紧迫且频繁的调整之中。

困惑三：

作为国家级重点出版物规划项目，《中国手工纸文库》在撰写开卷总序的时候，按照规范的说明要求，应该清楚地叙述分卷的标准与每一卷的覆盖范围，同时提供中国手工造纸业态及地点分布现

贵州省仁怀市五马镇
取缔手工造纸作坊的横幅
2009年4月

Banner of a handmade papermaking mill in Wuma Town of Renhuai City in Guizhou Province, saying “Handmade papermaking mills should be closed as encouraged by the local government”
April 2009

which also led to uncertainty in the planning of *Library of Chinese Handmade Paper* and that of each volume.

Problem Two:
China is currently under the process of rapid industrialization and urbanization. As a traditional manual technique, the industry of handmade papermaking is being confronted with pressures such as economic benefits, environmental protection, integrated operation, technological progress, consumption transfer, and many other important changes in industry and society. During nine years of field work, the project team found out that most handmade papermaking mills are on the verge of extinction, except a few gathering places of handmade paper production like Jingxian, Jiajiang, Fuyang, etc. Some handmade papermaking mills stopped production long before the team arrived or had just recently ceased production; others stopped production when the team paid a second visit to the mills. In some mills, only one old papermaker or an elderly couple were working, without any inheritor to learn their techniques... The whole picture of this industry is in great transition, which left our field work and sample collection scrambling with hasty and frequent changes.

Problem Three:
As a national key publication project, the preface of *Library of Chinese Handmade Paper* should clarify the standard and the scope of each volume according to the research plan. At the same time, general information such as the map with locations of Chinese handmade

状图等整体性信息。但由于前述的不确定性，开宗明义的工作只能等待田野调查全部完成或进行到尾声时再来弥补。当然，这样的流程一定程度上会给阅读者带来系统认知的先期缺失，以及项目组工作推进中的迷茫。尽管如此，作为拓荒性的中国手工造纸整体研究与田野调查就在这样的现状下全力推进着！

当然，我们的团队对《中国手工纸文库》的未来仍然满怀信心与憧憬，期待着通过项目组与国际国内支持群体的协同合作，尽最大努力实现尽可能完善的田野调查与分析研究，从而在我们这一代人手中为中国经典的非物质文化遗产样本——中国手工造纸技艺留下当代的全面记录与文化叙述，在中国非物质文化遗产基因库里绘制一份较为完整的当代手工纸文化记忆图谱。

汤书昆
2017年12月

papermaking industry should be provided. However, due to the uncertainty mentioned above, those tasks cannot be fulfilled, until all the field surveys have been completed or almost completed. Certainly, such a process will give rise to the obvious loss of readers' systematic comprehension and the team members' confusion during the following phases. Nevertheless, the pioneer research and field work of Chinese handmade paper have set out on the first step.

There is no doubt that, with confidence and anticipation, our team will make great efforts to perfect the field research and analysis as much as possible, counting on cooperation within the team, as well as help from domestic and international communities. It is our goal to keep a comprehensive record, a cultural narration of Chinese handmade paper craft as one sample of most classic intangible cultural heritage, to draw a comparatively complete map of contemporary handmade paper in the Chinese intangible cultural heritage gene library.

Tang Shukun
December 2017

# 编撰说明





1

关于类目的划分标准，《中国手工纸文库·安徽卷》（以下简称《安徽卷》）在充分考虑安徽地域当代手工造纸高度聚集于泾县一地，而且手工纸的历史传承品种相对丰富的特点后，决定不按地域分布划分类目，而是按照宣纸、书画纸、皮纸、竹纸、加工纸、工具划分第一级目类，形成“章”的类目单元，如第二章“宣纸”、第三章“书画纸”。章之下的二级类目以造纸企业或家庭纸坊为单元，形成“节”的类目，如第二章第一节“中国宣纸股份有限公司”、第四章第三节“潜山县星杰桑皮纸厂”。

2

《安徽卷》成书内容丰富，篇幅较大，从适宜读者阅读和装帧牢固角度考虑，将其分为上、中、下三卷。上卷内容为第一章“安徽省手工造纸概述”、第二章“宣纸”；中卷内容为第三章“书画纸”、第四章“皮纸”、第五章“竹纸”；下卷内容为第六章“加工纸”、第七章“工具”以及“附录”。

3

《安徽卷》第一章为概述，其格式与先期出版的《中国手工纸文库·云南卷》（以下简称《云南卷》）、《中国手工纸文库·贵州卷》（以下简称《贵州卷》）等类似。其余各章各节的标准撰写格式则因有手工纸业态高度密集的县级区域存在，所以与《云南卷》《贵州卷》所用的单一标准撰写格式不同，分为三类撰写标准格式。

第一类与《云南卷》《贵州卷》相近，适应一个县域内手工造纸厂坊不密集、品种相对单纯的业态分布。通常分为七个部分，即“××××纸的基础信息及分布”“××××纸生产的人文地理环境”“××××纸的历史与传承”“××××纸的生产工艺与技术分析”“××××纸的用途与销售情况”

## Introduction to the Writing Norms

1. Referring to the categorization standards, *Library of Chinese Handmade Paper: Anhui* will not be categorized based on location, but the paper types, i.e. Xuan Paper, Calligraphy and Painting Paper, Bast Paper, Bamboo Paper, Processed Paper and Tools, due to the fact that papermaking sites in the region cluster around Jingxian County, and the diverse paper types historically inherited in the area. Each category covers a whole chapter, e.g. Chapter II “Xuan Paper”, Chapter III “Calligraphy and Painting Paper”. Each chapter consists of sections based on different papermaking factories or family-based papermaking mills. For instance, first section of the second chapter is “China Xuan Paper Co., Ltd.”, and the third section of Chapter IV is “Xingjie Mulberry Bark Paper Factory in Qianshan County”.

2. Due to its rich content and great length, *Library of Chinese Handmade Paper: Anhui* is further divided into three sub-volumes (I, II, III) for convenience of the readers and bookbinding. *Anhui* I consists of Chapter I “Introduction to Handmade Paper in Anhui Province”, Chapter II “Xuan Paper”; *Anhui* II contains Chapter III “Calligraphy and Painting Paper”, Chapter IV “Bast Paper” and Chapter V “Bamboo Paper”; *Anhui* III is composed of two chapters, i.e. Chapter VI “Processed Paper”,Chapter VII “Tools”, and “Appendices”.

3. First chapter of *Library of Chinese Handmade Paper: Anhui* is introduction, which follows the volume format of *Yunnan* and *Guizhou*, which have already been released. Sections of other chapters follow three different writing norms, because of the concentrated distribution of county-level handmade papermaking practice, and this is different from two volumes that have been published.

First type of volume writing norm is similar to that of *Yunnan* and *Guizhou*: each section consists of seven sub-sections introducing various aspects of each kind of handmade paper, namely, Basic Information and Distribution, The Cultural and Geographic

"××××纸的品牌文化与习俗故事""××××纸的保护现状与发展思考"。如遇某一部分田野调查和文献资料均未能采集到信息，则将按照实事求是原则略去标准撰写格式的相应部分。

第二类主要针对泾县宣纸与书画纸企业以及少数加工纸企业的特征，手工造纸厂坊在一个小地区聚集度特别高，或者纸品非常丰富，不适合采用第一类撰写格式时采用。通常的格式及大致名称为："××××纸（纸厂）的基础信息与生产环境""××××纸（纸厂）的历史与传承情况""××××纸（纸厂）的代表纸品及其用途与技术分析""××××纸（纸厂）生产的原料、工艺与设备""××××纸（纸厂）的市场经营状况""××××纸（纸厂）的品牌文化与习俗故事""××××纸（纸厂）的业态传承现状与发展思考"。

第三类主要针对当代世界最大的手工造纸企业——中国宣纸股份有限公司，由于其从业人数多达1 300余人，工艺、产品、制度与文化的丰富性独具一格，因此专门设计了撰写类目形式，分为："中国宣纸股份有限公司的基础信息与生产环境""中国宣纸股份有限公司的历史与传承情况""中国宣纸股份有限公司的关键岗位和产量变更情况""'红星'宣纸制作技艺的基本形态""原料、辅料、人员配置、工具和用途""'红星'宣纸的分类与品种""'红星'宣纸的价格、销售、包装信息""社会名流品鉴'红星'宣纸的重要掌故""中国宣纸股份有限公司保护宣纸业态的措施"。

4

《安徽卷》专门安排一节讲述的手工纸的入选标准是：（1）项目组进行田野调查时仍在生产；（2）项目组田野调查时虽已不再生产，但保留着较完整的生产环境与设备，造纸技师仍能演示或讲述完整技艺和相关知识。

考虑到竹纸在安徽省历史上曾经是大宗民生产品，而其当代业态萎缩特别明显，处于几近消亡状态，因此对调查组所能够找到的很少的竹纸产地中的泾县竹纸放宽了"保留着较完整的生产环境与设备"这一项标准。

5

《安徽卷》调查涉及的造纸点均参照国家地图标准绘制两幅示意图：一幅为造纸点在安徽省和所属县的地理位置图，另一幅为由该县县城前往造纸点的路线图，但在具体出图时，部分节会将两图合一呈现。在标示地名时，均统一标示出



Environment, History and Inheritance, Papermaking Technique and Technical Analysis, Uses and Sales, Brand Culture and Stories, Preservation and Development. Omission is also acceptable if our fieldwork efforts and literature review fail to collect certain information. This writing norm applies to the handmade papermaking practice in the area where factories and papermaking mills are not dense, and the paper produced is of single variety.

The second writing norm is applied to Xuan paper, and calligraphy and painting paper factories in Jingxian County, and a few processed paper factories, which all cluster in a small area, and produce diverse paper types. In such chapter, sections are: Basic Information and Production Environment; History and Inheritance; Representative Paper and Its Uses and Technical Analysis; Raw Materials, Papermaking Techniques and Tools; Marketing Status; Brand Culture and Stories; Current Status of Business Inheritance and Thoughts on Development.

The third writing norm is applied to China Xuan Paper Co., Ltd., which boasts the largest handmade papermaking factory around the world. It harbors over 1,300 employees and unique papermaking techniques, products, and colorful management system and culture. In this chapter, sections are listed differently: Basic Information and Production Environment of China Xuan Paper Co., Ltd.; History and Inheritance of China Xuan Paper Co., Ltd.; Key Positions and Production Profile of China Xuan Paper Co., Ltd.; "Red Star" Xuan Papermaking Techniques; Types and Varieties of "Red Star" Xuan Paper; Celebrities and "Red Star" Xuan Paper; Preservation of Xuan Paper by China Xuan Paper Co., Ltd.

4. The handmade paper included in each section of this volume conforms to the following standards: firstly, it was still under production when the research group did their fieldwork. Secondly, the papermaking equipment and major sites were well preserved, and the handmade papermakers were still able to demonstrate the papermaking techniques and relevant knowledge, in case of ceased production.

县城、乡镇两级，乡镇下一级则直接标示造纸点所在村，而不再做行政村、自然村、村民组之区别。示意图上的行政区划名称及编制规则均依据中国地图出版社、国家基础地理信息中心的相关地图。



6

《安徽卷》原则上对每一个所调查的造纸厂坊的代表纸品，均在珍稀收藏版书中相应章节后附调查组实地采集的实物纸样。采样量足的造纸点代表纸品附全页纸样；由于各种限制因素，采样量不足的则附2/3、1/2、1/4或更小规格的纸样；个别因近年停产等导致未能获得纸样或采样严重不足的，则不附实物纸样。

7

《安徽卷》原则上对所有在章节中具体描述原料与工艺的代表纸品进行技术分析，包括实物纸样可以在书中呈现的类型，以及个别只有极少量纸样遗存，可以满足测试要求而无法在"珍稀收藏版"中附上实物纸样的类型。

全卷对所采集纸样进行的测试参考了中国宣纸的技术测试分析标准（GB/T 18739—2008），并根据安徽地域手工纸的多样性特色做了必要的调适。实测、计算了所有满足测试分析标示足量需求的已采样的手工纸中的宣纸类、书画纸类、皮纸类的厚度、定量、紧度、抗张力、抗张强度、撕裂度、湿强度、白（色）度、耐老化度下降、尘埃度、吸水性(数种熟宣未测该指标)、伸缩性、纤维长度和纤维宽度共14个指标；加工纸类的厚度、定量、紧度、抗张力、抗张强度、撕裂度、色度、吸水性共8个指标；竹纸类的厚度、定量、紧度、抗张力、抗张强度、色度、纤维长度和纤维宽度共8个指标。由于所采集的安徽省各类手工纸样的生产标准化程度不同，因而若干纸种纸品所测数据与机制纸、宣纸的标准存在一定差距。



8

测试指标说明及使用的测试设备如下：

（1） 厚度 ▸ 所测纸的厚度指标是指纸在两块测量板间受一定压力时直接

Because bamboo paper used to be mass produced in Anhui Province, while the practice shrank greatly or even is lingering on extinction in current days, the research team decided to omit the requirement of comparatively complete preservation of production environment and equipment.

5. For each handmade papermaking site, we draw two standard illustrations, i.e. distribution map and roadmap from the county center to the papermaking sites (in some sections, two figures are combined). We do not distinguish the administrative village, natural village or villagers' group, and we provide county name, town name and village name of each site based on standards released by Sinomaps Press and National Geomatics Center of China.

6. For each type of paper included in Special Edition, we attach a piece of paper sample (a full page, 2/3, 1/2 or 1/4 of a page, or even smaller if we do not have sufficient sample available) to the corresponding section. For some sections, no sample is attached for the shortage of sample paper (e.g. the papermakers had ceased production).

7. All the paper samples elaborated on in this volume, in terms of raw materials and papermaking techniques, were tested, including those attached to the special edition, or not attached to this volume due to scarce sample which only enough for technical analysis.

The test was based on the technical analysis standards of Chinese Xuan paper (GB/T 18739—2008), with modifications adopted according to the specific features of the handmade paper in Anhui Province. All paper with sufficient sample, such as Xuan paper, calligraphy and painting paper, bast paper, was tested in terms of 14 indicators, including thickness, mass per unit area, tightness, resistance force, tensile strength, tear resistance, wet strength, whiteness, ageing resistance, dirt count, absorption of water (several processed Xuan paper was not tested on the indicator), elasticity, fiber length and fiber width. Processed paper was tested in terms of 8 indicators, including thickness, mass per unit area, tightness, resistance force, tensile strength, tear resistance, whiteness,

测量得到的厚度。根据纸的厚薄不同，可采取多层指标测量、单层指标测量，以单层指标测量的结果表示纸的厚度，以mm为单位。

所用仪器▶长春市月明小型试验机有限责任公司JX-HI型纸张厚度仪、杭州品享科技有限公司PN-PT6厚度测定仪。

(2) 定量▶所测纸的定量指标是指单位面积纸的质量，通过测定试样的面积及质量，计算定量，以g/m²为单位。

所用仪器▶上海方瑞仪器有限公司3003电子天平。

(3) 紧度▶所测纸的紧度指标是指单位体积纸的质量，由同一试样的定量和厚度计算而得，以g/cm³为单位。

(4) 抗张力▶所测纸的抗张力指标是指在标准试验方法规定的条件下，纸断裂前所能承受的最大张力，以N为单位。

所用仪器▶杭州高新自动化仪器仪表公司DN-KZ电脑抗张力试验机、杭州品享科技有限公司PN-HT300卧式电脑拉力仪。

(5) 抗张强度▶所测纸的抗张强度指标一般用在抗张强度试验仪上所测出的抗张力除以样品宽度来表示，也称为纸的绝对抗张强度，以kN/m为单位。

《安徽卷》采用的是恒速加荷法，其原理是使用抗张强度试验仪在恒速加荷的条件下，把规定尺寸的纸样拉伸至撕裂，测其抗张力，计算出抗张强度。公式如下：

$$S=F/W$$

式中，$S$为试样的抗张强度（kN/m），$F$为试样的绝对抗张力（N），$W$为试样的宽度（mm）。

(6) 撕裂度▶所测纸张撕裂强度的一种量度，即在测定撕裂度的仪器上，拉开预先切开一小切口的纸达到一定长度时所需要的力，以mN为单位。

所用仪器▶长春市月明小型试验机有限责任公司ZSE-1000型纸张撕裂度测定仪、杭州品享科技有限公司PN-TT1000电脑纸张撕裂度测定仪。

(7) 湿强度▶所测纸张在水中浸润规定时间后，在润湿状态下测得的机械强度，以mN为单位。





and absorption of water. Bamboo paper was tested in terms of 8 indicators, including thickness, mass per unit area, tightness, resistance force, tensile strength, whiteness, fiber length and fiber width. Due to the various production standards involved in papermaking in Anhui Province, the data might vary from those standards of machine-made paper and Xuan paper.

8. Test indicators and devices:

(1) Thickness: the values obtained by using two measuring boards pressing the paper. In the measuring process, single layer or multiple layers of paper were employed depending on the thickness of the paper, and its measurement unit is mm. The thickness measuring instruments employed are produced by Yueming Small Testing Instrument Co., Ltd., Changchun City (specification: JX-HI) and Pinxiang Science and Technology Co., Ltd., Hangzhou City (specification: PN-PT6).

(2) Mass per unit area: the sample mass divided by area, with the measurement unit g/m². The measuring instrument employed is 3003 electronic balance produced by Shanghai Fangrui Instrument Co., Ltd.

(3) Tightness: mass of paper per volume unit, obtained by measuring the mass per unit area and thickness, with the measurement unit g/cm³.

(4) Tensile strength: the resistance of sample paper to a force tending to tear it apart, measured as the maximum tension the material can withstand without tearing. The resistance force testing instrument (specification: DN-KZ) is produced by Gaoxin Technology Company, Hangzhou City and PN-HT300 horizontal computer tensiometer by Pinxiang Science and Technology Co., Ltd., Hangzhou City.

(5) Unit tensile strength: the resistance of one unit sample paper to a force, with the measurement unit kN/m. In *Library of Chinese Handmade Paper: Anhui*, constant loading method was employed to measure the tensile strength. The sample's maximum resistance force against the constant loading was tested, then we divided the maximum force by the sample width. The formula is:

$$S=F/W$$

$S$ stands for tensile strength (kN/m) for each unit, $F$ is resistance force (N) and $W$ represents sample width (mm).

(6) Tear resistance: a measure of how well a piece of paper can

所用仪器▸长春市月明小型试验机有限责任公司ZSE-1000型纸张撕裂度测定仪、杭州品享科技有限公司PN-TT1000电脑纸张撕裂度测定仪。

(8) 白(色)度▸白度测试针对白色纸，色度测试针对其他颜色的纸。白度是指被测物体的表面在可见光区域内与完全白（标准白）的物体漫反射辐射能的大小的比值，用百分数来表示，即白色的程度。所测纸的白度指标是指在D65光源、漫射/垂射照明观测条件下，以纸对主波长475 nm蓝光的漫反射因数表示白度的测定结果。

所用仪器▸杭州纸邦仪器有限公司ZB-A色度测定仪、杭州品享科技有限公司PN-48A白度颜色测定仪。

(9) 耐老化度下降▸指所测纸张进行高温试验的温度环境变化后的参数及性能。本测试采用105 ℃高温恒温放置72小时后进行测试，以百分数（%）表示。

所用仪器▸上海一实仪器设备厂3GW-100型高温老化试验箱、杭州品享科技有限公司YNK/GW100-C50耐老化度测试箱。

(10) 尘埃度▸所测纸张单位面积上尘埃涉及的黑点、黄茎和双浆团个数。测试时按照标准要求计算出每一张试样正反面每组尘埃的个数，将4张试样合并计算，然后换算成每平方米的尘埃个数，计算结果取整数，以个/$m^2$为单位。

所用仪器▸杭州品享科技有限公司PN-PDT尘埃度测定仪。

(11) 吸水性▸所测纸张在水中能吸收水分的性质。测试时使用一条垂直悬挂的纸张试样，其下端浸入水中，测定一定时间后的纸张吸液高度，以mm为单位。

所用仪器▸四川长江造纸仪器有限责任公司J-CBY100型纸与纸板吸收性测定仪、杭州品享科技有限公司PN-KLM纸张吸水率测定仪。

(12) 伸缩性▸指所测纸张由于张力、潮湿，尺寸变大、变小的倾向性。分为浸湿伸缩性和风干伸缩性，以百分数（%）表示。

所用仪器▸50 cm × 50 cm × 20 cm长方体容器。





withstand the effects of tearing. It measures the strength the test specimen resists the growth of any cuts when under tension. The measurement unit is mN. Paper tear resistance testing instrument (specification: ZSE-1000) is produced by Yueming Small Testing Instrument Co., Ltd., Changchun City and computer paper tear resistance testing instrument (specification: PN-TT1000) produced by Pinxiang Science and Technology Co., Ltd., Hangzhou City.

(7) Wet strength: a measure of how well the paper can resist a force of rupture when the paper is soaked in the water for a set time. The measurement unit is mN. Paper tear resistance testing instrument (specification: ZSE-1000) is produced by Yueming Small Testing Instrument Co., Ltd., Changchun City and computer paper tear resistance testing instrument (specification: PN-TT1000) produced by Pinxiang Science and Technology Co., Ltd., Hangzhou City.

(8) Whiteness: degree of whiteness, represented by percentage, which is the ratio obtained by comparing the radiation diffusion value of the test object in visible region to that of the completely white (standard white) object. Whiteness test in our study employed D65 light source, with dominant wavelength 475nm of blue light, under the circumstances of diffuse reflection or vertical reflection. The whiteness testing instrument (specification: ZB-A) is produced by Zhibang Instrument Co., Ltd., Hangzhou City and whiteness tester (specification: PN-48A) produced by Pinxiang Science and Technology Co., Ltd., Hangzhou City respectively.

(9) Ageing Resistance: the performance and parameters of paper sample when put in high temperature. In our test, temperature is set 105 degrees centigrade, and the paper is put in the environment for 72 hours. It is measured in percentage(%) . The high temperature ageing test box (specification: 3GW-100) is produced by Yishi Testing Instrument Factory and ageing test box (specification: YNK/GW100-C50) produced by Pinxiang Science and Technology Co., Ltd., Hangzhou City.

(10) Dirt count: fine particles (black dots, yellow stems, fiber knots) in the test paper. It is measured by counting fine particles in every side of four pieces of paper sample, adding up and then calculate the number (integer only) of particles every square meter. It is measured by the number of particles/$m^2$. Dust tester (specification: PN-PDT) is produced by Pinxiang Science and Technology Co., Ltd., Hangzhou City.

（13）纤维长度/宽度▸所测纸的纤维长度/宽度是指从所测纸里取样，测其纸浆中纤维的自身长度/宽度，分别以mm和μm为单位。测试时，取少量纸样，用水湿润，用Herzberg试剂染色，制成显微镜试片，置于显微分析仪下采用10倍及20倍物镜进行观测，并显示相应纤维形态图各一幅。

所用仪器▸珠海华伦造纸科技有限公司生产的XWY-VI型纤维测量仪和XWY-VII型纤维测量仪。

9

《安徽卷》对每一种调查采集的纸样均采用透光摄影的方式制作成图像，以显示透光环境下的纸样纤维纹理影像，作为实物纸样的另一种表达方式。其制作过程为：先使用透光台显示纯白影像，作为拍摄手工纸纹理透光影像的背景底，然后将纸样铺平在透光台上进行拍摄。拍摄相机为佳能5DIII。

10

《安徽卷》引述的历史与当代文献均以当页脚注形式标注。所引文献原则上要求为一手文献来源，并按统一标准注释，如“[宋]罗愿.《新安志》整理与研究[M]. 萧建新，杨国宜，校. 合肥:黄山书社,2008:371.”“民国三年（1913年）泾县小岭曹氏编撰.曹氏宗谱[Z]. 自印本.”“魏兆淇.宣纸制造工业之调查:中央工业试验所工业调查报告之一[J]. 工业中心,1936(10):8.”等。

11

《安徽卷》所引述的田野调查信息原则上要求标示出调查信息的一手来源，如：“据访谈中刘同烟的介绍，星杰桑皮纸厂年产5 000多刀纸，年销售额约100万元”“按照访谈时沈维正的说法，以他为核心的这个团队专注造纸新技术的研发和传统技艺的保护”等。





(11) Absorption of water: it measures how sample paper absorbs water by dipping the paper sample vertically in water and testing the level of water. It is measured in mm. Paper and Paper Board Water Absorption Tester (specification: J-CBY100) is produced by Changjiang Papermaking Instrument Co., Ltd., Sichuan Province, and Water absorption tester (specification: PN-KLM) produced by Pinxiang Science and Technology Co., Ltd., Hangzhou City.
(12) Elasticity: continuum mechanics of paper that deform under stress or wet. It is measured in percentage(%), consists of two types, i.e. wet elasticity and dry elasticity. Testing with a rectangle container (50cm×50cm×20cm).
(13) Fiber length and width: analyzed by dying the moist paper sample with Herzberg reagent, and the fiber pictures were taken through ten times and twenty times objective lens of the microscope, with the measurement unit mm and μm. We used the fiber testing instruments (specifications: XWY-VI and XWY-VII) produced by Hualun Papermaking Technology Co., Ltd., Zhuhai City.

9. Each paper sample included in this volume was photographed against a luminous background, which vividly demonstrated the fiber veins of the sample. This is a different way to present the status of our paper sample. Each piece of paper sample was spread flat-out on the light table giving white light, and photographs were taken with Canon 5DIII camera.

10. All the quoted literature are original first-hand resources and the footnotes are used for documentation with a uniform standard. For instance, "[Song Dynasty] Luo Yuan. *Xin'an Records* [M]. Proofread by Xiao Jianxin and Yang Guoyi. Hefei: Huangshan Publishing House, 2008:371." and "*Genealogy of The Caos* [Z]. compiled by the Caos in Xiaoling Village of Jingxian County, 1913. Self-printed." and "Wei Zhaoqi. *Investigation of Xuan paper industry*: *One of the national industrial investigation report series* [J]. Industrial Center, 1936 (10):8" etc.

11. Sources of field investigation information were attached in this volume. For instance, "According to Liu Tongyan, annual output of Xingjie Mulberry Bark Factory exceeded 5,000 *dao* each year, with annual sales about one million RMB." "According to Shen Weizheng,

12

《安徽卷》所使用的摄影图片主体部分为调查组成员在实地调查时所拍摄的图片，也有项目组成员在既往田野工作中积累的图片，另有少量属撰稿过程中所采用的非项目组成员的摄影作品。由于项目组成员在完成全卷过程中形成的图片的著作权属集体著作权，且在调查过程中多位成员轮流拍摄或并行拍摄为工作常态，因而全卷对图片均不标示项目组成员作者。项目组成员既往积累的图片，以及非项目组成员拍摄的图片在图题文字或后记中特别说明，并承认其个人图片著作权。

13

考虑到《安徽卷》中文简体版的国际交流需要，编著者对全卷重要或提要性内容同步给出英文表述，以便英文读者结合照片和实物纸样领略全卷的基本语义。对于文中一些晦涩的古代文献，英文翻译采用意译的方式进行解读。英文内容包括：总序、编撰说明、目录、概述、图目、表目、术语、后记，以及所有章节的标题，全部图题、表题与实物纸样名。

"安徽省手工造纸概述"为全卷正文第一章，为保持与后续各章节体例一致，除保留章节英文标题及图表标题英文名外，全章的英文译文作为附录出现。

14

《安徽卷》的名词术语附录兼有术语表、中英文对照表和索引三重功能。其中收集了全卷中与手工纸有关的地理名、纸品名、原料与相关植物名、工艺技术和工具设备、历史文化等5类术语。各个类别的名词术语按术语的汉语拼音先后顺序排列。每条中文名词术语后都以英文直译，可以作中英文对照表使用，也可以当作名词索引使用。

he played a key role in the team which focused on papermaking techniques R&D, and preserving the traditional skills".

12. The majority of photographs included in the volume were taken by the researchers when they were doing fieldworks of the research. Others were taken by our researchers in even earlier fieldwork errands, or by the photographers who were not involved in our research. We do not give the names of the photographers in the book, because almost all our researchers are involved in the task and they agreed to share the intellectual property of the photos. Yet, as we have claimed in the epilogue or the caption, we officially admit the copyright of all the photographers, including those who are not our researchers.

13. For the purpose of international academic exchange, English version of some important chapters is provided, so that the English readers can have a basic understanding of the volume based on the English parts together with photos and samples. For the ancient literature which is hard to understand, free translation is employed to present the basic idea. English part includes Preface, Introduction to the Writing Norms, Contents, Introduction, Figures, Tables, Terminology, Epilogue, and section titles, figure and table captions and paper sample names.

Among them, "Introduction to Handmade Paper in Anhui Province" is the first chapter of the volume and its translation is appended in the appendix part, apart from the section titles and table and figure titles.

14. Terminology is appended in *Library of Chinese Handmade Paper: Anhui*, which covers five categories of places, paper names, raw materials and plants, techniques and tools, history and culture, etc., relevant to our handmade paper research. All the terms are listed following the alphabetical order of the first Chinese character. The Chinese and English parts in the Terminology can be used as check list and index.

# 目 录
# Contents

# 第三章
# 书画纸

# Chapter Ⅲ
# Calligraphy and Painting Paper

# 第一节
# 泾县载元堂工艺厂



安徽省
Anhui Province

宣城市
Xuancheng City

泾县
Jingxian County

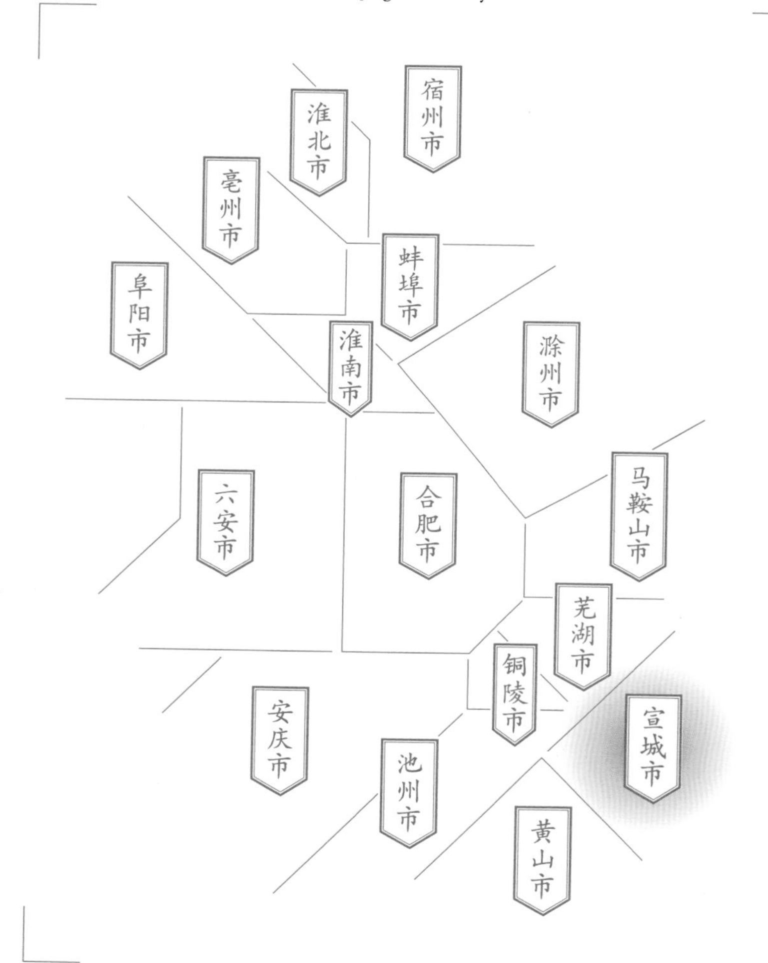




调查对象

泾川镇
泾县载元堂工艺厂
书画纸

## Section 1
## Zaiyuantang Craft Factory in Jingxian County

Subject

Calligraphy and Painting Paper of Zaiyuantang Craft Factory in Jingxian County in Jingchuan Town

# 一
# 泾县载元堂工艺厂的基础信息与生产环境

## 1
## Basic Information and Production Environment of Zaiyuantang Craft Factory in Jingxian County

泾县载元堂工艺厂创设于1988年，书画纸生产厂家，以“载元堂”作为系列纸产品的品牌。生产厂区位于泾县泾川镇的城西工业集中区（原园林村）内，紧邻S322省道，交通较为便利。地理坐标为：东经118° 21′ 58″，北纬30° 41′ 28″。

工厂依湖山而建，占地面积约5 600 m²，厂房面积约2 000 m²，建筑布局有中式园林庭院风格。厂房则依庭院三面环抱而建。根据工序、加工内容分为几大块，如书画纸包装和加工纸车间分布在庭院两边，原纸生产与销售则在庭院的前后位置。调查组于2015年7月24日与2016年4月25日先后两次对载元堂工艺厂进行了调查。

⊙1

⊙2

泾县载元堂工艺厂生产的纸品种较多，原纸有画仙纸、皮纸等；加工类纸品有册页、印谱、洒金纸、洒银纸、万年红、仿古纸等。2016年4月调查组调查时年产量为20 000多刀，年产值为600万元。该厂主要以画仙纸为代表，销量约占总销量的50%，产品主要出口日本；加工纸约占40%，以万年红、洒金纸和册页为主，主要销往台湾地区；大约10%份额的皮纸销往韩国。从调查数据看，载元堂工艺厂是泾县代表性的以出口外销为

⊙1 载元堂工艺厂大门 Gate of Zaiyuantang Craft Factory

⊙2 厂区外部环境 External view of the factory

# 泾县载元堂工艺厂位置示意图

路线图
泾县县城
↓
泾县载元堂工艺厂
Road map from Jingxian County centre
to Zaiyuantang Craft Factory in Jingxian County

# Location map of Zaiyuantang Craft Factory in Jingxian County

考察时间
2015年7月 / 2016年4月

Investigation Date
July 2015 / Apr. 2016

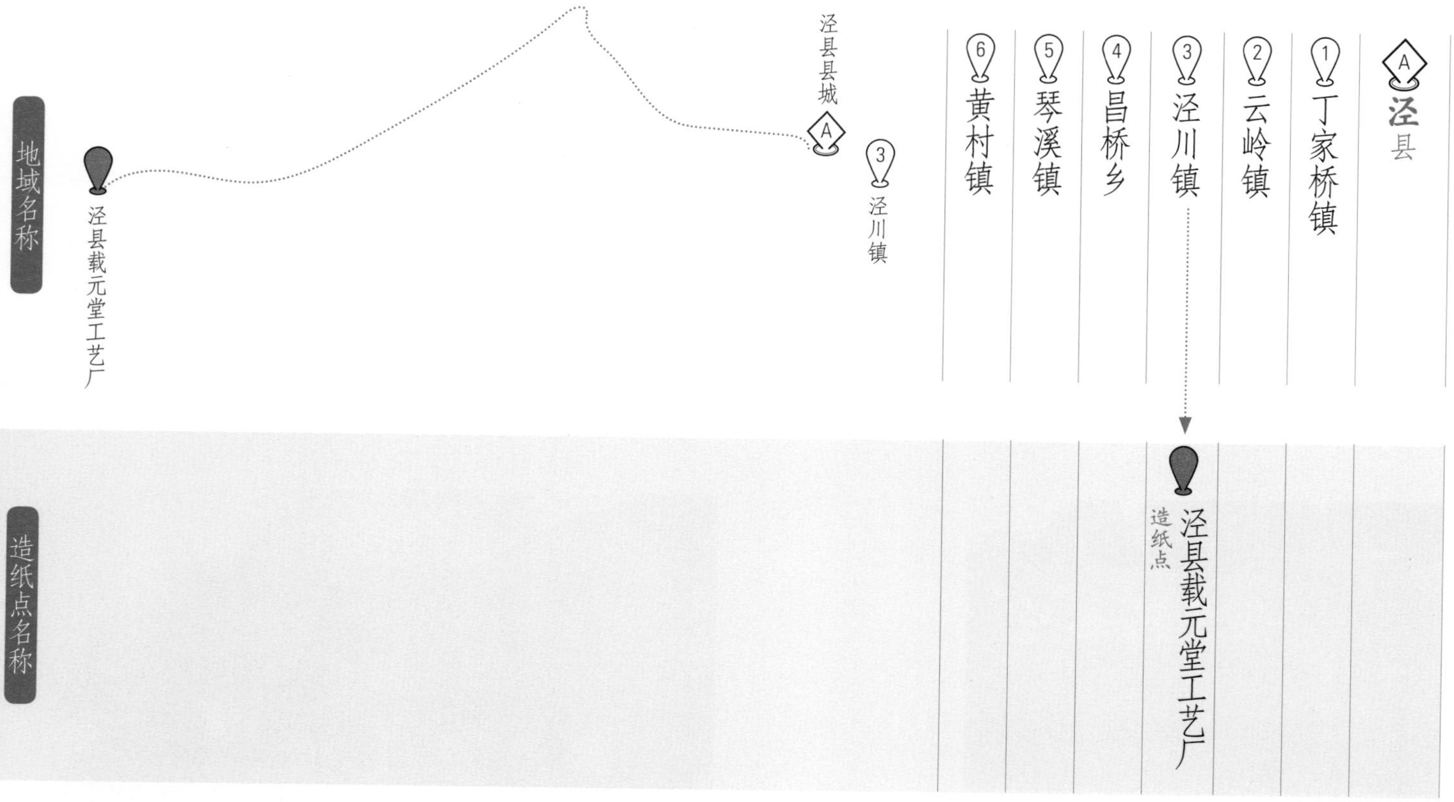

地域名称
泾县载元堂工艺厂
泾县县城
A
3
泾川镇
6 黄村镇
5 琴溪镇
4 昌桥乡
3 泾川镇
2 云岭镇
1 丁家桥镇
A 泾县
造纸点名称
造纸点
泾县载元堂工艺厂


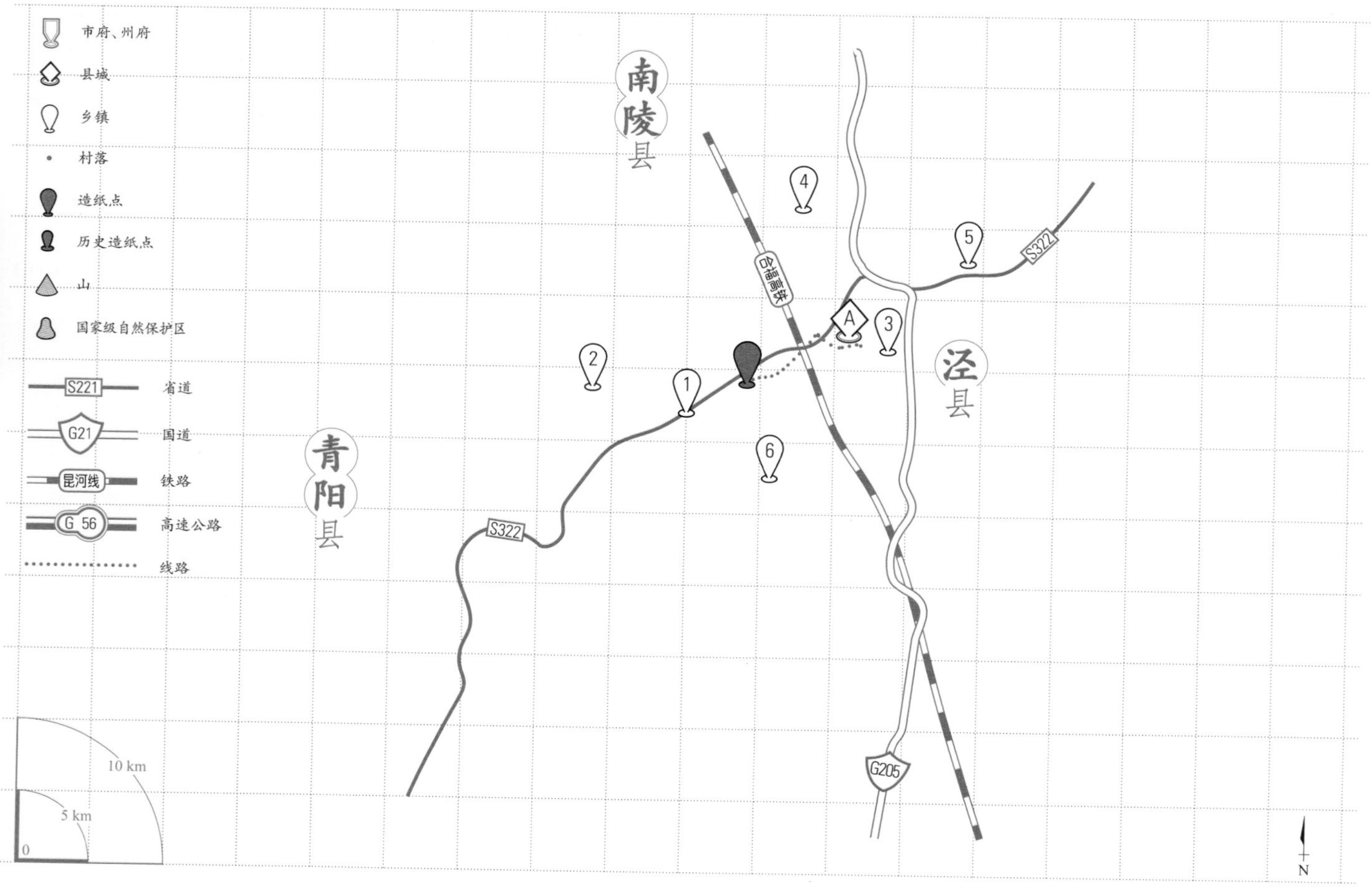

位置分布
市府、州府
县城
乡镇
村落
造纸点
历史造纸点
山
国家级自然保护区
S221 省道
G21 国道
昆河线 铁路
G 56 高速公路
线路
南陵县
青阳县
泾县
合福高铁
S322
G205
A
1
2
3
4
5
6
10 km
5 km
0
N

⊙1

主的书画纸企业。

截至2016年4月25日第二次调查时，所获泾县载元堂工艺厂的基础生产信息如下：有8台喷浆制纸设备、2个手工捞纸槽位、3个晒纸焙房，共有员工40名。载元堂工艺厂工人一般每天工作10小时，早上7点上班，下午5点下班；而捞纸工人工作时间最长，每天早上4点上班，下午4点下班。工人每周休息一天，月工作日25天左右，春节放假约15天。根据品种不同槽单位产量也有所不同，如喷浆槽单位产量为6～8刀，手工纸槽单位产量为20刀（四尺）、7刀（六尺）和7～8刀（吊帘）。

⊙2

⊙3

⊙1 厂区内部环境图 Internal view of the factory

⊙2 喷浆技艺 Pulp shooting technique

⊙3 手工捞纸技艺 Handmade papermaking technique

# 二
# 泾县载元堂工艺厂的历史与传承情况

# 2
# History and Inheritance of Zaiyuantang Craft Factory in Jingxian County

沈维正是载元堂工艺厂的创建者与法人代表，1963年生于泾县。访谈中，比较健谈的沈维正介绍了其与造纸相关的经历：

1982年沈维正进入百岭坑宣纸厂学习晒纸技艺，4个月后被派往安徽省博物馆学习宣纸加工（木刻水印、染色、装裱、册页制作）技艺。当年年底，沈维正回到百岭坑宣纸厂专门从事宣纸加工，后联合沈伯全（百岭坑宣纸厂采购员）和沈庭喜（园林牧场老四甲生产队队长）承包了百岭坑宣纸厂的加工纸车间——百岭轩，沈伯全与沈庭喜出资及负责销售，沈维正负责生产和技术管理，并担任生产副厂长。1985年，沈维正担任百岭坑宣纸厂副厂长，专门负责销售。

1987年，沈维正从百岭坑宣纸厂辞职，筹划建立“载元堂”；1988年4月，载元堂工艺厂正式成立，使用“元一”商标，“载元堂”商标到2008年才申请注册。载元堂工艺厂初创时，租用的是百岭坑宣纸厂的厂房，主要为其他宣纸厂家贴牌生产宣纸，也制作一些加工纸。这一业务格局持续了12年，直至2000年，沈维正才将厂房搬迁至现在位置，添置了生产流程设备。搬迁时，共投资了200余万元，建有4帘纸槽，包括2帘四尺槽、1帘八尺槽、1帘尺八屏槽。据沈维正的说法，由于现厂区水源不适合生产宣纸，所以形成了以书画纸生产为主的格局。

与泾县其他纸厂企业主有所不同的是，沈维正属于最早在泾县县城开店的个人。1991年，位于县城迎宾路的泾川书画院下属门市部“德厚斋”聘请沈维正出任总经理。当时的泾县县城只有3家销售文房四宝的企业：一是中国宣纸集团公司开设的“翰墨堂”；二是泾县宣笔厂开设的“泾县宣笔厂”门市部；三就是“德厚斋”。沈维正边管理自己的企业边揽下了“德厚斋”，让其妻负责门面业务，自己当“幕后”总经理。1993年，沈维正正式承包了“德厚斋”，仍交给其妻

⊙1

⊙1 接受访谈的沈维正 Interviewing Shen Weizheng

打理。2005年，由于需要照顾孩子上学和企业规模扩大，沈维正之妻难以兼顾，便将“德厚斋”关闭。

沈维正共有两个孩子，但调查中沈维正表示下一代都对造纸不感兴趣，为了保护和传承自己的造纸技艺，同时也尊重孩子们的选择，只好组建了家庭体系外的核心技艺、技术与营销团队。按照访谈时沈维正的说法，以他为核心的这个团队专注造纸新技术的研发和传统技艺的保护，不仅关注传承非物质文化遗产的造纸传统，也关注新的技术应用，努力做到在尊重传统基础上的“出新”。

⊙2 调查组成员访谈沈维正 Researchers interviewing Shen Weizheng





## 三 泾县载元堂工艺厂的代表纸品及其用途与技术分析

## 3 Representative Paper and Its Uses and Technical Analysis of Zaiyuantang Craft Factory in Jingxian County

### （一）代表纸品及其用途

2016年4月调查时，载元堂工艺厂主要有书画纸、加工纸、皮纸3个类别。其中，书画纸以画仙纸*（画仙纸是日本的书道用纸之一）最具特色，也是该厂的主要产品；加工纸主要包括“万年红”“洒金纸”和“册页”等。从年销量看，目前画仙纸占50%左右的份额，加工纸占40%，而皮纸约占10%。

据沈维正介绍，“载元堂”画仙纸主要出口

* 根据百度的检索，画仙纸有如下几种注释：其一，画仙纸就是中国的宣纸，“仙纸”等于“宣纸”，系中文在日语中使用的汉字；其二，画仙纸指绘画用的纸，“画仙”指绘画大师；其三，本意是“宣纸”，传到日本后，日本人开始仿造，也称之为“和画仙”。因此，综合来说，画仙纸是日本人对书画用纸的一种指称，其来源于宣纸，但今天已不仅指宣纸真纸，画仙纸与“半纸”是今天日本书道的大宗用纸，而雁皮纸、麻纸等则属于高级书道用纸。

至日本，与青檀皮配沙田稻草的宣纸不同，画仙纸的主要原料为龙须草、木浆、竹浆及其皮料，因其价格相对较低，故而适合大众性的书画需求，日本市场需求明显大于国内，“载元堂”画仙纸出口渠道较为稳定。画仙纸的规格为适应日本市场需求，除常规四尺、六尺、尺八屏外，还有60 cm × 180 cm、70 cm × 180 cm、70 cm × 205 cm、90 cm × 180 cm、90 cm × 210 cm、90 cm × 240 cm、120 cm × 240 cm、140 cm × 170 cm和140 cm × 205 cm等我国市场不多的规格；品种有薄口（日本的称呼，比一般的单宣要薄一点）、夹宣、二层和罗纹，主要用于山水画、写意画和书法。

## (二)
## 代表纸品的性能分析

测试小组对“载元堂”画仙纸所做的性能分析，主要包括厚度、定量、紧度、抗张力、抗张强度、撕裂度、湿强度、白度、耐老化度下降、尘埃度、吸水性、伸缩性、纤维长度和纤维宽度等。按相应要求，每一指标都重复测量若干次后求平均值，其中厚度抽取10个样本进行测试，定量抽取5个样本进行测试，抗张力（强度）抽取20个样本进行测试，撕裂度抽取10个样本进行测试，湿强度抽取20个样本进行测试，白度抽取10个样本进行测试，耐老化度下降抽取10个样本进行测试，尘埃度抽取4个样本进行测试，吸水性抽取10个样本进行测试，伸缩性抽取4个样本进行测试，纤维长度测试了200根纤维，纤维宽度测试了300根纤维。对“载元堂”画仙纸进行测试分析所得到的相关性能参数见表3.1。表中列出了各参数的最大值、最小值及测量若干次所得到的平均值或者计算结果。

由表3.1中的数据可知，“载元堂”画仙纸最厚约是最薄的1.11

⊙1

表3.1 “载元堂”画仙纸的相关性能参数
Table 3.1 Performance parameters of “Zaiyuantang” Huaxian paper

| 指标 | | 单位 | 最大值 | 最小值 | 平均值 | 结果 |
|---|---|---|---|---|---|---|
| 厚度 | | mm | 0.080 | 0.072 | 0.078 | 0.078 |
| 定量 | | g/m² | — | — | — | 30.1 |
| 紧度 | | g/cm³ | — | — | — | 0.386 |
| 抗张力 | 纵向 | N | 10.5 | 9.0 | 9.6 | 9.6 |
| | 横向 | N | 7.9 | 6.6 | 7.3 | 7.3 |
| 抗张强度 | | kN/m | — | — | — | 0.563 |
| 撕裂度 | 纵向 | mN | 320 | 250 | 282 | 282 |
| | 横向 | mN | 320 | 270 | 300 | 300 |
| 撕裂指数 | | mN · m²/g | — | — | — | 9.0 |
| 湿强度 | 纵向 | mN | 1 000 | 700 | 870 | 870 |
| | 横向 | mN | 500 | 350 | 410 | 410 |

⊙1 『载元堂』画仙纸
"Zaiyuantang" Huaxian paper

续表

| 指标 | | | 单位 | 最大值 | 最小值 | 平均值 | 结果 |
|---|---|---|---|---|---|---|---|
| 白度 | | | % | 76.4 | 75.9 | 76.3 | 76.3 |
| 耐老化度下降 | | | % | — | — | — | 2.7 |
| 尘埃度 | 黑点 | | 个/m$^2$ | — | — | — | 12 |
| | 黄茎 | | 个/m$^2$ | — | — | — | 4 |
| | 双浆团 | | 个/m$^2$ | — | — | — | 0 |
| 吸水性 | | | mm | — | — | — | 16 |
| 伸缩性 | 浸湿 | | % | — | — | — | 0.43 |
| | 风干 | | % | — | — | — | 0.78 |
| 纤维 | 皮 | 长度 | mm | 4.62 | 1.14 | 2.51 | 2.51 |
| | | 宽度 | μm | 18.0 | 5.0 | 10.0 | 10.0 |
| | 草 | 长度 | mm | 2.19 | 0.59 | 1.12 | 1.12 |
| | | 宽度 | μm | 20.0 | 3.0 | 7.0 | 7.0 |

倍，经计算，其相对标准偏差为0.003 01，纸张厚薄较为一致。所测“载元堂”画仙纸的平均定量为30.1 g/m$^2$。通过计算可知，“载元堂”画仙纸紧度为0.386 g/cm$^3$，抗张强度为0.563 kN/m。所测“载元堂”画仙纸撕裂指数为9.0 mN·m$^2$/g，撕裂度较大；湿强度纵横平均值为640 mN，湿强度较大。

所测“载元堂”画仙纸平均白度为76.3%，白度较高。白度最大值是最小值的1.007倍，相对标准偏差为0.150 55，白度差异相对较小。经过耐老化测试后，耐老化度下降2.7%。

所测“载元堂”画仙纸尘埃度指标中黑点为12个/m$^2$，黄茎为4个/m$^2$，双浆团为0个/m$^2$。吸水性纵横平均值为16 mm，纵横差为2.6 mm。伸缩性指标中浸湿后伸缩差为0.43%，风干后伸缩差为0.78%，说明“载元堂”画仙纸伸缩性差异不大。

“载元堂”画仙纸在10倍、20倍物镜下观测的纤维形态分别见图★1、图★2。所测“载元堂”画仙纸皮纤维长度：最长4.62 mm，最短1.14 mm，平均长度为2.51 mm；纤维宽度：最宽18.0 μm，最窄5.0 μm，平均宽度为10.0 μm。草纤维长度：最长2.19 mm，最短0.59 mm，平均长度为1.12 mm；纤维宽度：最宽20.0 μm，最窄3.0 μm，平均宽度为7.0 μm。“载元堂”画仙纸润墨效果见图⊙2。

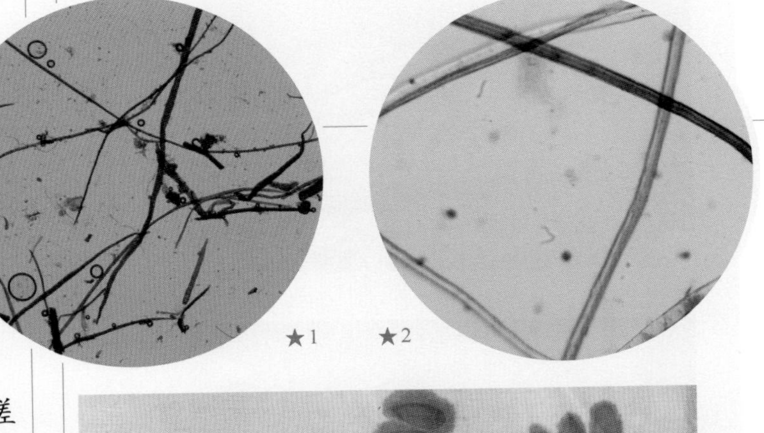

★1 “载元堂”画仙纸纤维形态图（10×） Fibers of "Zaiyuantang" Huaxian paper (10× objective)

★2 “载元堂”画仙纸纤维形态图（20×） Fibers of "Zaiyuantang" Huaxian paper (20× objective)

⊙2 “载元堂”画仙纸润墨效果 Writing performance of "Zaiyuantang" Huaxian paper

## 四
## "载元堂"书画纸生产的原料、工艺流程与工具设备

## 4
## Raw Materials, Papermaking Techniques and Tools of "Zaiyuantang" Calligraphy and Painting Paper

沈维正在访谈中向调查组成员介绍："载元堂"书画纸通过喷浆和手工捞制两种方式生产，其中喷浆主要生产画仙纸，以出口日本为主，而手工捞制的书画纸主要用作本厂加工纸所使用的原纸。在载元堂工艺厂未生产原纸之前，加工所需的纸都从泾县本地纸厂购买，自从开始生产原纸后，可自产自用，既节约了成本，也保证了原纸质量。

### (一)
### "载元堂"画仙纸生产的原料

#### 1. 主料

画仙纸的主要原料是龙须草、木浆、竹浆和皮料混合浆，其中皮料混合浆系将构皮、檀皮、楮皮、雁皮和三桠皮等根据不同客户要求混合。由于泾县不是构皮、楮皮、雁皮和三桠皮等皮料的丰产地，而且在这些皮料加工过程中还会产生大量的污水，对水源和环境都有所损害，因此考虑到皮料的加工成本和环保成本，沈维正表示：工厂自身没有购置皮料加工设备，几乎都是从外地直接购买经过蒸煮、漂白后的皮料，这些皮料到厂后经过清洗和挑拣即可打浆，不需要进行任何污水处理。沈维正介绍，构皮主要从广西或泰国购买，雁皮购自菲律宾，三桠皮来自我国的广西。

檀皮皮料是从泾县本地的原料生产商那里直接购买已漂白后的白皮。龙须草、木浆和竹浆均是直接购买的浆板，其中龙须草浆板主要来自河南，木浆浆板从加拿大进口，竹浆浆板来自四川。龙须草、木浆和竹浆浆板经过浸泡后便可以直接打浆了。

⊙1

⊙2

⊙1/2
购买来的皮料半成品
Semi-finished bark materials bought from elsewhere

2. 辅料

(1) 纸药。沈维正介绍，“载元堂”画仙纸使用的是化学纸药，是从日本进口的产品，但纸药的生产地是韩国，在日本贴牌销售。

(2) 水源。“载元堂”画仙纸使用的水是地下水。调查组实测其pH约为6.6，呈弱酸性。

⊙3

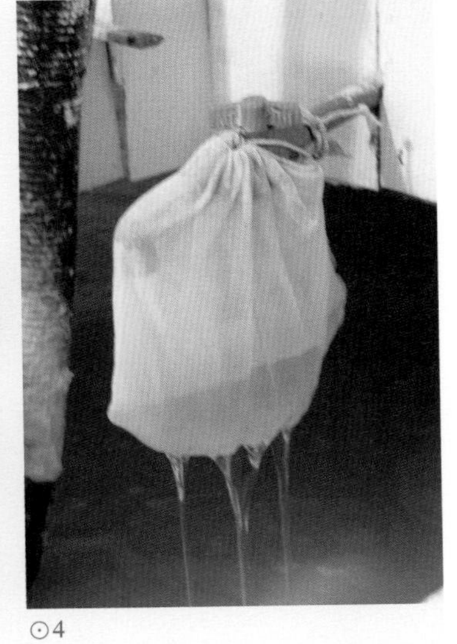

⊙4



## (二) “载元堂”画仙纸生产的工艺流程

调查组成员于2015年7月24日对“载元堂”画仙纸生产的工艺进行了第一次实地调查和访谈，2016年4月25日又进行了第二次补充调查，总结出画仙纸的主要生产工艺流程如下：

⊙5

| 壹 | 贰 | 叁 | 肆 | 伍 | 陆 | 柒 | 捌 | 玖 |
| --- | --- | --- | --- | --- | --- | --- | --- | --- |
| 浸泡、清洗 | 挑选 | 打浆 | 配浆 | 捞纸 | 压榨 | 晒纸 | 检验、剪纸 | 包装 |

### 壹 浸泡、清洗

1

⊙6

分别浸泡龙须草、木浆和竹浆浆板，一般浸泡24小时左右，目的是使浆板变软、充分吸收水分便于打浆；分别浸泡和清洗漂白后的檀皮、构皮、楮皮、雁皮和三桠皮皮料。

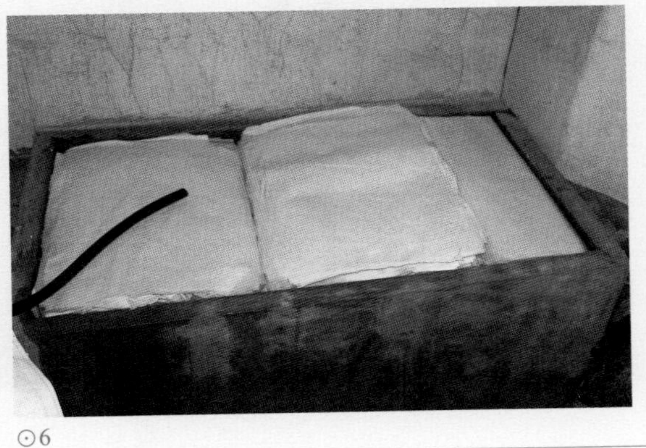

⊙6

### 贰 挑选

2

在浸泡龙须草、木浆和竹浆浆板的过程中，用钳子去除黑点和杂物。清洗后的皮料要进行人工挑选，选出未漂白的皮料、树枝和其他杂物，以保证皮料纯净。

⊙3 购买的龙须草、木浆和竹浆浆板 Eulaliopsis binata, wood pulp, and bamboo pulp board bought from elsewhere

⊙4 进口分张剂 Imported mucilage for separating the paper layers

⊙5 调查组成员在车间探询工艺 A researcher inquiring papermaking procedures

⊙6 浸泡浆板 Soaking the pulp board

叁

## 打 浆

3 ⊙7

对浸泡后的浆板和皮料分别直接进行打浆。

肆

## 配 浆

4

将龙须草、木浆、竹浆浆料与皮浆按比例混合后再进行打浆，其配比比例根据画仙纸的品种和客户的需要决定。

⊙7



伍

## 捞 纸

5 ⊙8⊙9

喷浆书画纸纸槽一边装有一个封闭的管道循环装置，该装置在每帘槽边均设有喷浆口，控制开关在捞纸工的脚边。每个喷浆口只需一个捞纸工操作。操作时，捞纸工推动装有滑轨的帘床到喷浆口，用脚打开开关，运动中的浆料从喷浆口喷出，浆料布满整张帘子后，操作工松开开关，喷浆停止。在这一过程中，捞纸工需要掌握好纸的厚薄，随后将帘床拉回槽沿，将吸附有湿纸的帘子揭走，放入身后的纸板上。放完纸后，将帘子揭起放回帘床，开始下一张纸的操作，如此往复。这种喷浆帘床下面带有滑轨，捞纸工可以借助这个滑轨前后滑动帘床，较为省力。一位捞纸工平均每天能捞6～8刀喷浆画仙纸。

⊙8

⊙9

⊙7 打浆 Beating the pulp

⊙8 捞纸 Papermaking

⊙9 喷浆捞纸车间 Workshop for pulp shooting and papermaking

## 陆
## 压榨

6 ⊙10⊙11

每天捞完纸后，将捞好的湿纸放在一边自然沥水。等水沥到一定程度后，就将纸板整体整块抬走，按4~5个槽集中进行扳榨。将最上层的纸帖盖上纸板，顶上千斤顶进行压榨。扳榨时间有长有短，天热则短，天冷则稍长，完全看操作工的掌握程度。纸帖榨干后送入晒纸车间。

⊙10

⊙11

## 柒
## 晒纸

7 ⊙12~⊙15

与其他书画纸不同的是，画仙纸压榨后的纸帖不需要放到焙房进行烘干，而是直接进入晒纸的流程。画仙纸晒纸流程大体分为以下五步：

(1) 将压榨好的纸帖放置一晚，第二天晒纸前在纸帖的四周浇水，俗称“浇帖”。浇帖的目的是使干燥的纸帖边缘吸附水分，易于晒纸时分张。

(2) 用切刀对浇帖后的纸帖进行人工切边，俗称“杀额”，目的是使额头整齐，便于揭下。

(3) 将杀额后的纸帖抬到焙房的晒纸架上，由晒纸工用额枪（也叫“擀棍”）轻轻划过纸边来松纸，俗称“做边”。

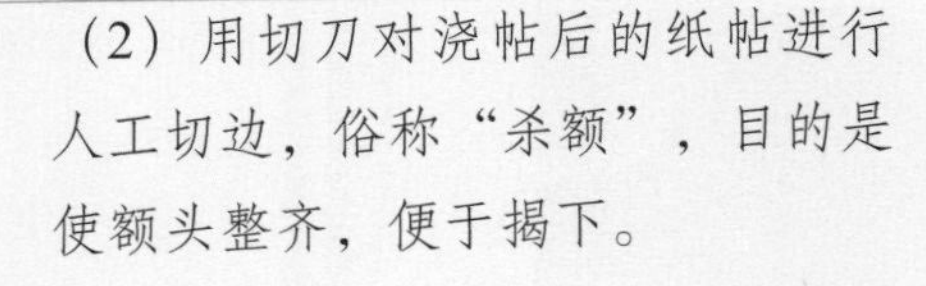

⊙13

⊙14

⊙15

⊙12

(4) 用手指尖将纸帖上方的纸边扭松，并卷起纸角，纸贴边，尤其上面一条边最重要，俗称“做额”。

(5) 将纸帖上的纸一张一张地揭离下来，又一张一张地依次刷到烘焙上。将纸贴上烘焙时，需要用刷子来回刷几下将纸贴平整。由于烘焙温度高，几十秒后，刷上去的湿纸就因水分蒸发而迅速干透，晒纸工即可从铁焙上揭取下干纸。

⊙10/11 工人在压榨纸帖 A worker pressing the paper pile
⊙12 浇帖 Watering the paper pile
⊙13 杀额 Trimming the deckle edges
⊙14 做边 Processing the edge of paper pile
⊙15 晒纸 Drying the paper

## 捌 检验、剪纸

8 ⊙16～⊙18

将晒好的纸张运到剪纸车间进行裁剪。剪纸前需要对纸张进行检查，如发现纸张有破损或其他瑕疵则弃之不用。将检验好的纸张放在纸台上，根据重量不同分别归类。裁剪时纸张要用压纸石压紧，用特制大剪刀进行裁边，要求纸张四边裁剪整齐。一般情况下，四尺的画仙纸不进行裁边，而是检验后根据客户需要用切纸机切齐边或裁成半切，每次能切9～10刀；四尺以上规格的画仙纸需要进行人工裁剪，每次裁剪50张。

⊙16

⊙17

⊙18

## 玖 包装

9 ⊙19

裁剪之后，四尺的画仙纸每100张分成1刀，四尺以上规格的每50张1刀，用牛皮纸包好后，按照品种和重量的不同，分别装箱成为待售成品。沈维正介绍，画仙纸一般不用透明塑料袋包装，使用的牛皮纸包袋是从泾县当地的商店购买的。

⊙19

⊙16 检验纸张 Checking the paper

⊙17 工人操作切纸机 Workers operating the paper cutting machine

⊙18 剪纸工在登记槽单位基本数据 A worker registering basic data of papermaking trough size

⊙19 待售的画仙纸产品 Huaxian paper products for selling

## (三)

## “载元堂”画仙纸生产的工具设备

### 壹 打浆机 1

打浆机是用来打碎龙须草、竹浆、木浆浆板和各种皮料的设备。

⊙20

⊙21

⊙22

### 贰 纸槽 2

纸槽是用来盛浆料的设施，用水泥浇筑而成。“载元堂”有8个半自动喷浆槽（其中5个四尺、1个尺八屏和2个六尺）、2个手工一改二纸槽和1个吊帘槽。

⊙23

⊙20 打浆机 Beating machine

⊙21 喷浆纸槽 Pulp shooting trough

⊙22 手工纸槽 Handmade papermaking trough

⊙23 吊帘纸槽 Trough with movable papermaking screen

## 叁 纸 帘 3

纸帘是抄纸的重要工具，由竹子编织而成，表面光滑平整，帘纹细密。据沈维正介绍，纸帘都是从制造商那里直接购买的，泾县丁家桥镇后山村和云岭镇靠山村都是生产质地精良纸帘的地方。实测时，根据产品规格的不同，纸帘的大小也不同。其中，喷浆四尺画仙纸纸帘长152 cm，宽87 cm；六尺纸帘长200 cm，宽120 cm；尺八屏纸帘长250 cm，宽73 cm。

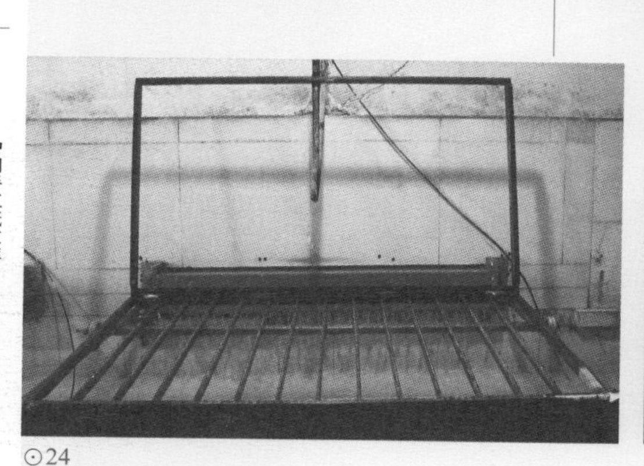

⊙24

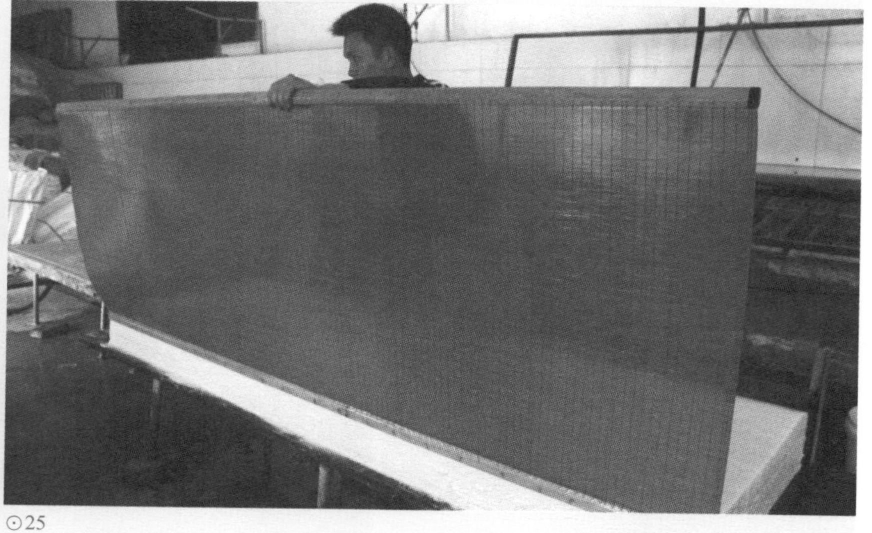

⊙25

## 肆 定位桩 4

为了使刚捞出的湿纸能堆放整齐，每个槽的纸板上贴近纸边的地方都固定两根木桩作为定位桩。在放帘前，只需把纸帘的额头靠着定位桩轻轻放下，纸帘呈圆筒形倒扣在纸板上，湿纸便能整齐地堆放在木板上了。

⊙26

## 伍 帖 架 5

木制，主要用来浇帖和杀额。晒纸前需要将纸帖放在帖架上进行人工浇帖和切边。

⊙27

⊙24 纸帘 Papermaking screen

⊙25 浆帘床 Screen frame for pulp shooting

⊙26 定位桩 Fixed timber pile

⊙27 帖架 Frame for supporting the paper pile

陆
## 烘焙
6

烘焙的外观像一堵独立的“墙”，上方和一边留有空隙，由钢板制成，中间贮水，将水加热升温后晒纸。载元堂工艺厂通过环保锅炉燃烧生物颗粒来加热水。烘焙表面温度保持在60～80 ℃，一张湿纸贴上去几十秒即可烘干揭下。

⊙28

柒
## 棒槌
7

棒槌又叫木槌，长约40 cm，晒纸前松纸时用来槌打纸帖，使纸张变松而易于分张。

捌
## 额枪
8

长约20 cm，晒纸前做边时用来松纸，便于晒纸时分张。

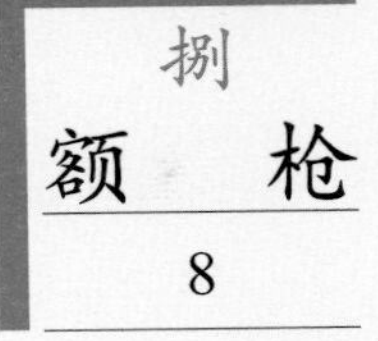
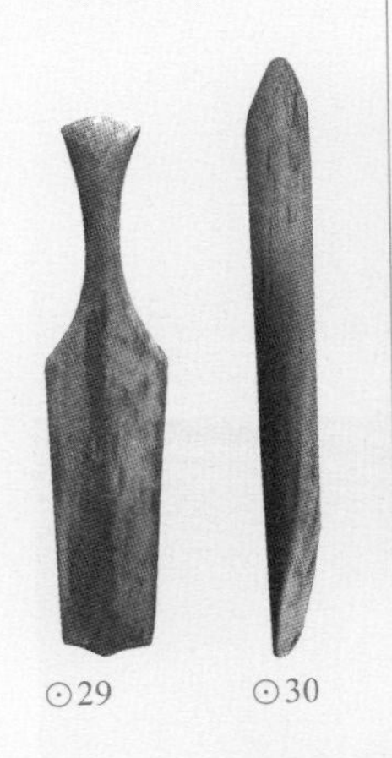
⊙29　⊙30

玖
## 刷子
9

长约50 cm，宽约14 cm，刷毛由松针制成。晒纸时，用刷子将湿纸刷上烘焙。

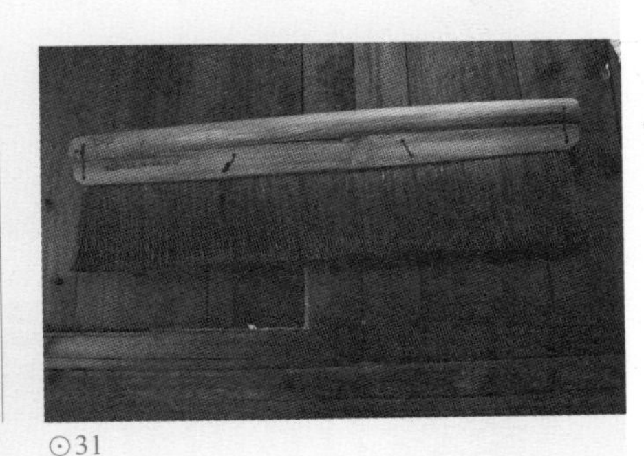
⊙31

拾
## 剪刀
10

长约33 cm，用于裁纸，将纸边沿着直线依次裁下。剪刀来源同纸帘一样，也是从本县丁家桥镇后山村直接购买的。

⊙32

拾壹
## 压纸石
11

压纸石虽只是普通的石头，但在造纸中有两次被用到。将晒好的纸张整齐地堆放在纸板上，用压纸石来固定纸张；剪纸过程中，用压纸石压紧纸张以便于剪纸工裁剪纸边。

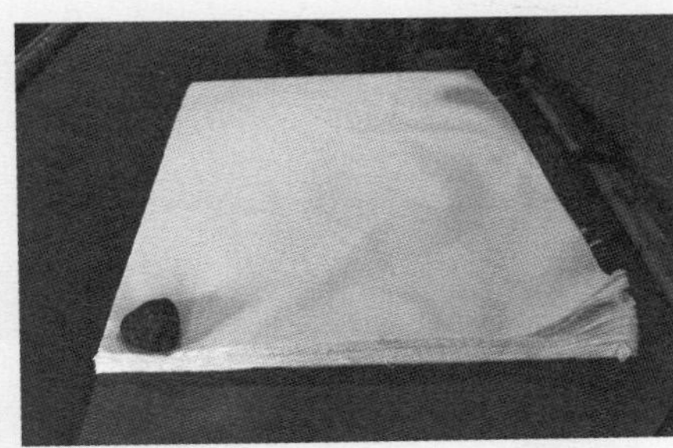
⊙33

⊙34

⊙28 工人在清洗烘焙表面 Workers cleaning the surface of the drying wall
⊙29 棒槌 Wooden hammer
⊙30 额枪 Stick for separating the paper layers
⊙31 刷子 Brush
⊙32 大剪刀 Shears
⊙33 晒纸过程中的压纸石 Stone for pressing the paper in the procedure of drying the paper
⊙34 剪纸过程中的压纸石 Stone for pressing the paper in the procedure of cutting the paper

# 五 泾县载元堂工艺厂的市场经营状况

# 5 Marketing Status of Zaiyuantang Xuan Paper Factory in Jingxian County

安徽省泾县载元堂工艺厂产品价目表

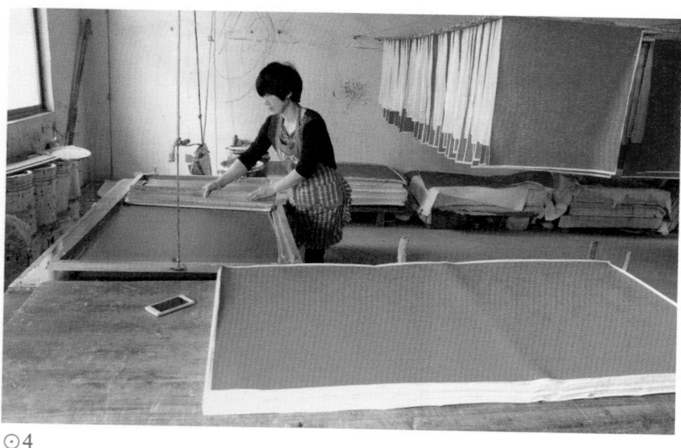

⊙1 纸库局部 A part of paper warehouse

⊙2 载元堂工艺厂2015年产品价目表 Product price list of Zaiyuantang Craft Factory in 2015

⊙3 制作中的『万年红』春联纸 Making "Wannianhong", red paper for writing spring couplets

⊙4 制作中的洒金『万年红』 Making "Wannianhong", red paper with golden dots

调查中沈维正介绍：从2014年开始，到2016年4月底入厂调查时止，国内书画纸市场一直处于降温状态，而载元堂工艺厂的国内销售额一直维持在200万～300万元，能稳住的主要积极措施是由合伙人在广州、杭州、上海等中心消费城市开设销售点，由合伙人代管，实行销售和生产分离的模式。

出口外销渠道是“载元堂”纸产品的重点，占年营业额的60%以上，年出口额500万元左右。由于从事出口贸易较早，“载元堂”品牌在日本、韩国等境外市场信誉度较高，通常是客户直接将订单发至公司邮箱，也有部分客户和市场通过贸易公司联系。载元堂工艺厂不直接参与报关等手续，分别委托上海、合肥等地不同的贸易公司代理出口。截至调查时的2016年4月，载元堂工艺厂产品主要出口到日本，约占境外销售总额的2/3，韩国和中国台湾约占境外销售的1/3。通过与沈维正及具体管理人的详细交流获知具体销售品种和价格信息如下：

### （1）画仙纸

日本对画仙纸的需求量较大，载元堂根据客户需求生产多规格的画仙纸，并通过编号如zy-1和zy-2等（zy是“载元”二字的汉语拼音缩写）进行管理，目前在销的共有25种。在发货中，对相同货号的画仙纸则根据纸张厚度的不同分成不同的品种，常见的如薄口、夹宣、二层、罗纹等，售价为100～400元/刀（四尺）、200～800元/刀（六尺）；货号zy-1（二层）为画仙纸产品中的代表，由于其单张纸厚度是其他纸种的2倍，故售价也是同规格中最高的，四尺、六尺和尺八屏2015年的售价分别为386元/刀、772元/刀和695元/刀。

### （2）加工纸

“万年红”是“载元堂”加工纸产品中的代表纸品，主要销往台湾地区。“万年红”是春联

⊙5

⊙6

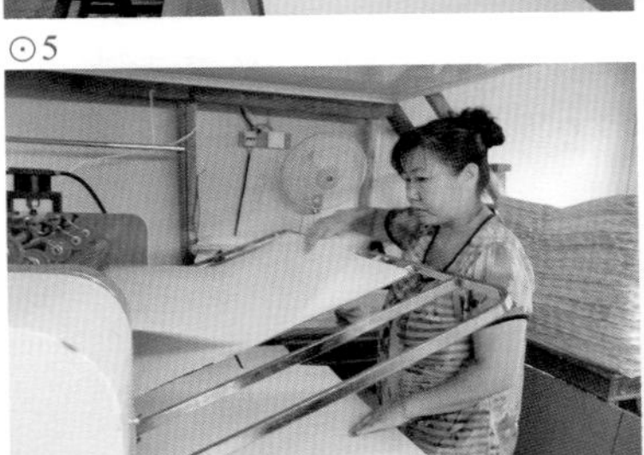
⊙7

⊙8

⊙5/8 不同工种忙碌的工人 Busy workers in different steps of papermaking

所需的纸材，有较为严格的发货时间，一般在第一年的春节之前发第二年要用的春节用纸，收货方会请专门的书法老师书写对联内容，成本则是20元以上一副（一对纸）。

载元堂工艺厂有25种不同规格的册页，价格在6～150元/本区间，其中62 cm×42 cm为其规格最大的册页，价格为122元/本；8 cm×5 cm的袖珍版册页价格为6元/本，是规格最小、价格最低的册页产品。

### （3）皮纸

皮纸主要有雁皮纸、连史纸、半纸、半色皮纸、黑壳皮纸和云龙纸等品种。在皮纸销售中，雁皮纸价格最高，2015年调查时，70 cm×140 cm规格的为310元/刀，相同规格的连史纸、本色皮纸、黑壳皮纸和本色云龙纸价格分别为180元/刀、250元/刀、190元/刀和200元/刀。

## 六 泾县载元堂工艺厂的品牌文化与民俗故事

## 6 Brand Culture and Stories of Zaiyuantang Craft Factory in Jingxian County

### （一）“载元堂”名称的由来

关于“载元堂”名称的由来，调查时沈维正回顾了当初的一段轶事：1989年，纸厂创立之初，沈维正不知该取何名，便请到著名画家、时任上海美术馆馆长的方增先为公司取名并题词。方增先了解了纸厂的情况后，挥墨写下“安徽省泾县载元堂”，并解释道：“‘载’为承载之意，‘元’近音‘原’，初意取最原始、最初的意思，是考虑到你们在泾县属于最早专业从事宣纸加工的纸厂。”同时，方增先又表示：“希望纸厂未来越做越好，成为同类企业中规模最大的公司，而‘元’含有最大的意思。”方增先的题字已被装裱好挂在沈维正的办公室墙壁上。

### （二）日韩客户"厉害"的下单方式

调查中，沈维正感叹不已地说："日韩客户对纸的要求非常细致。"由于纸浆配比不同，纸的润墨效果和层次感也不尽相同，日韩客户通常通过寄来几张绘写过的纸样品表达自己对产品的诉求。"看到每次寄来的纸样，首先需研究如何配比纸浆，做出试样后需找人试验，检查是否与客户寄来纸样上的润墨效果相同，如不同还需重新制造。"不仅如此，产品包装也是日韩客户注意的环节，"他们会对纸箱材质、大小、外包裹纸层数和厚度都提出详细要求"。由此，在载元堂形成了一个"尺寸精准、重量相符、整齐划一、材质优良"的原则。

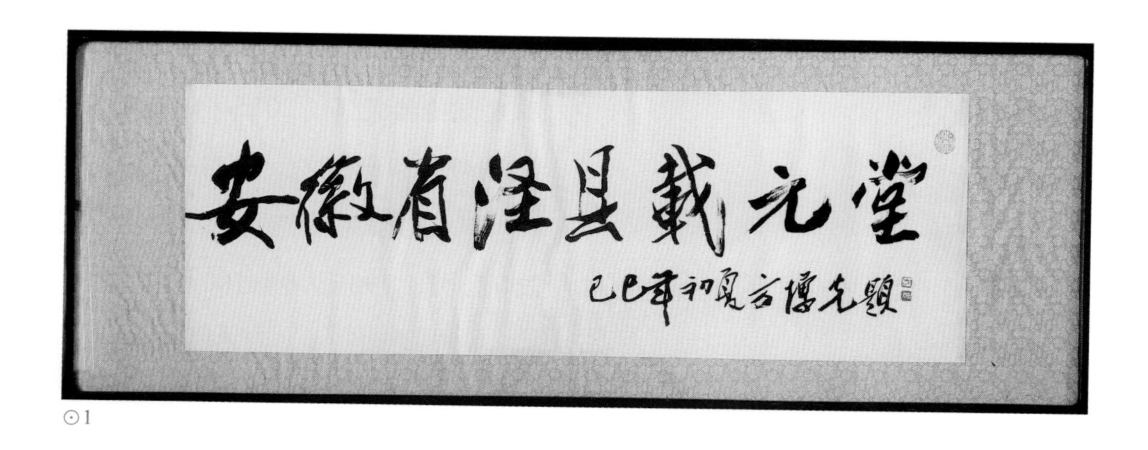

⊙1

## 七 泾县载元堂工艺厂的传承现状与发展思考

## 7 Current Status of Business Inheritance and Thoughts on Development of Zaiyuantang Craft Factory in Jingxian County

"载元堂"虽然产业规模不大，但是60%以上的出口销量是泾县同行业中较为典型的代表。在泾县，较有规模的纸厂多以国内销售为主，而"载元堂"则看中了国外的市场潜力，成为一家以出口外销为主的企业。面对2015年以来国内书画用纸市场下行及竞争激烈的态势，沈维正认为：着眼于国际市场对企业发展还是有利的。随着中国国际地位的提升，越来越多的国家希望了解中国文化。中国画和书法作为中国传统文化的代表形式，成为国外了解中国文化的窗口，而宣纸和书画纸作为其必备的工具也必然在海外国家和地区日益拓展开来，就像日本和韩国那样。

但是出口外销也存在较大的风险。沈维正的经营体会是：韩国近几年经济下滑，相应的纸产品订购量也呈下降趋势，但"载元堂"的多国/

⊙1 方增先『安徽省泾县载元堂』题字 Autograph of "Zaiyuantang in Jingxian County of Anhui Province" by Fang Zengxian

⊙2

⊙3

地区外销模式，尽管受到韩国市场影响，但总体形势还算乐观，由此提醒造纸企业走外向型经济"不能完全吊死在一棵树上"。

据沈维正介绍，关于书画纸产品的选择，对传统性和价格两种要素，大部分国内消费者倾向于价格，而国外消费者则更看重商品的传统性及所代表的品质感。像日本客户就看重商品的传统性，例如要求洒金纸必须用手工洒金。对于注重传统性的企业来说，国外客户的偏好有利于其发展，因此在面对发展出现疲态的泾县手工造纸行业，注重出口的载元堂工艺厂依然保持稳定销量。

访谈中沈维正特别强调独特性问题。沈维正表示：出口外销的纸品一定不能与别家重复，因为国外市场特别重视独特性。独特性是一个企业专有技术和特例产品的表现。对于各生产环节都对技术有要求的手工造纸，如何在某一环节或者某些环节进行创新是手工纸厂当前发展必须思考的关键问题。独特性不仅关系生存，也是纸厂品牌脱颖而出的保证。据沈维正自己的说法，目前泾县出口的洒金纸基本上产自"载元堂"，尤其是出口日本的。

⊙4

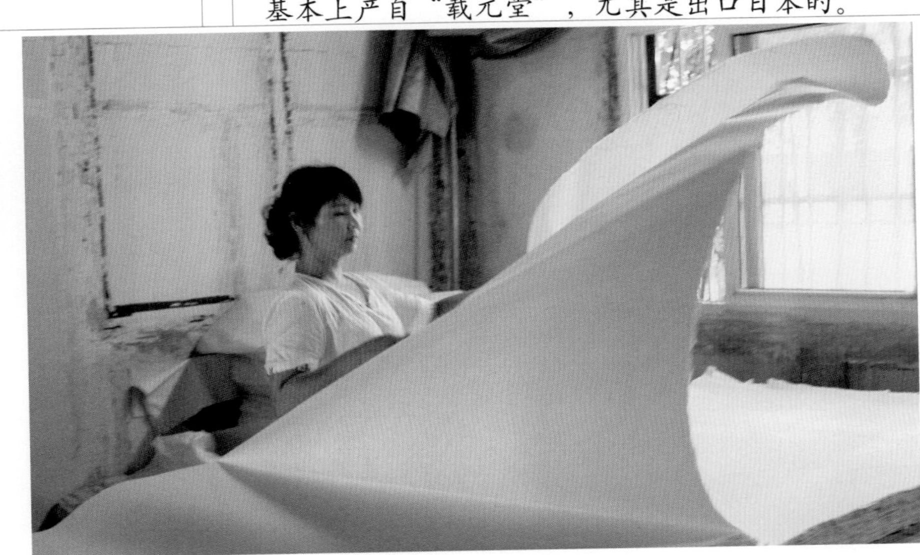

⊙5

⊙2
沈维正向调查组成员介绍纸样
Shen Weizheng introducing the paper sample to a researcher

⊙3
沈维正向调查组成员展示产品包装
Shen Weizheng showing the product packaging to the researchers

⊙4
载元堂工艺厂外景
External view of Zaiyuantang Craft Factory

⊙5
技艺娴熟的剪纸女工
A skilled female worker cutting the paper

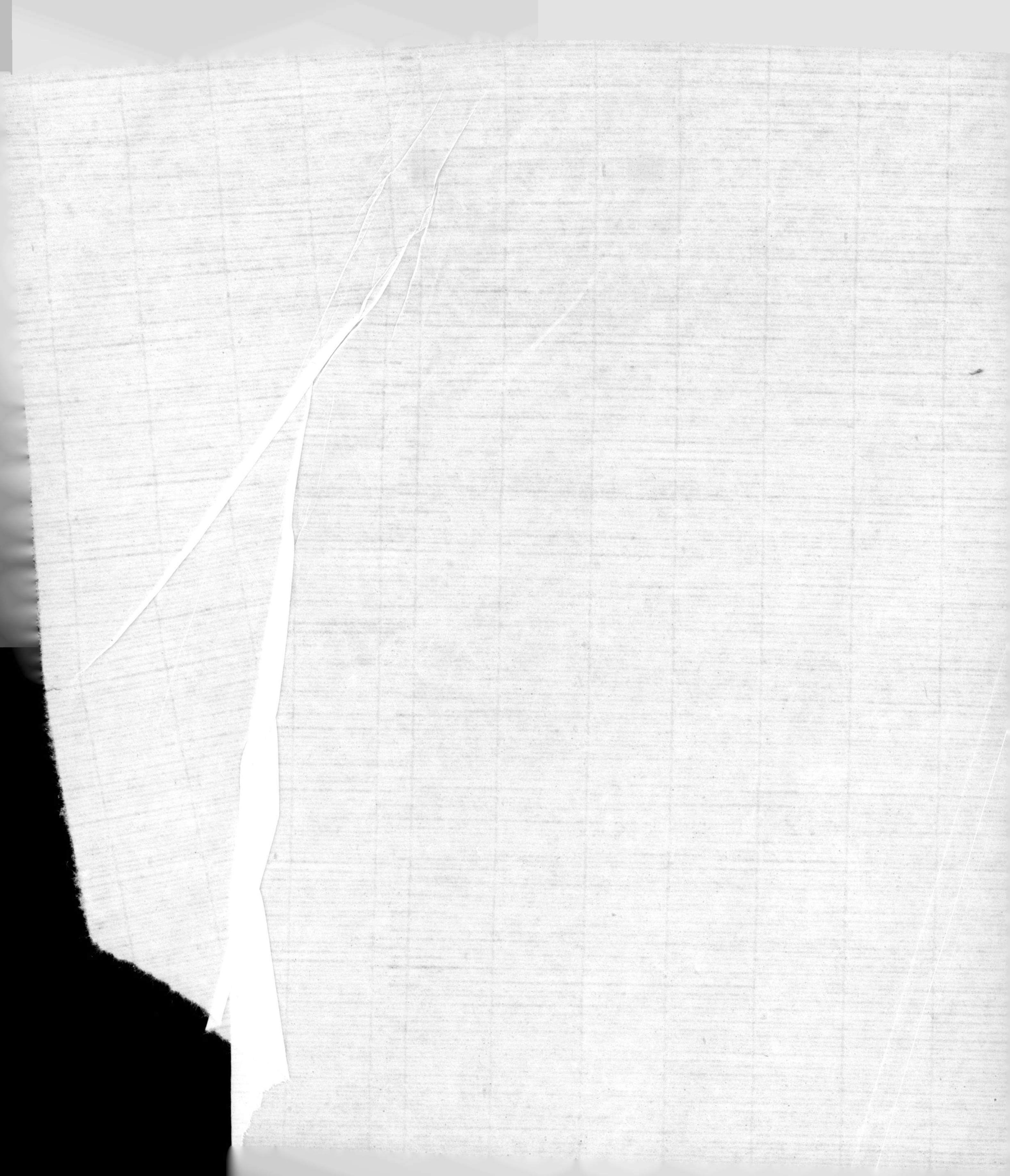

# 泾县载元堂工艺厂 画仙纸

Huaxian Paper of Zaiyuantang Craft Factory in Jingxian County

『载元堂』画仙纸透光摄影图

A photo of "Zaiyuantang" Huaxian paper seen through the light

# 第二节

# 泾县小岭强坑宣纸厂



安徽省
Anhui Province

宣城市
Xuancheng City

泾县
Jingxian County

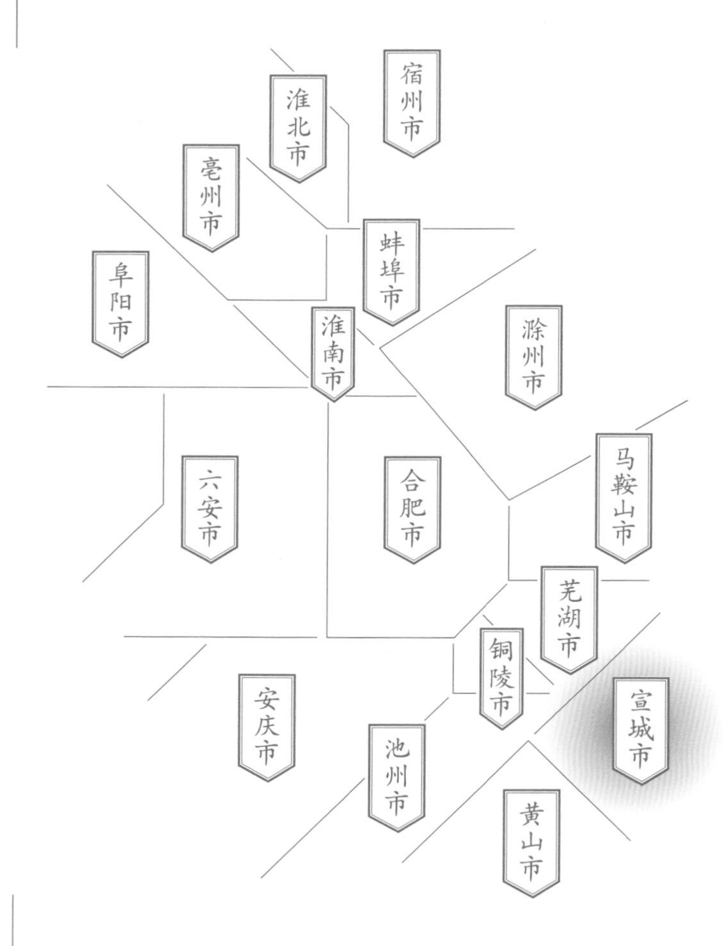


**调查对象**

丁家桥镇
泾县小岭强坑宣纸厂
书画纸

Section 2

## Xiaoling Qiangkeng Xuan Paper Factory in Jingxian County

**Subject**

Calligraphy and Painting Paper
of Xiaoling Qiangkeng Xuan Paper Factory
in Jingxian County
in Dingjiaqiao Town

## 一
## 泾县小岭强坑宣纸厂的基础信息与生产环境

### 1
### Basic Information and Production Environment of Xiaoling Qiangkeng Xuan Paper Factory in Jingxian County

泾县小岭强坑宣纸厂创办于1996年，坐落于泾县丁家桥镇小岭行政村西山村民组，地理坐标为：东经118° 18′ 52″，北纬30° 41′ 15″。通过登录查阅中国全国企业信用信息公示系统知：泾县小岭强坑宣纸厂于2012年登记，经营性质是个体工商户，厂长曹友泉，以“曹友泉”书画纸制造、加工与销售为主营业务范围。

2015年8月5日、2016年4月25日，调查组成员两次前往小岭强坑宣纸厂进行田野调查，通过现场考察与曹友泉的叙述，所获得的基础信息如下：泾县小岭强坑宣纸厂现有员工20人左右、生产槽位4帘，年产以龙须草浆板为主原料的书画纸18 000刀左右，2014年销售总量2万多刀，销售额200多万元。

西山，泾县当地人又称“上西山”，与方家山、汪义坑临近，是明清时期宣纸的重要原产地，也是宣纸原料制作的主产地之一。2015年夏天的调查数据显示，西山村民组共有200多口人，70多户人家，其中有3户从事造纸。20世纪70年代著名的集体所有制泾县小岭宣纸厂曾在此创办西山车间。20世纪80年代直至21世纪初，西山村为泾县小岭宣纸厂总部所在地，其宣纸生产规模一度较大。泾县小岭宣纸厂总部转让给私有资本后，购买小岭宣纸厂总部的朱志勇在此基础上创办了曹鸿记纸业有限公司，成为第二批宣纸地理标志保护企业。在小岭宣纸厂的西山车间原址附近，也诞生了小岭强坑宣纸厂等书画纸、宣纸生产厂家。

⊙1

⊙2

⊙3

⊙1/2 通往西山村的乡村公路 Country road to Xishan Village

⊙3 强坑宣纸厂内景之一 Internal view of Qiangkeng Xuan Paper Factory

路线图
泾县县城
↓
泾县小岭强坑宣纸厂
Road map from Jingxian County centre to Xiaoling Qiangkeng Xuan Paper Factory of Jingxian County

# 泾县小岭强坑宣纸厂位置示意图

# Location map of Xiaoling Qiangkeng Xuan Paper Factory in Jingxian County

考察时间
2015年8月 / 2016年4月

Investigation Date
Aug. 2015 / Apr. 2016

地域名称

泾县小岭强坑宣纸厂
泾县县城
丁家桥镇

A 泾县
1 丁家桥镇
2 云岭镇
3 泾川镇
4 昌桥乡
5 琴溪镇
6 黄村镇

造纸点名称

泾县小岭强坑宣纸厂 造纸点

位置分布

市府、州府
县城
乡镇
村落
造纸点
历史造纸点
山
国家级自然保护区
S221 省道
G21 国道
昆河线 铁路
G56 高速公路
线路

南陵县
泾县
青阳县
S322
G205
合福高铁
10 km
5 km
0
N

## 二
## 泾县小岭强坑宣纸厂的历史与传承情况

## 2
## History and Inheritance of Xiaoling Qiangkeng Xuan Paper Factory in Jingxian County

2015年8月入厂调查时，据曹友泉的回忆，在创办强坑宣纸厂之前，他与父亲曹炳集都在小岭宣纸厂工作。曹炳集在20世纪70年代至80年代中期先在小岭原料社从事宣纸原料加工工作，后在小岭宣纸厂的西山车间从事造纸工作，1994年开始在小岭宣纸厂担任车间主任；曹友泉在小岭宣纸厂工作期间，没有直接从事与宣纸生产和加工工艺有关的工种，而是在厂里当电工。1996年，小岭宣纸厂授权给小岭村当地具备条件的小厂贴牌生产“红旗”宣纸，曹炳集父子认为这是一个好机会，便在当年即快速创办了私人企业——泾县小岭强坑宣纸厂。

小岭强坑宣纸厂一直由曹友泉担任企业法人代表。创办初期，开有2帘纸槽，有十几个生产工人。这些早期的生产工人都在小岭宣纸厂学过造纸，大多因不能转为正式工而到私营小厂工作。因曹炳集既有一线造纸和原料加工经历，又有管理宣纸生产的经验，所以厂里生产全由曹炳集负责，曹友泉则主要负责经营外联事务。从1997年开始，小岭宣纸厂的“红旗”宣纸产品的生产经营开始衰落，宣纸厂于2000年倒闭，原小岭宣纸厂里的工人又有陆续来到强坑宣纸厂工作的，到调查时的2016年有20人左右的规模。

曹友泉，1970年出生于泾县小岭，兄弟姐妹4人，系小岭曹氏造纸世家出身。曹友泉家人初工作时都在本村的小岭宣纸厂，小岭宣纸厂倒闭前后，基本已回到自家创办的厂里工作。曹友泉的侄子曹凯从2010年开始经营网店，销售“曹友泉”书

⊙1

⊙2

⊙3

⊙1
强坑宣纸厂内景之二
Another internal view of Qiangkeng Xuan Paper Factory

⊙2
曹炳集工作过的小岭宣纸厂西山车间（分厂）
Xishan Workshop (branch factory) of Xiaoling Xuan Paper Factory where Cao Bingji once worked

⊙3
侄子曹凯（左）与曹友泉（右）
Cao Youquan (right) and his nephew Cao Kai (left)

⊙1

画纸。曹友泉的独生女儿曹圆，大学毕业后回到小岭的西山村，从事“曹友泉”书画纸网络销售，创办了名叫“墨淋”的网店。由于有年轻一辈较早介入网上经营，因此小岭强坑宣纸厂是泾县当地较早从事网络销售的书画纸企业。父亲曹炳集1937年生人，已于2012年去世。曹友泉大哥曹金祥1963年生人，在厂里从事捞纸工作；二哥曹康泉1968年生人，从事造纸工作；大姐曹琴方1966年生人，从事晒纸工作；妹妹曹秋云1973年生人，在北京琉璃厂开直营店，销售“曹友泉”书画纸。

⊙2

## 三 “曹友泉”书画纸的代表纸品及其用途与技术分析

## 3 Representative Paper and Its Uses and Technical Analysis of “Cao Youquan” Calligraphy and Painting Paper

### （一）代表纸品及其用途

据2015年8月5日的现场调查得知：小岭强坑宣纸厂纸品种类较多，规格各异，覆盖了书画纸、宣纸和加工纸三大类，规格有四尺、六尺、八尺、尺八屏等，商标全部是以人名注册的“曹友泉”品牌。据曹友泉的介绍，目前强坑宣纸厂对外销售中市场需求最大的是书画纸，销量相对比较好的纸品是高级书画纸，用于书法和国画创作。2015年，四尺高级书画纸在直营店的售价大

⊙1 『曹友泉宣纸』（书画纸）淘宝网店
Taobao online store of “Cao Youquan Xuan Paper” (calligraphy and painting paper)

⊙2 记录曹友泉介绍的纸厂历史
Recording the history of the paper factory that Cao Youquan introduced

致在170元／刀，四尺普通书画纸售价在100元／刀左右，四尺宣纸的售价在400元／刀左右，与四尺相比，六尺宣纸的售价要高出近一倍。

## （二）代表纸品的性能分析

测试小组对采样自泾县小岭强坑宣纸厂生产的“曹友泉”高级书画纸所做的性能分析，主要包括厚度、定量、紧度、抗张力、抗张强度、撕裂度、湿强度、白度、耐老化度下降、尘埃度、吸水性、伸缩性、纤维长度和纤维宽度等。按相应要求，每一指标都需重复测量若干次后求平均值，其中厚度抽取10个样本进行测试，定量抽取5个样本进行测试，抗张力（强度）抽取20个样本进行测试，撕裂度抽取10个样本进行测试，湿强度抽取20个样本进行测试，白度抽取10个样本进行测试，耐老化度下降抽取10个样本进行测试，尘埃度抽取4个样本进行测试，吸水性抽取10个样本进行测试，伸缩性抽取4个样本进行测试，纤维长度测试了200根纤维，纤维宽度测试了300根纤维。对“曹友泉”高级书画纸进行测试分析所得到的相关性能参数见表3.2。表中列出了各参数的最大值、最小值及测量若干次所得到的平均值或者计算结果。

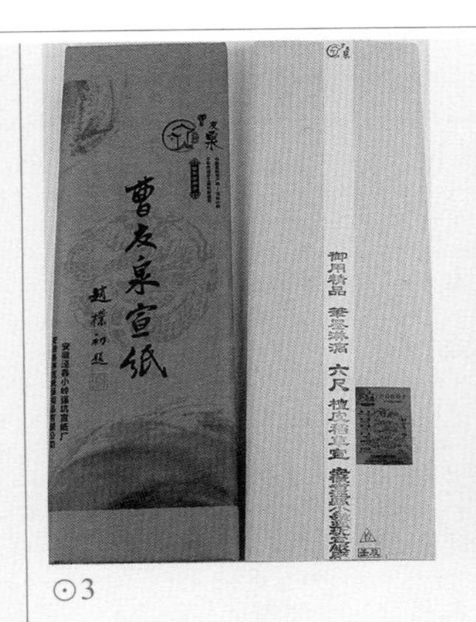

⊙3

⊙4

⊙3 檀皮稻草宣 Xuan Paper made with *Pteroceltis tatarinowii* Maxim. bark and straw

⊙4 檀皮稻草精品宣 High-quality Xuan paper made with *Pteroceltis tatarinowii* Maxim. bark and straw

表3.2 “曹友泉”高级书画纸相关性能参数
Table 3.2 Performance parameters of “Cao Youquan” advanced calligraphy and painting paper

| 指标 | | 单位 | 最大值 | 最小值 | 平均值 | 结果 |
|---|---|---|---|---|---|---|
| 厚度 | | mm | 0.091 | 0.081 | 0.088 | 0.088 |
| 定量 | | g/m$^2$ | — | — | — | 31.3 |
| 紧度 | | g/cm$^3$ | — | — | — | 0.360 |
| 抗张力 | 纵向 | N | 17.7 | 13.8 | 15.1 | 15.1 |
| | 横向 | N | 12.8 | 9.6 | 11.1 | 11.1 |
| 抗张强度 | | kN/m | — | — | — | 0.873 |
| 撕裂度 | 纵向 | mN | 290 | 250 | 261 | 261 |
| | 横向 | mN | 330 | 300 | 313 | 313 |
| 撕裂指数 | | mN·m$^2$/g | — | — | — | 9.1 |
| 湿强度 | 纵向 | mN | 500 | 300 | 410 | 410 |
| | 横向 | mN | 250 | 150 | 200 | 200 |
| 白度 | | % | 71.7 | 71.3 | 71.5 | 71.5 |
| 耐老化度下降 | | % | — | — | — | 3.6 |

续表

| 指标 | | 单位 | | 最大值 | 最小值 | 平均值 | 结果 |
|---|---|---|---|---|---|---|---|
| 尘埃度 | | 黑点 | 个/m$^2$ | — | — | — | 40 |
| | | 黄茎 | 个/m$^2$ | — | — | — | 12 |
| | | 双浆团 | 个/m$^2$ | — | — | — | 0 |
| 吸水性 | | | mm | — | — | — | 16 |
| 伸缩性 | | 浸湿 | % | — | — | — | 0.90 |
| | | 风干 | % | — | — | — | 1.23 |
| 纤维 | 皮 | 长度 | mm | 4.01 | 0.69 | 1.82 | 1.82 |
| | | 宽度 | μm | 19.0 | 1.0 | 8.0 | 8.0 |
| | 草 | 长度 | mm | 1.82 | 0.12 | 0.89 | 0.89 |
| | | 宽度 | μm | 9.0 | 1.0 | 4.0 | 4.0 |

由表3.2中的数据可知，“曹友泉”高级书画纸最厚约是最薄的1.123倍，经计算，其相对标准偏差为0.003，纸张厚薄较为一致。所测“曹友泉”高级书画纸的平均定量为31.3 g/m$^2$。通过计算可知，“曹友泉”高级书画纸紧度为0.360 g/cm$^3$。抗张强度为0.873 kN/m，抗张强度值较大。所测“曹友泉”高级书画纸撕裂指数为9.1 mN · m$^2$/g，撕裂度较大；湿强度纵横平均值为305 mN，湿强度较大。

所测“曹友泉”高级书画纸平均白度为71.5%，白度较高，白度最大值是最小值的1.006倍，相对标准偏差为0.132，白度差异相对较小。经过耐老化测试后，耐老化度下降3.6%。

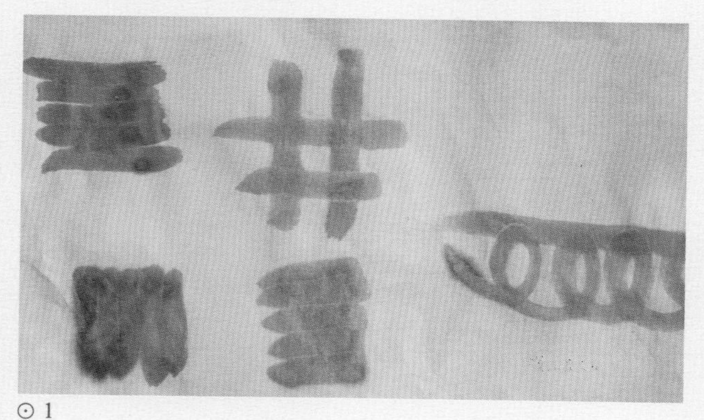

⊙1

所测“曹友泉”高级书画纸尘埃度指标中黑点为40个/m$^2$，黄茎为12个/m$^2$，双浆团为0个/m$^2$。吸水性纵横平均值为16 mm，纵横差为4.6 mm。伸缩性指标中浸湿后伸缩差为0.90%，风干后伸缩差为1.23%，说明“曹友泉”高级书画纸伸缩性差异不大。

★1　★2

“曹友泉”高级书画纸在10倍、20倍物镜下观测的纤维形态分别见图★1、图★2。所测“曹友泉”高级书画纸皮纤维长度：最长4.01 mm，最短0.69 mm，平均长度为1.82 mm；纤维宽度：最宽19.0 μm，最窄1.0 μm，平均宽度为8.0 μm。草纤维长度：最长1.82 mm，最短0.12 mm，平均长度为0.89 mm；纤维宽度：最宽9.0 μm，最窄1.0 μm，平均宽度为4.0 μm。“曹友泉”高级书画纸润墨效果见图⊙1。

性能分析

★1
『曹友泉』高级书画纸纤维形态图（10×）
Fibers of "Cao Youquan" advanced calligraphy and painting paper (10× objective)

★2
『曹友泉』高级书画纸纤维形态图（20×）
Fibers of "Cao Youquan" advanced calligraphy and painting paper (20× objective)

⊙1
『曹友泉』高级书画纸润墨效果
Writing performance of "Cao Youquan" advanced calligraphy and painting paper

## 四
## “曹友泉”高级书画纸生产的原料、工艺流程与工具设备

4
Raw Materials, Papermaking Techniques and Tools of “Cao Youquan” Advanced Calligraphy and Painting Paper

### (一)
### “曹友泉”高级书画纸生产的原料

#### 1. 主料

调查与访谈中了解到：小岭强坑宣纸厂的高级书画纸生产的主要原料为龙须草浆板、稻草浆、青檀皮浆。为提高纸张的墨线效果，还会加上适量的填充料碳酸钙粉。由于泾县当地极少产龙须草，强坑宣纸厂所需的龙须草浆主要从泾县本地的经销商处购买，而经销商则整体从湖北与河南交界的地区如十堰及陕西等地的纸浆板厂采购。由于青檀皮料加工过程中污染较为严重，考虑到环保成本和经济成本，强坑宣纸厂直接从当地青檀皮加工厂购买漂白后的青檀皮，回厂加工处理后使用；处理稻草浆的成本较高，污染较重，强坑宣纸厂也从本县经销商处购买。根据曹友泉访谈中给出的数据，强坑宣纸厂2015年采购龙须草浆板的价格达到11 000元/t，漂白后的半成品皮料收购价是20元/kg。

⊙2

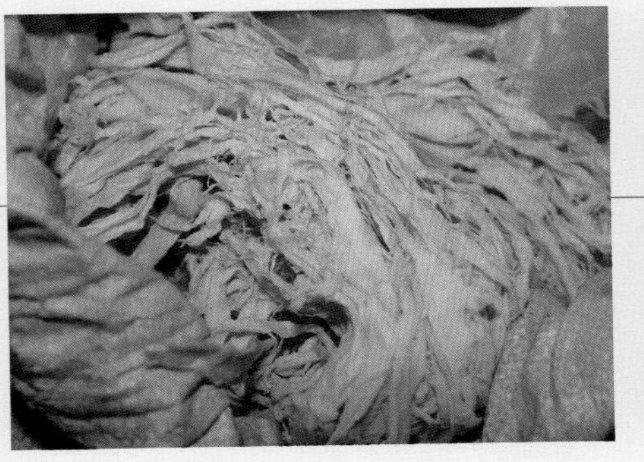
⊙3

⊙4

#### 2. 辅料

（1）纸药。强坑宣纸厂使用的纸药全部是化学分张剂聚丙烯酰胺（简称PAM），为水溶性高分子聚合物，具有良好的絮凝性，可用来替代泾县传统的杨桃藤汁液的纸药功能，起到使纸浆纤维在水中悬浮及纤维均匀的作用，均从泾县当地的供应商处采购。

⊙2 混合后的浆料 Mixed pulp materials

⊙3/4 漂白后的檀皮原料 Bleached *Pteroceltis tatarinowii* Maxim. bark

(2) 水源。小岭境内有10多条溪流，多是从山上茂密森林中的岩石砂砾间经过过滤而延接下来的山泉水，水质洁净，杂质含量少，直接影响着纸的洁净度与白皙度。据曹友泉介绍，强坑宣纸厂生产所需用水来自厂区溪流旁边挖掘的深井，主要是山泉溪流沉淀下的表层水，通过掘井抽取直接使用。经调查人员现场取样测试，其生产用水pH约为6.93，基本呈中性。

⊙1

⊙2

## (二) “曹友泉”高级书画纸生产的工艺流程

据曹友泉的描述，综合调查组2015年8月5日和2016年4月25日在强坑宣纸厂对生产工艺的实地调查，概述高级书画纸生产的工艺流程为：

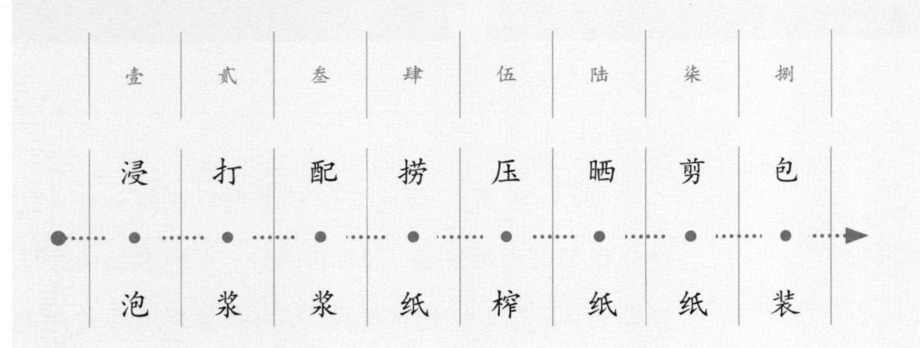

| 壹 | 贰 | 叁 | 肆 | 伍 | 陆 | 柒 | 捌 |
|---|---|---|---|---|---|---|---|
| 浸泡 | 打浆 | 配浆 | 捞纸 | 压榨 | 晒纸 | 剪纸 | 包装 |

⊙3

### 壹 浸泡

1

⊙3

根据所生产纸品品种规格要求，首先需要将龙须草浆板提前一天进行浸泡，一般浸泡一天一夜即可，浆板在充分吸收水分后软化，便于后期打浆，同时在浸泡过程中挑选出浆板中的杂质；浸泡檀皮料半成品，将浸泡后的皮料进行清洗、挑选，去除其中的杂质以备打浆。

⊙1 厂区旁的山泉 Mountain spring alongside the factory

⊙2 强坑宣纸厂的水井 Well of Qiangkeng Xuan Paper Factory

⊙3 浸泡浆板 Soaking the pulp board

贰

## 打　　浆

2

浸泡好的龙须草浆板和清洗后的皮料可直接进行打浆。据曹友泉介绍，强坑宣纸厂是将龙须草浆和檀皮料混合在一起打浆的，先打龙须草浆大概20分钟，然后将适量的檀皮料放入，混合打十几分钟，直至打成熟料。

叁

## 配　　浆

3　⊙4⊙5

根据产品品种不同，按照不同配比掺入龙须草浆与皮浆。强坑宣纸厂的书画纸分为普通书画纸和高级书画纸，普通书画纸的皮浆浆料含量一般在10%以下，甚至不加皮浆浆料，只使用龙须草浆；高级书画纸的檀皮浆含量在25%～30%。

⊙4

⊙5

肆

## 捞　　纸

4　⊙6～⊙9

强坑宣纸厂书画纸的捞纸工艺为：掌帘、抬帘两个捞纸工分别站在纸槽两头，其动作要领与泾县捞制宣纸动作基本上一样。不同的是，捞制“一改二”书画纸在提帘上档时，需将纸帘在帘床上拖动，便于提帘上档时能取纸帘的中间部位，以掌握平衡。2015年8月5日调查组入厂调查时，强坑宣纸厂拥有4个槽位，当天则只有1帘一改二的槽在生产。2016年4月25日，调查组成员对强坑宣纸厂进行回访时，有2帘一改二的槽在生产。据曹友泉介绍，强坑宣纸厂工人一天捞纸的定量是：四尺一改二的槽一天的单位产量大约是20刀纸，单张四尺、六尺的产量也基本在日均10刀左右。

⊙6

⊙7

⊙8

⊙9

⊙4 已配好的浆料 Mixed pulp materials

⊙5 放浆 Adding pulp

⊙6/9 捞纸工序的主要环节 Main steps of papermaking procedure

伍

## 压榨

5

⊙10⊙11

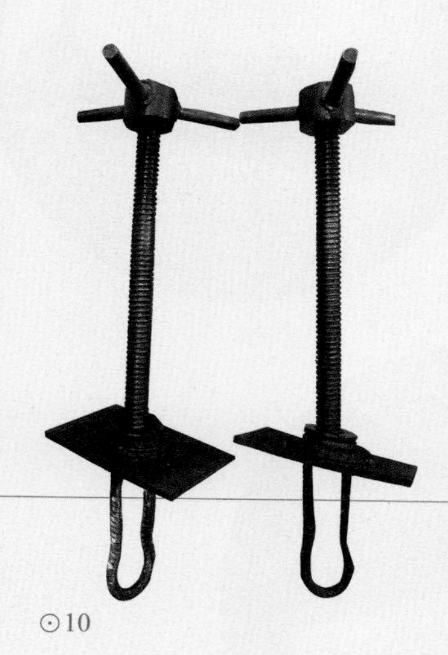

⊙10

⊙11

强坑宣纸厂的压榨方式分两种：一种是使用螺旋杆；另一种是使用千斤顶。无论使用何种工具扳榨，均需要先盖盖纸帘，再盖纸板，沥上几分钟的水后，再上榨杆。榨杆上或架千斤顶或上螺旋杆，压榨时力度均需从小到大缓慢进行，压榨过快会压坏纸张。直到纸帖不再滴水时，压榨才告完成。据曹友泉介绍，使用千斤顶压榨一般需要8～9小时，而使用螺旋杆通常只需一个多小时。

陆

## 晒纸

6

⊙12⊙13

强坑宣纸厂普通书画纸在晒纸之前，先用菜刀将纸帖的额头切掉一小边，一般以1～2 cm为宜；在晒制高级书画纸之前，需将纸帖放在纸焙上炕干，然后浇水润帖，将润好帖的额头切去一条边。此动作称为“切额”。将切好额的纸帖放在纸架上，用鞭帖板打纸帖整面，让

⊙12

⊙10 螺旋杆 Pressing device

⊙11 压榨 Pressing the paper

⊙12 切额 Trimming the deckle edges

纸帖发松后，用手将纸帖的四边往外翻，再用额枪将纸边打松便可晒纸了。晒纸时，先从掐角开始起头，将单张湿纸从纸帖上揭下后，用刷把将纸张贴在纸焙上，晒完整焙后逐张揭下，平放在纸桌上。

⊙13

## 柒 剪纸 ⊙14

### 7

首先将晒好的纸逐张进行检验，筛选出合格的纸，对有瑕疵或杂质的纸进行剔除回笼再打浆；合格的纸按照一定数量整理好，一般50张为一个刀口，压上石头，操作工用特制大剪刀，持平剪刀一气呵成地完成剪纸。

⊙14

## 捌 包装

### 8 ⊙15⊙16

剪好后的纸按照100张/刀分装好，再加盖"曹友泉"书画纸的商标、厂名、纸张品种以及尺寸的印章，包装完毕后运入贮纸仓库。

⊙15

⊙16

⊙13 晒纸 Drying the paper

⊙14 检验 Checking the paper

⊙15 加盖印章 Stamping the paper

⊙16 包装好的书画纸 Packaged calligraphy and painting paper

## (三)

## “曹友泉”书画纸生产的工具设备

### 壹 捞纸槽 1

即捞纸盛装浆料的水池，多用水泥浇筑。实测强坑宣纸厂捞纸槽尺寸规格为：长350 cm，宽185 cm，高80 cm。

⊙1

⊙2

### 贰 纸帘 2

用于捞纸，用苦竹丝多根排列，以丝线贯穿其中而编连成一个整体，然后涂上生漆，滤干即成纸帘。实测强坑宣纸厂一改二的纸帘尺寸规格为：长312 cm，宽90 cm。用于放置纸帘的帘床尺寸规格为：长315 cm，宽92 cm。

### 叁 铁架 3

用于拖运榨干的纸帖。实测强坑宣纸厂所用铁架尺寸规格为：长303 cm，宽170 cm。

⊙3

### 肆 切帖刀 4

晒纸前用来切帖，钢制。

⊙4

⊙1 捞纸槽 Papermaking trough

⊙2 挂在墙上的纸帘 Papermaking screen hanging on the wall

⊙3 铁架 Iron frame for carrying the paper

⊙4 切帖刀 Knife for cutting the paper pile

## 伍 松毛刷 5

晒纸时将纸刷上晒纸墙，刷柄为木制，刷毛为松毛。实测强坑宣纸厂所用的刷子长48 cm，带刷毛一共宽10 cm。

## 陆 刷　夹 6

在松毛刷用完闲置时或晒纸工每天下班后，用刷夹夹住毛刷的松毛部分，既起到保护作用，又防止松毛变弯。实测强坑宣纸厂所用的刷夹尺寸为：55 cm ×7 cm。

## 柒 剪　刀 7

用来剪纸的工具，剪刀口为钢制，其余部分为铁制，由泾县本地后山村一带为宣纸行业特制。实测强坑宣纸厂所用的大剪刀长33 cm。

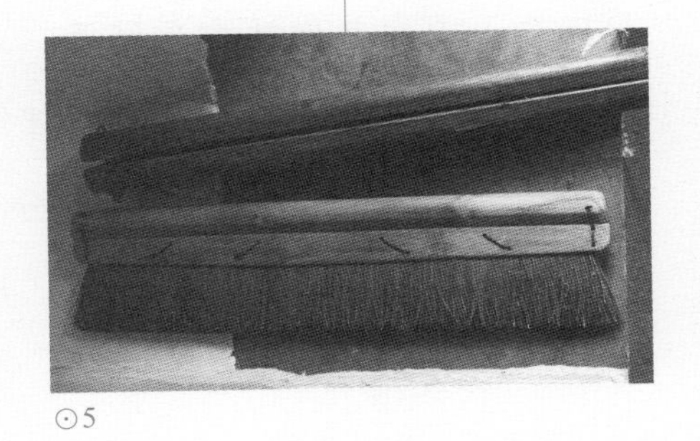

⊙5

⊙6

## 捌 木　尺 8

剪纸时用来测量尺寸的竹制工具。实测强坑宣纸厂所用的木尺全长为100 cm。

⊙7

## 玖 鞭帖板 9

晒纸前用来鞭打纸帖的竹制工具。实测强坑宣纸厂所用的鞭帖板上段长50 cm，下段长80 cm 。

⊙8

⊙5 刷夹和松毛刷 Brush holder and pine needle brush

⊙6 大剪刀 Shears

⊙7 压纸木尺 Wooden ruler for pressing the paper

⊙8 鞭帖板 Bamboo whip

# 五 泾县小岭强坑宣纸厂的市场经营状况

# 5 Marketing Status of Xiaoling Qiangkeng Xuan Paper Factory in Jingxian County

2015年8月和2016年4月两次调查期间，曹友泉反映的综合信息显示，受到宣纸市场行情大幅波动的影响，2015年以造书画纸为主的强坑宣纸厂整体销售效益有明显下滑，但调查组未能获得详细对比数据。

强坑宣纸厂采取“直营店+代理商+网店”的融合销售模式。曹友泉在北京琉璃厂西街投资经营一家直营店，店铺名称为“徽宝堂”，由其妹妹、妹夫经营管理；在泾县丁家桥镇经营一家“曹友泉”品牌宣纸店，销售的产品主要是强坑宣纸厂生产的纸品。除了北京和丁家桥镇开有直营店外，国内其他地方如陕西、山东、云南、广东、湖北、上海等省市也有代理经销商，至2016年4月，代理商有十几家，基本分布在省会城市，所有代理商拿货价格严格统一。网络平台销售基本依靠其女儿曹圆和侄子曹凯。

2014～2015年，小岭强坑宣纸厂年销量在20 000刀左右，销售额200多万元，其中50%来源于直营店，40%来源于代理商，10%来源于网络销售。访谈中曹友泉诉苦说：“手工产品的制作成本高，利润率越来越低，这几年手工纸的利润率保持在8%～10%。成本则主要来自原料和工人工资，工资主要支付给晒纸、捞纸工人，2015年一个工人每月工资在6 000元左右。工人通常一天工作10小时，一个月工作25天左右，除去节假日、员工请假和恶劣天气等因素，一年工作10～11个月。”

⊙1 访谈现场 Interviewing a papermaker

⊙2 『曹友泉宣纸』天猫旗舰店截屏 Screen shot of "Cao Youquan Xuan Paper" in Tmall flagship store

# 六
# 泾县小岭强坑宣纸厂的传承现状与发展思考

6
Current Status of Business Inheritance and Thoughts on Development of Xiaoling Qiangkeng Xuan Paper Factory in Jingxian County

⊙3

⊙4

## 1. 传承现状

泾县小岭强坑宣纸厂作为中国当代书画纸经典聚集地——丁家桥镇众多书画纸厂家的一个代表，其创始人曹炳集、曹友泉父子，以及家庭兄弟姐妹均有从原集体所有制企业——泾县小岭宣纸厂从业而后创办经营家族企业的经历，具有时代与地域特征较为鲜明的技艺传习轨迹。

⊙3 工厂周围的檀树 *Pterocelis tatarinowii* trees around the factory

⊙4 小岭强坑宣纸厂的库房 Warehouse of Xiaoling Qiangkeng Xuan Paper Factory

从当前传承来看，强坑宣纸厂处于较正常的状态：（1）曹炳集、曹友泉、曹凯和曹媛三代人仍全职从事手工纸的生产、管理、销售等，核心传承脉系并未出现青黄不接的断层；（2）较早开拓了网上营销渠道，建立了“直营店+代理商+网店”的融合销售模式，使家族第三代的力量和志趣获得了良好的激发；（3）维系了一批从原小岭宣纸厂学艺成长、具有丰富造纸工艺经验的技术工人在生产一线，技艺的生产性保护有较正常的团队支撑。

## 2. 发展思考：挑战与探索中的问题

关于发展中面临的问题，在访谈中曹友泉比较关注的有两点：

（1）强坑宣纸厂正探索集中力量发展龙须草浆板+青檀皮的“曹友泉”高级书画纸，这既与青檀皮+沙田稻草的宣纸拉开了距离，又与纯龙须草浆板（或添加机械废纸边角料等更劣等的材料）类的中低端书画纸拉开了距离。从积极的方面看，这似乎正好打入了宣纸与中低端书画纸的空白地带，价格与使用性能都有表达和营销优势；但从可能遇到的负面压力看，高端宣纸因处于行业低谷，材料、工艺与价格下行的趋势已部分形成，而传统中低端的书画纸因夹江、眉山等地较大型机械书画纸厂的出现而被迫往材料、工艺、性能、价格上行的求生通道走，两边夹击也可能很快会使其失去优势空间。

（2）强坑宣纸厂目前平均利润率为8%～10%，由于面临行业低谷，下降趋势更明显一些，但维持具有丰富造纸工艺经验的技术工人的成本则上升压力很大。强坑宣纸厂一直强调自己是用“古法”生产高级书画纸的，虽然其依据并不明确，但这一说法包含了明确的书画纸工艺不断优化的诉求和努力，因此，人工成本上升的挑战还在增大。对如何化解这一矛盾，强坑宣纸厂一时也无办法。

# 泾县小岭强坑宣纸厂 书画纸

Calligraphy and Painting Paper of Xiaoling Qiangkeng Xuan Paper Factory in Jingxian County

『曹友泉』书画纸透光摄影图

A photo of "Cao Youquan" calligraphy and painting paper seen through the light

# 第三节

# 泾县雄鹿宣纸厂

安徽省
Anhui Province

宣城市
Xuancheng City

泾县
Jingxian County

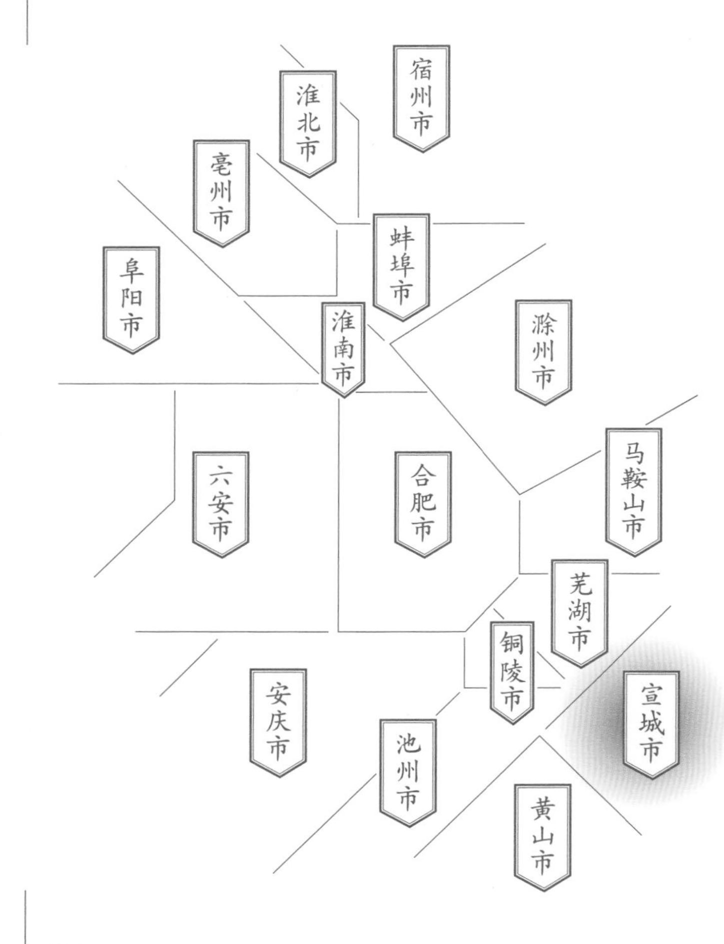




**调查对象**

丁家桥镇
泾县雄鹿宣纸厂
书画纸

## Section 3
## Xionglu Xuan Paper Factory in Jingxian County

**Subject**

Calligraphy and Painting Paper
of Xionglu Xuan Paper Factory in Jingxian County
in Dingjiaqiao Town

# 一 泾县雄鹿宣纸厂的基础信息与生产环境

## 1 Basic Information and Production Environment of Xionglu Xuan Paper Factory in Jingxian County

泾县雄鹿宣纸厂位于泾县丁家桥镇李园行政村五组甲3号，S322省道丁渡段北侧，地理坐标为：东经118° 19′ 14″，北纬30° 39′ 25″。2015年12月12日与2016年4月18日，调查组成员先后两次到雄鹿宣纸厂进行调查，综合两次调查的变动情况，归纳雄鹿宣纸厂的基础生产信息如下：有员工22人，手工纸生产槽位4个，其中，有2帘四尺槽、1帘八尺槽、1帘一改二槽。2016年4月18日第二次入厂调查时，现场有2帘纸槽在生产。

雄鹿宣纸厂主要生产书画纸和宣纸两大类产品，以书画纸为主，商标为“雄鹿”。2014～2015年，纸厂书画纸平均每月每个槽产量为500刀，宣纸平均每月每个槽产量为300刀，书画纸年产量约为15 000刀，宣纸年产量约为6 000刀。

2016年6月，通过检索得到泾县丁家桥镇李园村网站的信息如下：李园行政村地处丁家桥镇北部，青弋江北岸，距县城10 km，S322省道穿境而过。2004年由原来的枫坑、包村、周村、李园、丁渡5个自然村合并成立李园村，全村总面积11.79 km$^2$，辖28个村民组，人口约3 500人。李园村是泾县以至中国以宣纸、书画纸生产加工为主导产业的特色文化产业聚集村落，有宣纸、书画纸加工企业100余家，年产值百万元以上的企业有7家。

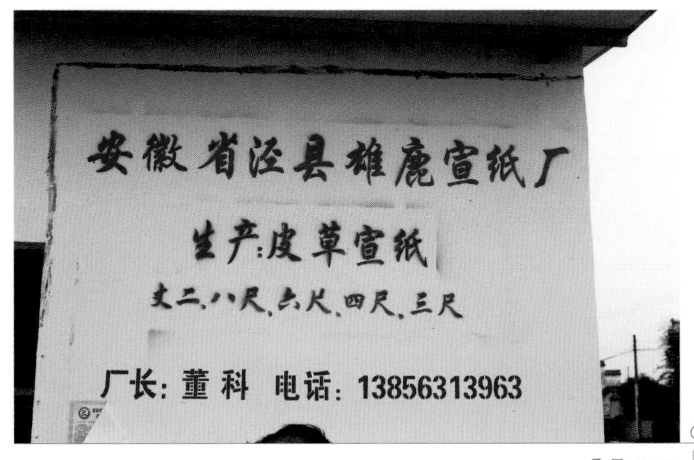


⊙1 雄鹿宣纸厂的路边宣传标牌 Billboard of Xionglu Xuan Paper Factory by the road

路线图
泾县县城
↓
泾县雄鹿宣纸厂
Road map from Jingxian County centre to Xionglu Xuan Paper Factory in Jingxian County

# 泾县雄鹿宣纸厂位置示意图

# Location map of Xionglu Xuan Paper Factory in Jingxian County

考察时间
2015年12月 / 2016年4月

Investigation Date
Dec. 2015 / Apr. 2016

地域名称

泾县县城
丁家桥镇
泾县雄鹿宣纸厂

泾县
1 丁家桥镇
2 云岭镇
3 泾川镇
4 昌桥乡
5 琴溪镇
6 黄村镇

造纸点名称

造纸点 泾县雄鹿宣纸厂

位置分布

市府、州府
县城
乡镇
村落
造纸点
历史造纸点
山
国家级自然保护区
S221 省道
G21 国道
昆河线 铁路
G 56 高速公路
线路

南陵县
青阳县
泾县
合福高铁
S322
G205
10 km
5 km
0
N

## 二 泾县雄鹿宣纸厂的历史与传承情况

## 2 History and Inheritance of Xionglu Xuan Paper Factory in Jingxian County

调查组成员2015年12月12日对雄鹿宣纸厂企业法人董科进行了第一次访谈。据董科的介绍，纸厂1997年建立，当时叫“金鹿宣纸厂”。初建时共有3个槽，其中1个六尺单槽、2个四尺一改二槽，启动资金花了4万～5万元，员工18人左右。2007年，因“金鹿”已被其他人注册，纸厂被迫更名为“雄鹿宣纸厂”，产品标记也由最初的“金鹿”改为“雄鹿”。

金鹿宣纸厂成立之初主要生产以龙须草浆板为主原料的书画纸，随着对市场需求认识的加深，董科发现即便是中高端书画纸也无法取代宣纸的消费功能，便从2000年左右开始在厂里增加生产宣纸的生产线，原料采用从外地购买的粗加工过的稻草浆，进行碾草、洗漂后，加上青檀皮浆制成宣纸。

董科1964年出生于泾县，高中毕业时，因身体残疾，两次体检不合格而错失了上大学的机会。1987年他进入泾县丁桥乡新办私营宣纸厂——包家宣纸厂（明星宣纸厂的前身）做行政管理工作。1989年，他在泾县开设酱制品厂。因饮食习惯的改变和酱制品投入风险增大，加上自己曾在宣纸厂工作过以及母亲一直在金星宣纸厂当主办会计等因素，在母亲的鼓动下，董科于1997年转行创办了自己的书画纸厂——金鹿宣纸厂。

值得关注的是，董科的祖辈和父辈均没有造纸经历和经验，在没有上辈经验传承的情况下，纸厂由他一手经营起来。通过勤奋的钻研，董科成为泾县手工造纸行业有一定口碑的“纸专家”。董科有一个儿子，2015年26岁，在常州开宣纸文房四宝店，代销雄鹿宣纸厂生产的纸品。

⊙1

⊙1 调查组成员访谈董科 A researcher interviewing Dong Ke

# 三
# 泾县雄鹿宣纸厂的代表纸品及其用途与技术分析

# 3
# Representative Paper and Its Uses and Technical Analysis of Xionglu Xuan Paper Factory in Jingxian County

## （一）
## 代表纸品及其用途

雄鹿宣纸厂生产的宣纸和书画纸均标以“雄鹿”。宣纸品种按原料可分为特净、净皮和棉料三大类，规格有三尺、四尺、六尺、八尺、丈二等；书画纸类产品有普通书画纸和中档书画纸两类，尺寸包括四尺、六尺等。调查时，据董科自己的描述，“雄鹿”宣纸润墨和定墨效果好，受到诸多画家的青睐，厂里因此常年会有为书画家定期生产高端宣纸的订单。书画纸从低端到中端的品种较多，以满足不同层级消费者的需求为宗旨。

⊙1

⊙1 雄鹿宣纸厂产品图 A photo of product of Xionglu Xuan Paper Factory

## （二）
## 代表纸品的性能分析

测试小组对雄鹿宣纸厂所产的书画纸做了性能分析，主要包括厚度、定量、紧度、抗张力、抗张强度、撕裂度、湿强度、白度、耐老化度下降、尘埃度、吸水性、伸缩性、纤维长度和纤维宽度等。按相应要求，每一指标都重复测量若干次后求平均值，其中厚度抽取10个样本进行测试，定量抽取5个样本进行测试，抗张力（强度）抽取20个样本进行测试，撕裂度抽取10个样本进行测试，湿强度抽取10个样本进行测试，白度抽取10个样本进行测试，耐老化度下降抽取10个样本进行测试，尘埃度抽取4个样本进行测试，吸水性抽取10个样本进行测试，伸缩性抽取4个样本进行测试，纤维长度测试了200根纤维，纤维宽度测试了300根纤维。对雄鹿宣纸厂书画纸进行测试分析所得到的相关性能参数见表3.3。表中列出了各参数的最大值、最小值及测量若干次所得到的平均值或者计算结果。

表3.3 雄鹿宣纸厂书画纸相关性能参数
Table 3.3 Performance parameters of calligraphy and painting paper in Xionglu Xuan Paper Factory

| 指标 | | | 单位 | 最大值 | 最小值 | 平均值 | 结果 |
|---|---|---|---|---|---|---|---|
| 厚度 | | | mm | 0.100 | 0.080 | 0.090 | 0.090 |
| 定量 | | | g/m² | — | — | — | 29.4 |
| 紧度 | | | g/cm³ | — | — | — | 0.327 |
| 抗张力 | 纵向 | | N | 18.0 | 12.8 | 15.2 | 15.2 |
| | 横向 | | N | 11.0 | 8.0 | 9.8 | 9.8 |
| 抗张强度 | | | kN/m | — | — | — | 0.833 |
| 撕裂度 | 纵向 | | mN | 250 | 230 | 240 | 240 |
| | 横向 | | mN | 340 | 310 | 332 | 332 |
| 撕裂指数 | | | mN · m²/g | — | — | — | 8.6 |
| 湿强度 | 纵向 | | mN | 2 300 | 1 830 | 2 118 | 2 118 |
| | 横向 | | mN | 1 170 | 1 080 | 1 128 | 1 128 |
| 白度 | | | % | 67.5 | 66.9 | 67.2 | 67.2 |
| 耐老化度下降 | | | % | — | — | — | 3.5 |
| 尘埃度 | 黑点 | | 个/m² | — | — | — | 96 |
| | 黄茎 | | 个/m² | — | — | — | 48 |
| | 双浆团 | | 个/m² | — | — | — | 0 |
| 吸水性 | | | mm | — | — | — | 12 |
| 伸缩性 | 浸湿 | | % | — | — | — | 0.40 |
| | 风干 | | % | — | — | — | 0.68 |
| 纤维 | 皮 | 长度 | mm | 3.45 | 0.59 | 1.78 | 1.78 |
| | | 宽度 | μm | 23.0 | 1.0 | 9.0 | 9.0 |
| | 草 | 长度 | mm | 1.67 | 0.36 | 0.87 | 0.87 |
| | | 宽度 | μm | 9.0 | 1.0 | 4.0 | 4.0 |

★1

★2

由表3.3中的数据可知，雄鹿宣纸厂书画纸最厚约是最薄的1.25倍，经计算，其相对标准偏差为0.007，纸张厚薄很均匀。所测雄鹿宣纸厂书画纸平均定量为29.4 g/m²。通过计算可知，雄鹿宣纸厂书画纸紧度为0.327 g/cm³。抗张强度为0.833 kN/m，抗张强度值较大。所测雄鹿宣纸厂书画纸撕裂指数为8.6 mN · m²/g，撕裂度较大；湿强度纵横平均值为1 623 mN，湿强度较大。

所测雄鹿宣纸厂书画纸平均白度为67.2%，白度较高。白度最大值是最小值的1.009倍，相对标准偏差为0.194，白度差异相对较小。经过耐老化测试后，耐老化度下降3.5%。

所测雄鹿宣纸厂书画纸尘埃度指标中黑点为96个/m²，黄茎为48个/m²，双浆团为0个/m²。吸水性纵横平均值为12 mm，纵横差为0.6 mm。伸缩性指标中浸湿后伸缩差为0.40%，风干后伸缩差为0.68%，说明雄鹿宣纸厂书画纸伸缩性差异不大。

雄鹿宣纸厂书画纸在10倍、20倍物镜下观测的纤维形态分别见图★1、图★2。所测雄鹿宣纸厂书画纸皮纤维长度：最长3.45 mm，最短

★1 雄鹿宣纸厂书画纸纤维形态图(10×)
Fibers of calligraphy and painting paper in Xionglu Xuan Paper Factory (10× objective)

★2 雄鹿宣纸厂书画纸纤维形态图(20×)
Fibers of calligraphy and painting paper in Xionglu Xuan Paper Factory (20× objective)

0.59 mm，平均长度为1.78 mm；纤维宽度：最宽23.0 μm，最窄1.0 μm，平均宽度为9.0 μm。草纤维长度：最长1.67 mm，最短0.36 mm，平均长度为0.87 mm；纤维宽度：最宽9.0 μm，最窄1.0 μm，平均宽度为4.0 μm。雄鹿宣纸厂书画纸润墨效果见图⊙1。

⊙1

⊙1
雄鹿宣纸厂书画纸润墨效果
Writing performance of calligraphy and painting paper in Xionglu Xuan Paper Factory

## 四 “雄鹿”书画纸生产的原料、工艺流程与工具设备

## 4 Raw Materials, Papermaking Techniques and Tools of “Xionglu” Calligraphy and Painting Paper

### （一）“雄鹿”书画纸生产的原料

#### 1. 主料

龙须草是“雄鹿”书画纸的主要原料之一，均从泾县本地的经销商处购买。董科介绍说，因泾县当地不产龙须草，雄鹿宣纸厂使用的泾县经销商经营的龙须草浆板都是从河南等地进的货，2014～2015年各企业在经销商处提货价为11 000元/t。

⊙2

（2）水源。井水和山泉水。据董科介绍，雄鹿宣纸厂清洗浆料的时候一般用井水，而捞纸时使用山泉水。之所以使用两种水源，是因为山泉水有时不够用，非关键环节就用井水代替山泉水。雄鹿宣纸厂所用井水也是地表水，成分跟山泉水差异并不大。调查组成员实测井水pH为6.5，山泉水pH为6，后者更偏酸性一点。

⊙3

⊙4

“雄鹿”书画纸根据品类需要时常会加入一定量的青檀皮料，主要用于中高档书画纸的生产。因雄鹿宣纸厂没有环保治污设备，所需的青檀皮料都在泾县专事檀皮加工的原料生产户处购买。“雄鹿”书画纸所用青檀皮料均是加工后的皮料半成品，2014年前后皮料半成品价格约为16元/kg，价格随市场行情调整，出浆率在19%～25%。

雄鹿宣纸厂根据市场需要，也生产青檀皮料与稻草浆料混合的纸。生产这种纸所需的稻草并非泾县本地的原料，其草浆从河南采购，据访谈中董科的表述，雄鹿宣纸厂将河南购来的稻草浆料通过碾料、洗漂等工艺处理后，再根据宣纸棉料等三大类别的原料配比量与青檀皮料浆混合，可生产出与标准宣纸中稻草与檀皮同样含量的高端书画用纸。雄鹿宣纸厂将此纸作为“雄鹿”宣纸销售。

## 2. 辅料

（1）纸药。调查组入厂时了解到的信息是：雄鹿宣纸厂用的纸药全部为化学纸药聚丙烯酰胺，2015年价格约为20元/kg。

⊙5

⊙2 浸泡前的龙须草浆板
Pulp board of *Eulaliopsis binata* before soaking

⊙3 收购的稻草浆原料
Pulp materials of straw bought from elsewhere

⊙4 工人在调制化学纸药
A worker modulating chemical papermaking mucilage

⊙5 雄鹿宣纸厂从地下井取水
Xionglu Xuan Paper Factory getting water from a underground well

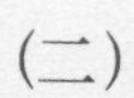

## (二)

## “雄鹿”书画纸生产的工艺流程

调查组于2015年12月12日和2016年4月18日对“雄鹿”书画纸生产工艺进行了实地调查和工艺访谈，总结其主要制作流程如下：

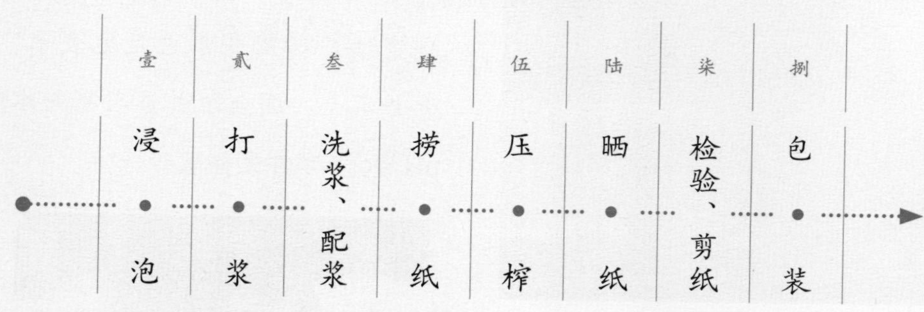

| 壹 | 贰 | 叁 | 肆 | 伍 | 陆 | 柒 | 捌 |
| --- | --- | --- | --- | --- | --- | --- | --- |
| 浸泡 | 打浆 | 洗浆、配浆 | 捞纸 | 压榨 | 晒纸 | 检验、剪纸 | 包装 |

### 壹 浸泡 1

购买的龙须草浆板需要在浸泡池浸泡3天以上，使其充分吸收水分软化。

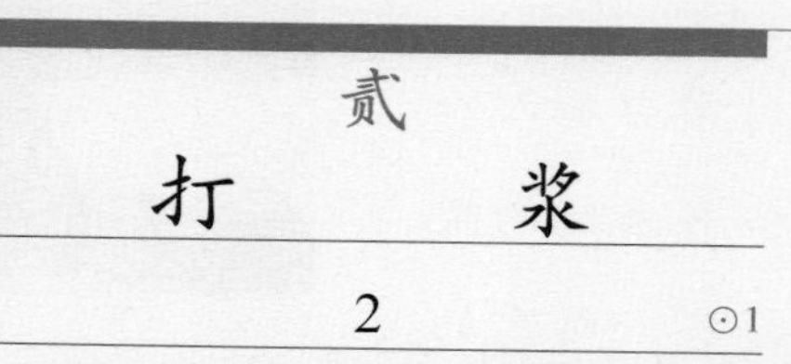

### 贰 打浆 2

⊙1

浸泡后的龙须草浆板需要放入打浆池打浆，使浆板充分打碎均匀；生产中档书画纸时，还需要对檀皮料进行打浆。

⊙1

### 叁 洗浆、配浆 3

⊙2

打好的浆料用浆泵抽取至圆筒筛进行洗料。圆筒筛是一种圆柱体洗料装置，四周裹有丝网布，通过圆筒筛的旋转能除去浆料中的污水和残留的化学成分。生产中档书画纸时，洗浆后需要对龙须草浆料和檀皮浆料进行混合配比，其中主体是龙须草浆料，檀皮料比例则根据客户需要添加。

⊙2

⊙1 打浆 Beating the pulp

⊙2 清洗后的浆料 Pulp materials after cleaning

## 肆 捞纸

### 4 ⊙3～⊙5

“雄鹿”书画纸均采用手工捞制方式。掌帘、抬帘两个捞纸工分站在纸槽两头，先将纸帘放上帘架，然后两人用帘尺夹紧纸帘，将帘床全部放入槽中，左右轻晃各一次，端平后立即抬出水面，将帘上多余的浆料由缝隙滤出，帘上的浆料就形成了一张湿纸。再将纸帘从帘床上拉出一段，右手拎起纸帘，左手托住纸帘的下端，将附着在纸帘上的湿纸以圆筒状方式倒覆在纸槽旁的纸板上。每捞一张湿纸，抬帘工按一下计数器。其动作要领与捞制宣纸时一样。不同的是捞制一改二书画纸在提帘上档时，需在帘床上拖动纸帘，以便于提帘上档时能取纸帘的中间部位，掌握平衡。捞低档书画纸时，清洗后的纸浆可直接放入打浆机中打浆；捞中档书画纸时，龙须草浆和檀皮浆需要在打浆机中分别打浆后再按需配浆，打匀后放入纸槽捞纸，捞纸前加入化学纸药。据董科的说法，该厂一帘槽每天能捞四尺书画纸16～17刀。

⊙3

⊙4

⊙5

## 伍 压榨

### 5 ⊙6

捞纸工下班后，由帮槽工先后将盖纸帘、盖纸板先后盖在纸帖上，等水沥到一定程度后，将榨杆放在盖纸板上，安上千斤顶后逐步加压。待纸帖上的水分逐渐被挤榨干后，湿纸会慢慢变硬成为一块纸帖，这时要逐渐加大压榨力度，压到纸帖不再出水时为止。

⊙6

⊙3 捞纸 Papermaking

⊙4 提帘放纸 Turning the papermaking screen upside down on the board

⊙5 计数器 Counting apparatus

⊙6 压榨 Pressing the paper

陆

## 晒纸

6 ⊙7～⊙9

“雄鹿”普通书画纸在晒纸之前，先用菜刀将纸帖的额头切掉一小边，一般以1～2 cm为宜；在晒制掺有青檀皮和稻草浆的纸之前，需将纸帖放在纸焙上炕干，然后浇水润帖，再将润好帖的额头切去一条边。称之“切额”。将切好额的纸帖放在纸架上，用鞭帖板击打纸帖整面，让纸帖发松后，用手将纸帖的四边往外翻，再用额枪将纸边打松便可晒纸了。晒纸时，先从掐角开始起头，将单张湿纸从纸帖上揭下后，用刷把将纸张贴在纸焙上，晒完整焙后逐张揭下，平放在纸桌上。

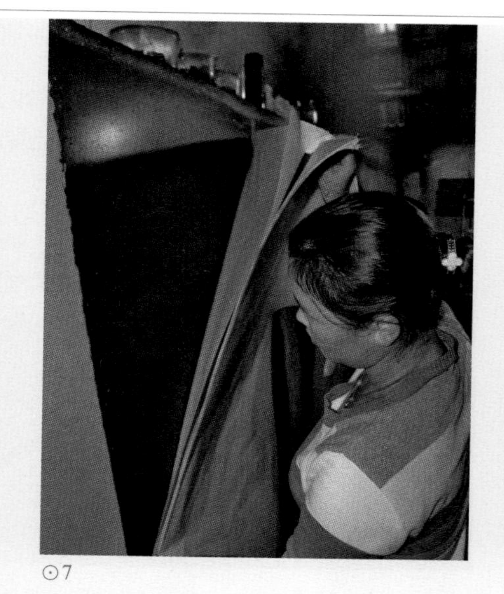
⊙7

⊙8

⊙9



柒

## 检验、剪纸

7 ⊙10⊙11

晒好的纸需要逐张进行检验，将不合格的纸取出。以每块纸帖为单位，全部检验好后放在检纸桌上，根据规格不同分别归类。裁剪时纸张要用压纸石压紧，用特制大剪刀进行裁边，要求纸张四边裁剪整齐。

⊙10

⊙11

捌

## 包装

8 ⊙12

剪好的纸按100张/刀分好，再加盖“雄鹿”印章，将盖章后的纸张进行装箱。

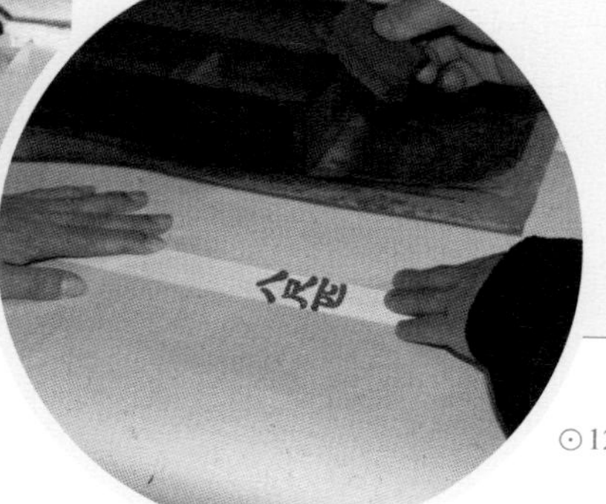
⊙12

⊙7 揭纸 Separating the paper layers

⊙8 晒纸上墙 Drying the paper on a wall

⊙9 收纸 Taking back the paper

⊙10 检验纸张 Checking the paper

⊙11 剪纸 Cutting the paper

⊙12 加盖印章 Stamping the paper

## (三)
## “雄鹿”书画纸生产的工具设备

### 壹 打浆机 1

用于制作龙须草浆和檀皮浆的电动式器具。

### 贰 圆筒筛 2

用于清洗浆料、脱去污水和残留化学物质的工具。

### 肆 纸槽 4

盛浆设施，用水泥浇筑而成。实测雄鹿宣纸厂手工捞纸所用的四尺单槽尺寸为：长197 cm，宽 195 cm，高86 cm；八尺纸槽尺寸为：长297 cm，宽215 cm，高86 cm；一改二纸槽尺寸为：长320 cm，宽190 cm，高86 cm。

### 叁 浆泵 3

用于抽取打浆、清洗后的浆料入槽的电动工具。

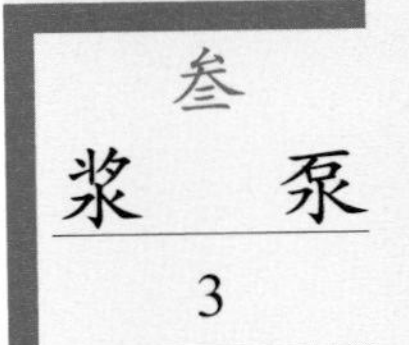

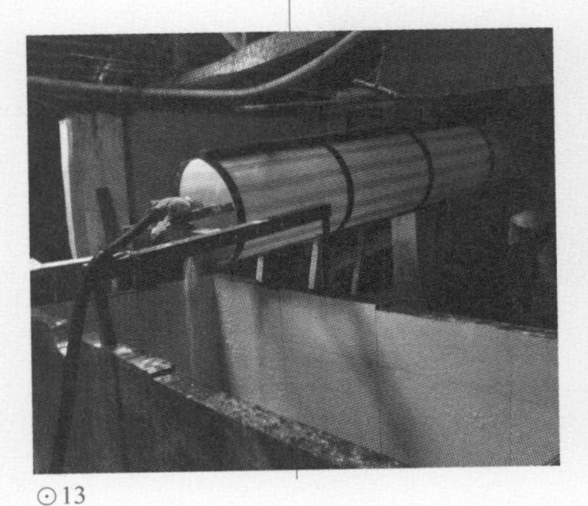

⊙13

⊙14

### 伍 纸榨 5

用来压榨捞好的湿纸帖的设施。

### 陆 纸帘 6

用于捞纸，由竹丝编织而成，表面光滑平整，帘纹细而密集。

### 柒 帘床 7

捞纸时用来放置纸帘的木制托架。

⊙15

### 捌 千斤顶 8

压榨纸帖时用来增加力度的机械。

⊙17

⊙16

⊙13 圆筒筛 Cylinder filter
⊙14 纸槽 Papermaking trough
⊙15 纸榨 Tools for pressing the paper
⊙16 帘床 Frame for supporting the papermaking screen
⊙17 千斤顶 Lifting jack

### 玖 纸焙 9

晒纸设施，由两块长方形钢板焊接而成，中间贮水，蒸汽加热，双面墙，可以两边晒纸。

### 拾 刷把 10

晒纸时用刷把将纸刷上晒纸墙，刷柄为木制，刷毛为松针。

⊙18

### 拾壹 压纸石 11

在造纸过程中两次被用到。晒纸时，将晒好的纸张整齐堆放在纸板上，用压纸石来固定纸张；在剪纸过程中，压纸石用来压紧纸张便于剪纸工裁剪纸边。



## 五 泾县雄鹿宣纸厂的市场经营状况

## 5 Marketing Status of Xionglu Xuan Paper Factory in Jingxian County

⊙18 纸刷 Bush

泾县雄鹿宣纸厂成立之初主要生产龙须草浆板原料的书画纸，2000年左右才开始生产宣纸，形成了宣纸与书画纸生产相结合的现有体系。据访谈中董科对经营特征的说法，纸厂采取灵活生产的方式，宣纸和书画纸会根据市场需求来快速动态调整生产量。同时，进行高端和中低端产品市场分割，强调实现产品的多样化和差异化，以保持满足不同人群使用需求的能力为宗旨。交流中董科表示：作为一家小型民营手工造纸企业，纸厂规模虽然不大，年销售额也不高，但纸品的质量一直很稳定，宣纸和书画纸都受到客户的喜爱。

从市场销售看，雄鹿宣纸厂销路主要包括三块：一是国内经销商体系，已经有较广的分布，全国各个省份基本都有“雄鹿”宣纸和书画纸的经销商，多分布在省会城市，每年有相当一部分产品发至经销商。二是网上销售部分，除了董科的儿子在常州开了网店代销厂里的产品外，他的弟弟董建农也开了淘宝店销售“雄鹿”产品。此外，泾县当地开的网上淘宝店也会从雄鹿宣纸厂拿货配销。三是高端定制纸渠道，据董科介绍，“雄鹿”宣纸因其良好的绘画润墨效果而受到多家画院的青睐，如近几年纸厂每年均会定制生产北京凤凰岭书画院、清华美院等机构专用的高端宣纸。

从售价来看，截至调查时的2016年4月，“雄鹿”特种净皮四尺单宣市场价为560元/刀，六尺市场价为1 120元/刀。“雄鹿”普通书画纸四尺市场价为95元/刀；中档书画纸四尺市场价有145元/刀、180元/刀、220元/刀等数种，价格高低主要取决于加入皮料的比例大小。

2014～2015年，雄鹿宣纸厂书画纸平均每月每个槽产量为500刀，宣纸平均每月每个槽产量为300刀，若按一年10个月的生产期计算，书画纸年产量约为15 000刀，宣纸年产量约为6 000刀。

⊙19

⊙20

⊙19
调查组成员观察董科试纸
A researcher watching Dong Ke testing the paper

⊙20
剪纸车间及成品暂存室
Paper cutting workshop and temporary warehouse for paper products

## 六
## 泾县雄鹿宣纸厂的品牌故事

## 6
## Brand Stories of Xionglu Xuan Paper Factory in Jingxian County

据雄鹿宣纸厂厂长董科介绍，1997年，因母亲在金星宣纸厂工作及受附近办纸厂成功案例的影响，他决定开办自己的纸厂。当时，启动资金不足，在外工作的兄弟姊妹得知后，纷纷解囊相助。企业正式成立后，兄弟姊妹又提议：“企业如要想生存长远，第一要有高质量的产品，第二要有一个叫得响的牌子。”董科便以“金鹿”为厂名，后因该商标已被别的厂家注册，2007年改名“雄鹿”，取“雄睨天下，逐鹿中原”之意。





## 七
## 泾县雄鹿宣纸厂的传承现状与发展思考

## 7
## Current Status of Business Inheritance and Thoughts on Development of Xionglu Xuan Paper Factory in Jingxian County

调查组通过访谈得知的现状信息是：雄鹿宣纸厂因其产品分类目标的定位精准、品质稳定，以及追求随市场快速调整生产量与品种的经营方式，市场销路一直较好，2014～2016年，在整个宣纸和手工书画纸行业严重产能过剩加上礼品纸渠道销售断崖式萎缩的不景气情况下，纸厂未出现产品滞销和库存积压。

在提到宣纸传统技艺传承和发展问题时，董科提出了3个观点：

（1）宣纸制作技艺流传至今已有很多年的历史，但宣纸行业相比机械纸行业技术与工艺不断的革新还有明显不足，显得保守性很强。在实际生产过程中，固然需要对其进行生产性保护和传承，但行业也迫切需要在传承中有改良和创新。董科举例说，现代的纸浆净化设备就比古法的要

⊙1

进步，不仅省时省力，净化效果也好很多。因此，提倡在传承传统技艺的基础上改良设备和工具是可取的，需要坚持继承传统和发展创新并举。

(2) 调查时董科特别提出：虽然宣纸已申报成为国家级非物质文化遗产，但国家和当地政府对其传承和保护力度以及普惠度不足，对行业内量大面广的手工造纸企业扶持资金也不到位。

(3) 最后董科提到宣纸的原料如稻草的质量与六七年前或再早前已无法相比，这与种植过程中持续大量使用化肥不无关系，认为这已成为相当大的核心原料发展瓶颈。

雄鹿宣纸厂产品系列覆盖改良型高端书画纸、层级丰富的中端与低端书画纸，立足有效满足各类书画用纸人群的使用需求，实现了产品市场的差异化分割。同时，纸厂采取灵活的生产方式，特别强调根据市场需求来调节生产，减小了公司生产成本的消耗，强化了市场经营的风险防范意识，这对其他厂家或具有经营路径选择上的启示。

⊙2

⊙1 产品库 Paper warehouse

⊙2 整理剪好的纸张 Sorting the processed paper

# 泾县雄鹿宣纸厂
# 书画纸

Calligraphy and Painting Paper
of Xionglu Xuan Paper Factory
in Jingxian County

『雄鹿』书画纸透光摄影图

A photo of "Xionglu" calligraphy and painting paper seen through the light

# 第四节
# 泾县紫光宣纸书画社

安徽省
Anhui Province

宣城市
Xuancheng City

泾县
Jingxian County

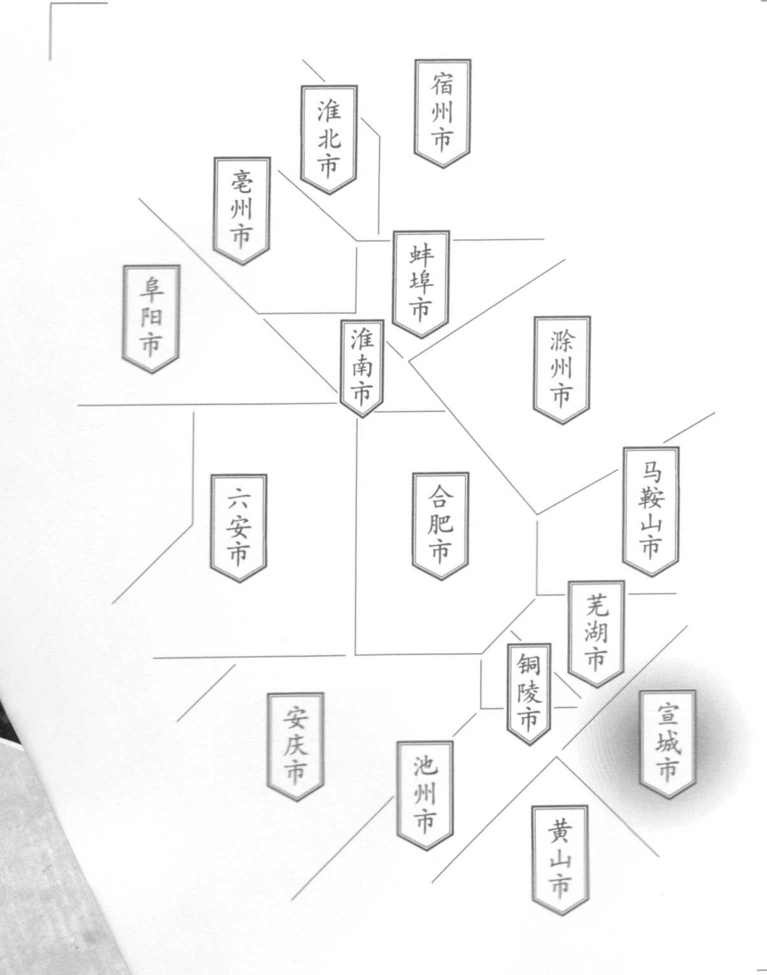


Section 4
g Calligraphy
aper Agency
an County

**Subject**

Calligraphy and Painting Paper of Ziguang Calligraphy and Painting Xuan Paper Agency in Jingxian County in Dingjiaqiao Town



调
查

## 一 泾县紫光宣纸书画社的基础信息与生产环境

## 1 Basic Information and Production Environment of Ziguang Calligraphy and Painting Xuan Paper Agency in Jingxian County

泾县紫光宣纸书画社坐落于“宣纸之乡”丁家桥镇后山行政村上阳村民组，地理坐标为：东经118° 19′ 2″，北纬30° 37′ 51″。纸厂大门紧邻丁（丁家桥）黄（黄村）县域公路西侧，交通较为便利。紫光宣纸书画社始创于1998年，占地面积6 800 m$^2$，注册商标为“绿杨宝”。

2015年7月23日和2016年4月27日，调查组成员两次前往紫光宣纸书画社调查访谈，获知的基础生产信息如下：紫光宣纸书画社现有员工120人左右，“绿杨宝”书画纸均采用喷浆捞纸工艺，共有槽位20个，其中六尺槽2个、尺八屏槽2个、四尺槽16个。2015年7月23日调查组入厂调查当天，共有12个槽在生产，其中2个六尺槽、9个四尺槽和1个尺八屏槽。四尺槽、六尺槽每人每天能捞6刀纸，尺八屏槽每人每天能捞5～6刀纸。

⊙1

⊙2

⊙1
丁黄公路
Ding-Huang Highway

⊙2
紫光宣纸书画社宣传栏
Billboard of Ziguang Calligraphy and Painting Xuan Paper Agency

路线图
泾县县城
↓
泾县紫光宣纸书画社
Road map from Jingxian County centre to Ziguang Calligraphy and Painting Xuan Paper Agency in Jingxian County

# 泾县紫光宣纸书画社位置示意图

# Location map of Ziguang Calligraphy and Painting Xuan Paper Agency in Jingxian County

考察时间
2015年7月 / 2016年4月

Investigation Date
July 2015 / Apr. 2016

地域名称

泾县县城
丁家桥镇
泾县紫光宣纸书画社

A 泾县
1 丁家桥镇
2 云岭镇
3 泾川镇
4 昌桥乡
5 琴溪镇
6 黄村镇

造纸点名称

泾县紫光宣纸书画社 造纸点

位置分布

市府、州府
县城
乡镇
村落
造纸点
历史造纸点
山
国家级自然保护区
S221 省道
G21 国道
昆河线 铁路
G56 高速公路
线路

南陵县
青阳县
泾县
S322
G205
合福高铁

10 km
5 km
0
N

## 二 泾县紫光宣纸书画社的历史与传承情况

## 2 History and Inheritance of Ziguang Calligraphy and Painting Xuan Paper Agency in Jingxian County

通过田野调查和泾县地方纸业文献的研究，获知的综合信息如下：泾县紫光宣纸书画社的创始人与负责人吴报景，1970年出生于丁家桥镇后山村，1985年初中毕业即到泾县宣纸二厂学习晒纸，师从其兄张必清（吴报景原姓张，从小过继入吴家，故改姓吴），学习6个月后，正逢泾县小岭宣纸厂氧－碱制浆新工艺生产线招人，吴报景遂进入小岭宣纸厂从事晒纸工作。1988年，因小岭宣纸厂是集体企业，正式工比临时工收入高，吴报景是农村户口，转正无望，而紧邻的乡办企业金竹坑宣纸厂（金星宣纸厂前身）没有正式工、临时工的等级差，加上离家近，便转到金竹坑宣纸厂继续晒纸。

1992年，吴报景萌生了创业办厂的念头，但当时小岭宣纸厂的师傅告诫他：徽商的成功在于先学徒，再营销，最后才能当大老板。于是，从1993年开始，吴报景只身到全国各地学习跑销售，西到西安，北赴京城，南下深圳，东去上海，几乎跑遍了大半个中国，掌握了市场一线的宣纸销售状况和需求。1998年，吴报景回乡在后山村搭盖了7间简陋的厂房，开始筹建紫光宣纸书画社。

⊙1

1999年到2000年12月，吴报景在妹婿朱水兵位于北京琉璃厂附近的"双旗宣纸厂驻北京直销处"从事销售工作。为打好市场基础，吴报景于2000 年12月到2002年7月，在西安先帮弟弟张三清开设了泾县紫光宣纸书画社销售处。2002年7月，吴报景再回到泾县，在丁家桥镇后山行政村的汪家店正式注册创办泾县紫光宣纸书画社。开厂生产时设2帘槽，专门生产书画纸；有工人16人，均为原小岭宣纸厂西山车间的熟练纸工。

随着生产规模的扩大，汪家店村的厂房已无法容纳。2005年7月，吴报景将紫光宣纸书画社的生产厂区搬迁到丁家桥镇后山行政村上阳村民组，开设了10个喷浆书画纸槽、2个吊帘（1个四尺、1个六尺）工位。搬迁新建厂房共投资40余万元，其中征地 5万元、厂房建设和设备35万元。

⊙1 资深宣纸工艺师朱正海访谈吴报景（右）
Zhu Zhenghai, the senior Xuan paper craftsman, interviewing Wu Baojing (right)

2006年，公司注册了“绿杨宝”商标。

紫光宣纸书画社的产品以自有“绿杨宝”品牌投放市场后，逐步有了出口到日本、韩国的外销订单。2008年，紫光宣纸书画社进一步扩大了生产规模，增设了10个喷浆槽，取消了吊帘生产。2010年，紫光宣纸书画社又增设了5个手工纸槽，主要生产北京三希堂艺术院订购的书画专用纸。2015年，随着市场萎缩，紫光宣纸书画社逐步减少生产，至调查阶段只保持了12个喷浆槽、2个手工捞纸槽（1个四尺、1个六尺）的生产。2015年底，吴报景与陕西省周至县起良村村民合作，开发了仿古汉纸品种。

据访谈中吴报景的自述，其原姓小岭村张家是宣纸世家，外公张春先原在泾县宣纸厂从事晒纸工作，退休后在百岭坑宣纸厂当晒纸车间主任直到76岁。吴报景有两个女儿，调查时大女儿吴璇（1993年出生）在西安学习古画修复，小女儿吴可馨（2003年出生）还在读中学，两个孩子都没有学习过造纸技艺，也未参与纸厂的生产和管理。

调查时，吴报景还担任着泾县中国宣纸协会副会长。

⊙1

⊙1
在社区办公室访谈吴报景
Interviewing Wu Baojing in the office of the agency



## 三 泾县紫光宣纸书画社的代表纸品及其用途与技术分析

## 3 Representative Paper and Its Uses and Technical Analysis of Ziguang Calligraphy and Painting Xuan Paper Agency in Jingxian County

### （一）代表纸品及其用途

调查中了解到的纸品信息如下：紫光宣纸书画社主要生产“绿杨宝”系列宣纸、书画纸和加工纸三大类。“绿杨宝”系列宣纸产品包括书画专用纸、花鸟专用纸、山水专用纸、古法檀皮宣、特制绿杨宣、绿杨色宣；系列加工纸包括精品生宣册页、精品锦盒手卷、绿杨生宣卡纸、卡纸扇开内泥金、卡纸扇开外泥金、矾宣、印谱、对联等。其中，以“绿杨宝”书画纸——“绿杨宣”最具有代表性。

“绿杨宝”产品销售以国内为主，因其价格相对低廉，书画使用效果较佳，所以市场需求量很大。“绿杨宝”品牌中的“绿杨宣”纸张规格多样，主要有三尺、四尺、五尺、六尺、尺八屏等。从主流用途来说，“绿杨宣”主要用于书画爱好人群的绘画、书法练习。

## (二)
## 代表纸品的性能分析

测试小组对采样自紫光宣纸书画社生产的“绿杨宝”书画纸所做的性能分析，主要包括厚度、定量、紧度、抗张力、抗张强度、撕裂度、湿强度、白度、耐老化度下降、尘埃度、吸水性、伸缩性、纤维长度和纤维宽度等。按相应要求，每一指标都重复测量若干次后求平均值，其中厚度抽取10个样本进行测试，定量抽取5个样本进行测试，抗张力（强度）抽取20个样本进行测试，撕裂度抽取10个样本进行测试，湿强度抽取20个样本进行测试，白度抽取10个样本进行测试，耐老化下降抽取10个样本进行测试，尘埃度抽取4个样本进行测试，吸水性抽取10个样本进行测试，伸缩性抽取4个样本进行测试，纤维长度测试了200根纤维，纤维宽度测试了300根纤维。对“绿杨宝”书画纸进行测试分析所得到的相关性能参数见表3.4。表中列出了各参数的最大值、最小值及测量若干次所得到的平均值或者计算结果。

表3.4 “绿杨宝”书画纸相关性能参数
Table 3.4 Performance parameters of “Lvyangbao” calligraphy and painting paper

| 指标 | | 单位 | 最大值 | 最小值 | 平均值 | 结果 |
|---|---|---|---|---|---|---|
| 厚度 | | mm | 0.120 | 0.100 | 0.106 | 0.106 |
| 定量 | | $g/m^2$ | — | — | — | 38.1 |
| 紧度 | | $g/cm^3$ | — | — | — | 0.359 |
| 抗张力 | 纵向 | N | 19.5 | 17.3 | 18.2 | 18.2 |
| | 横向 | N | 14.2 | 11.4 | 12.8 | 12.8 |
| 抗张强度 | | kN/m | | | | 1.033 |
| 撕裂度 | 纵向 | mN | 470 | 380 | 420 | 420 |
| | 横向 | mN | 510 | 430 | 460 | 460 |
| 撕裂指数 | | $mN \cdot m^2/g$ | — | — | — | 11.3 |
| 湿强度 | 纵向 | mN | 1 120 | 1 050 | 1 086 | 1 086 |
| | 横向 | mN | 830 | 800 | 812 | 812 |
| 白度 | | % | 73.9 | 73.6 | 73.8 | 73.8 |
| 耐老化度下降 | | % | — | — | — | 2.5 |
| 尘埃度 | 黑点 | 个/$m^2$ | — | — | — | 40 |
| | 黄茎 | 个/$m^2$ | — | — | — | 12 |
| | 双浆团 | 个/$m^2$ | — | — | — | 0 |
| 吸水性 | | mm | — | — | — | 15 |
| 伸缩性 | 浸湿 | % | — | — | — | 0.50 |
| | 风干 | % | — | — | — | 0.75 |

续表

| 指标 | | | 单位 | 最大值 | 最小值 | 平均值 | 结果 |
|---|---|---|---|---|---|---|---|
| 纤维 | 皮 | 长度 | mm | 5.57 | 1.55 | 2.73 | 2.73 |
| | | 宽度 | μm | 27.0 | 5.0 | 11.0 | 11.0 |
| | 草 | 长度 | mm | 3.03 | 0.45 | 1.19 | 1.19 |
| | | 宽度 | μm | 27.0 | 2.0 | 7.0 | 7.0 |

由表3.4中的数据可知，“绿杨宝”书画纸最厚约是最薄的1.200倍，经计算，其相对标准偏差为0.006，纸张厚薄较为一致。所测“绿杨宝”书画纸的平均定量为38.1 g/m$^2$。通过计算可知，“绿杨宝”书画纸紧度为0.359 g/cm$^3$。抗张强度为1.033 kN/m，抗张强度值较小。所测“绿杨宝”书画纸撕裂指数为11.3 mN·m$^2$/g，撕裂度较大；湿强度纵横平均值为949 mN，湿强度较大。

所测“绿杨宝”书画纸平均白度为73.8%，白度较高，是由于其加工过程中有漂白工序。白度最大值是最小值的1.004倍，相对标准偏差为0.088，白度差异相对较小。经过耐老化测试后，耐老化度下降2.5%。

所测“绿杨宝”书画纸尘埃度指标中黑点为40个/m$^2$，黄茎为12个/m$^2$，双浆团为0个/m$^2$。吸水性纵横平均值为15 mm，纵横差为1.6 mm。伸缩性指标中浸湿后伸缩差为0.50%，风干后伸缩差为0.75%，说明“绿杨宝”书画纸伸缩性差异不大。

“绿杨宝”书画纸在10倍、20倍物镜下观测的纤维形态分别见图★1、图★2。所测“绿杨宝”书画纸皮纤维长度：最长5.57 mm，最短1.55 mm，平均长度为2.73 mm；纤维宽度：最宽27.0 μm，最窄5.0 μm，平均宽度为11.0 μm。草纤维长度：最长3.03 mm，最短0.45 mm，平均长度为1.19 mm；纤维宽度：最宽27.0 μm，最窄2.0 μm，平均宽度为7.0 μm。“绿杨宝”书画纸润墨效果见图⊙1。

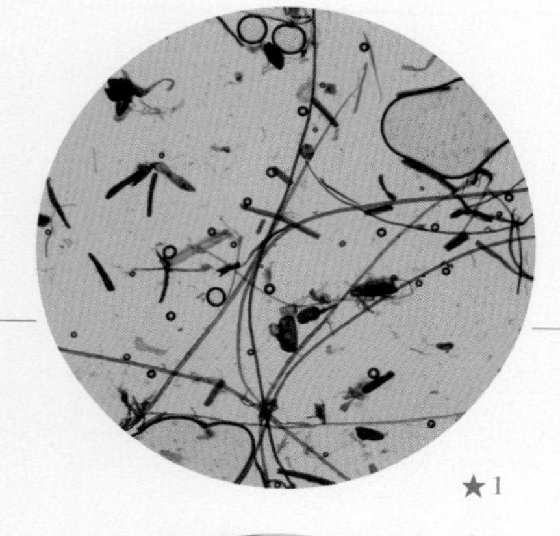

★1

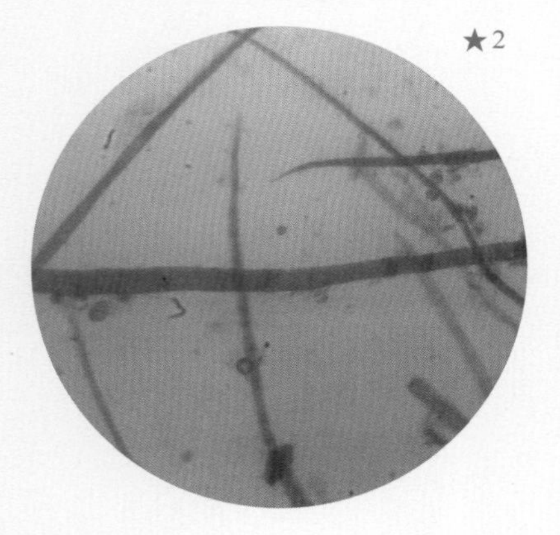

★2

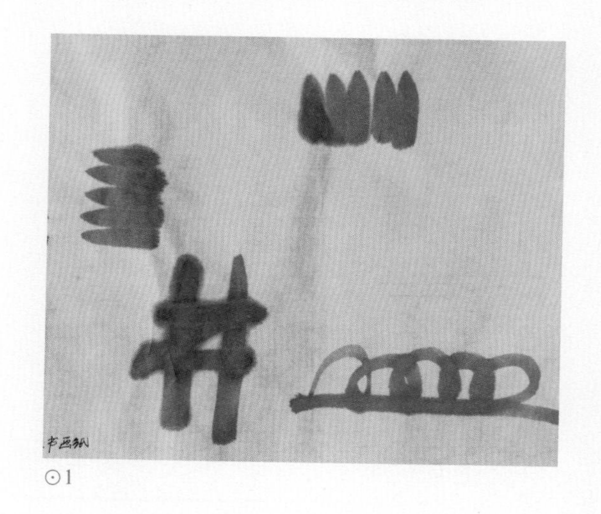


⊙1

★1 『绿杨宝』书画纸纤维形态图(10×) Fibers of "Lvyangbao" calligraphy and painting paper (10× objective)

★2 『绿杨宝』书画纸纤维形态图(20×) Fibers of "Lvyangbao" calligraphy and painting paper (20× objective)

⊙1 『绿杨宝』书画纸润墨效果 Writing performance of "Lvyangbao" calligraphy and painting paper

# 四 “绿杨宝”书画纸生产的原料、工艺流程与工具设备

## 4 Raw Materials, Papermaking Techniques and Tools of “Lvyangbao” Calligraphy and Painting Paper

调查时调查组成员对泾县紫光宣纸书画社现场生产管理负责人张世坤进行了详细访谈，张世坤介绍的工艺信息如下：紫光宣纸书画社生产“绿杨宝”书画纸时均通过半自动喷浆方式来捞纸，喷浆技艺和设备从台湾地区购入，制作“绿杨宣”的主要原料是龙须草和檀皮，其中龙须草是从外地直接购买的浆板，浆板经过浸泡后便可以打浆；檀皮是在泾县当地收购的已加工好的浆料。

⊙2

⊙3

### (一) “绿杨宝”书画纸生产的原料

#### 1. 主料

“绿杨宝”书画纸作为紫光宣纸书画社“绿杨宝”品牌的特色主打品种，主要原料是龙须草和檀皮。值得注意的是，泾县当地极少产龙须草，而且龙须草的加工要求高，加工过程中会产生大量污水，因此多从外地直接购买浆板。紫光宣纸书画社的龙须草浆板来源地是河南，檀皮是直接从加工户处购买的成熟青檀皮浆料。如生产绿杨色纸，还需要在混合浆料中加入一定量的化学染料，化学染料系从泾县当地的文房四宝商店购买。

#### 2. 辅料

(1) 纸药。紫光宣纸书画社“绿杨宝”书画纸以喷浆生产工艺为主，使用的化学纸药聚丙烯酰胺均在泾县当地的经销商处购买。

(2) 水源。紫光宣纸书画社“绿杨宝”书画纸使用的水从自有的水井中抽取。调查组成员实测的用水pH接近7，基本呈中性。

⊙2 正在讲解的张世坤 Zhang Shikun introducing papermaking techniques

⊙3 购自河南的龙须草浆板 Pulp board of *Eulaliopsis binata* from Henan Province

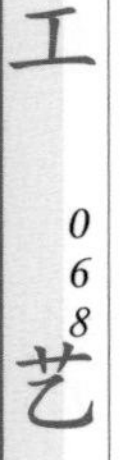

## （二）

## “绿杨宝”书画纸生产的工艺流程

### 壹 浸泡

1

需要将购买的龙须草浆板在浸泡池浸泡24小时左右，使浆板变软、充分吸收水分以便于打浆。

### 贰 制浆

2

制浆工序主要为龙须草制浆，非常简便，将浸泡后的浆板运至打浆机即可打浆。檀皮制浆工序相对繁琐，包括浸泡、蒸煮、清洗、拣黑皮、漂白、清洗、榨干、拣白皮、打浆等工序。因紫光宣纸书画社直接从加工户处购买成熟的青檀皮浆料，所以省略了檀皮制浆工序。

### 叁 配浆

3

将打浆后的龙须草浆料和购进的青檀皮浆料按照一定比例进行混合。根据张世坤的介绍，“绿杨宝”书画纸混合浆中龙须草浆占60%，檀皮浆占40%。

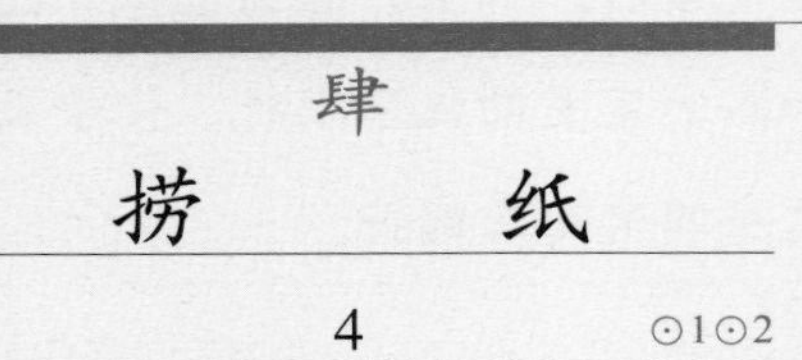

### 肆 捞纸

4　⊙1⊙2

“绿杨宝”书画纸采用半自动喷浆捞纸。半自动喷浆捞纸的原理和过程主要为：混合浆料在半自动喷浆机中会形成循环水流，运输管将浆料传输到各个半自动喷浆口，半自动喷浆帘架下面带有滑轨，可以前后滑动，捞纸比手工捞纸省力。在各个半自动喷浆池外侧有一个阀门作为控制开关。每个喷浆口只需一个捞纸工人。捞纸时，捞纸工将帘子滑向自动喷浆口，用脚踏阀门，浆料从喷浆口喷出时捞纸工用纸帘接住浆料，前后抖动后倒掉余水，再将纸帘从帘床上揭起，将所捞的湿纸反扣到纸板上。一个捞纸工每天能捞6刀左右的四尺“绿杨宝”书画纸。

⊙1 半自动喷浆捞纸 Semi-automatic pulp shooting papermaking

⊙2 提帘放纸 Turning the papermaking screen upside down on the board

伍

## 压 榨

5 ⊙3⊙4

捞纸工下班后，由帮槽工对当天的湿纸帖进行压榨。因湿纸含水多而强度降低，易破，需先盖上盖纸帘、纸板，等水分滤去部分后，再将2～3帘槽一榨抬到一边集中压榨，顺次加上榨杠、千斤顶，逐步加力，直到纸帖不再出水时为止。

陆

## 晒 纸

6 ⊙5⊙6

用切刀将压榨后的纸帖的额头切除一条边，然后将纸帖抬到晒纸房的晒纸架上，晒纸工用短木棍（称为“额枪”）将纸帖额头（上边）、两边划松（称为“做边”）。晒纸时，晒纸工用手指尖点沾一张纸角，将纸整张揭离下来，用刷把刷贴在纸焙上，等整条焙晒满后，依次将干纸揭下，放在纸桌上理好。

柒

## 检验、剪纸

7 ⊙7

对晒好的纸逐张进行检查，如发现纸张破损或其他瑕疵则弃之不用，然后将检验合格的纸放在检验台上，用压纸石压紧纸，由剪纸工用特制大剪刀将纸边裁剪整齐。四尺“绿杨宣”书画纸每次裁剪50张。

⊙3

⊙5

⊙7

⊙4

⊙6

捌

## 包 装

8 ⊙8～⊙10

纸裁剪好后，按每100张分成1刀，用包装纸包好后放入仓库，等待销售和发货。

⊙8

⊙9

⊙10

⊙3/4 室内、室外的压榨 Indoor and outdoor pressing device

⊙5/6 晒纸 Drying the paper

⊙7 检验纸张 Checking the paper

⊙8 待包装的纸 Paper to be packed

⊙9/10 包装好的『绿杨宣』成品 Packaged "Lvyangxuan" paper products

## (三)
## "绿杨宝"书画纸生产的工具设备

### 壹 打浆机 1

用来打散龙须草浆板和混合浆料的设备。

⊙1

⊙2

⊙3

### 贰 纸　槽 2

盛浆料的设施，用水泥浇筑而成。紫光宣纸书画社共有20个半自动喷浆槽。实测紫光宣纸书画社四尺喷浆槽尺寸为：长152 cm，宽127 cm；六尺喷浆槽尺寸为：长200 cm，宽133 cm；尺八屏喷浆槽尺寸为：长250 cm，宽127 cm。

### 叁 纸　帘 3

抄纸的主要工具，由竹子编织而成，表面光滑平整，帘纹细密。根据"绿杨宝"产品规格的不同，纸帘的大小也不同。实测紫光宣纸书画社四尺纸帘尺寸为：长148 cm，宽83 cm；六尺纸帘尺寸为：长200 cm，宽108 cm；尺八屏纸帘尺寸为：长245 cm，宽70 cm。

⊙4

⊙1/2 打浆机 Beating machine

⊙3 纸槽 Papermaking trough

⊙4 纸帘 Papermaking screen

肆

## 帖　架

4

主要用来浇帖和切边，木制。晒纸前需要将纸帖放在帖架上进行浇帖和切边。

⊙5

伍

## 晒纸焙

5

晒纸焙两边由钢板制成，中空贮水，通过加热使水快速升温。晒纸焙表面温度保持在70 ℃左右，一张纸晒上去很快即可烘干。

⊙6

陆

## 额　枪

6

也称“擀棍”，晒纸前做边时用来松纸。紫光宣纸书画社所用额枪长约20 cm。

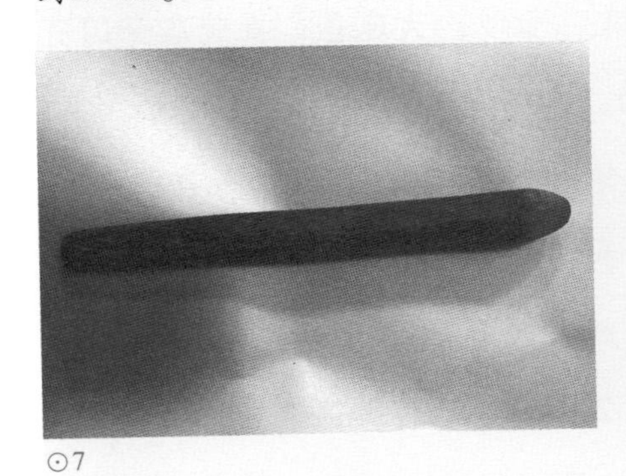

⊙7

柒

## 刷　子

7

紫光宣纸书画社所用刷子尺寸为：长约48 cm，宽约13 cm。刷毛由松针制成。

⊙8

捌

## 压纸石

8

在剪纸过程中，剪纸工人需要用压纸石来压紧纸帖，便于裁剪纸张。

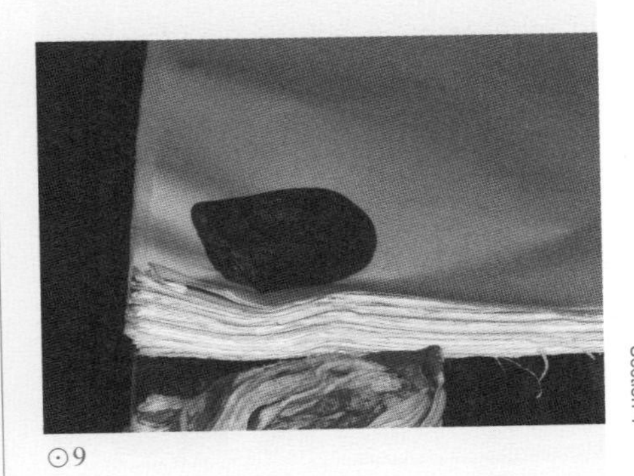

⊙9

⊙5 帖架 Frame for supporting the paper pile

⊙6 晒纸焙 Wall for drying the paper

⊙7 额枪 Tool for patting the paper

⊙8 刷子 Brush

⊙9 压纸石 Stone for pressing the paper

## 五 泾县紫光宣纸书画社的市场经营状况

## 5 Marketing Status of Ziguang Calligraphy and Painting Xuan Paper Agency in Jingxian County

根据2015年7月调查时获取的紫光宣纸书画社纸品销售价目表，其产品包括三尺、四尺、五尺、六尺、八尺对开、尺八屏、六尺对开、四开对开、特六尺等。以四尺和六尺常规品种来说，书法专用纸四尺市场价为222元/刀，六尺为443元/刀；花鸟专用纸四尺市场价为256元/刀，六尺为509元/刀；山水专用纸四尺市场价为280元/刀，六尺为560元/刀；特制“绿杨宝”书画纸四尺市场价为322元/刀，六尺为646元/刀；特制贡宣四尺市场价为373元/刀，六尺为743元/刀；绿杨色宣四尺市场价为305元/刀，六尺为610元/刀。“绿杨宝”系列加工纸产品，种类丰富繁多，规格多样，其中精品生宣册页（规格：35 cm × 50 cm）市场价为623元/本，精品锦盒手卷（规格：0.32 m × 10 m）市场价为553元/盒，绿杨生宣卡纸（规格：38 cm × 44.5 cm）市场价为28元/枚，绿杨宣卡纸（规格：39 cm × 39 cm）市场价为23元/枚，绿杨生宣国画长卷纸（规格：33 cm × 30 m）市场价为287元/卷。

绿扬宝® 安徽泾县紫光宣纸书画社绿杨系列产品价格表

| 品名 \ 规格 | 三尺 | 四尺 | 五尺 | 六尺 | 八尺对开 | 尺八屏 | 六尺对开 | 四尺对开 | 特六尺 | 八尺匹 | 丈二匹 |
|---|---|---|---|---|---|---|---|---|---|---|---|
|  | 53×100 | 69×138 | 84×153 | 97×180 | 69×240 | 53×234 | 48×180 | 35×138 | 69×180 | 124×248 | 146×368 |
| A级净皮 | 116 | 252 | 354 | 504 | 542 | 504 | 252 | 116 | 474 |  |  |
| A级特净 | 139 | 278 | 402 | 556 | 610 | 556 | 278 | 139 | 547 | 3596 | 6289 |
| 书法专用纸 | 211 | 415 | 619 | 829 | 933 | 829 | 415 | 211 | 702 |  |  |
| 花鸟专用纸 | 243 | 478 | 719 | 956 | 1070 | 956 | 478 | 243 | 815 |  |  |
| 山水专用纸 | 259 | 524 | 757 | 1048 | 1138 | 1048 | 524 | 259 | 897 |  |  |
| 特制绿杨宣 | 304 | 602 | 907 | 1209 | 1360 | 1209 | 602 | 304 | 1030 |  |  |
| 特制贡宣 | 346 | 699 | 1048 | 1391 | 1563 | 1391 | 699 | 346 | 1178 | 8650 | 16848 |
| 绿杨白纸带纹 | 270 | 538 | 815 | 1075 | 1219 | 1075 | 538 | 270 | 915 |  |  |
| 绿杨色纸带纹 | 321 | 638 | 965 | 1277 | 1445 | 1277 | 638 | 321 | 1086 |  |  |
| 澄心堂色宣 | 180 | 298 |  | 596 | 586 | 596 | 298 | 180 | 439 |  |  |
| 澄心堂色纸带纹 | 216 | 336 |  | 672 | 960 | 672 | 336 | 216 | 768 |  |  |
| 紫光点墨 |  | 1560 |  | 3120 |  |  |  |  | 1975 |  |  |
| 韵墨 |  | 960 |  | 1920 |  |  |  |  | 3336 |  |  |
| 冰翼 |  | 720 |  | 1440 |  |  |  |  |  |  |  |
| 凝霜 |  | 480 |  | 960 |  |  |  |  |  |  |  |

备注:以上价格含国内普通运费,不含税。款到发货(1刀=100张)　E-mail: zgxzs@163.com　网址:www.ahzgxz.com

厂址：安徽省泾县丁家桥镇　社长：吴报景　电话:0563-5703086　13866964893　安徽泾县紫光宣纸书画社

⊙1

从销售渠道看，调查组通过访谈得知：紫光宣纸书画社目前以经销商和直营店为主。其中，经销商全国约有130家，分布于各大省会和发达副省级城市，如北京、广州、石家庄、兰州、深圳、郑州、成都、宁波、长沙、大连等；直营店全国有7家，分布于陕西、山西、河南、甘肃等省份。

除了产品内销的常规渠道布局体系外，也有一定的销往国外和专项定制的产品渠道。据吴报景介绍，紫光宣纸书画社创办初期主要以出口日本和韩国为主，2006～2007年每年出口约30万元，占到当时年销售额的50%。2008年由于扩大生产，年销售额达到150万元；2010年年销售额达到600万元，其中北京三希堂书画院的定制纸品销售额为200万元；2015年销售额为500万元。近5年平均毛利润率为15%～20%，纯利润率为10%左右。

⊙2

⊙1 紫光宣纸书画社价目表
Price list of Ziguang Calligraphy and Painting Xuan Paper Agency

⊙2 紫光宣纸书画社西安店
Store of Ziguang Calligraphy and Painting Xuan Paper Agency in Xi'an City

据访谈中张世坤介绍，出口的产品叫“画仙纸”，主要销往韩国。这种纸属于定制纸，原料主要是龙须草、三桠树皮和纸边，几种原料按一定比例混合制成纸。韩国人主要用“画仙纸”画山水、花鸟、写意画，也用于书法。紫光宣纸书画社每年集中出口两次左右，出口量根据订单要求而定。

调查组入厂调查时，发现紫光宣纸书画社建有自己的官方网站，主要用以宣传推广，网络销售则以周边网店商家上门拿纸销售为主。

⊙3





## 六 紫光宣纸书画社的品牌文化与民俗故事

## 6 Brand Culture and Stories of Ziguang Calligraphy and Painting Xuan Paper Agency

### 1. “绿杨宝”品牌的来历

紫光宣纸书画社的核心概念是“紫光”。为什么会取一个看起来与“紫光”寓意完全不搭的“绿杨宝”作为产品的品牌呢？面对调查组成员提出的疑问，吴报景回忆起当年的故事：大约在2004～2005年，吴报景在考察厦门市场时发现，当地有个叫绿阳的造纸厂生产出来的纸在台湾地区书画市场上供不应求，而且台湾地区有绿杨集团造纸的技艺。吴报景觉得“紫光”作为产品品牌不如“绿杨”寓意和口碑积累好，因此取了“绿杨宝”作为书画用纸的品牌。2006年正式注册后，“绿杨宣”成为紫光宣纸书画社“龙须草+青檀皮”特制纸的流行品牌。书法家孟繁禧专为“绿杨宣”题诗一首：“绿杨绵薄如蝉翼，皴擦点染尽相宜。龙意文舫添妙品，疾书浅画堪称奇。”

⊙3 出口韩国打包好的『画仙纸』 Packaged “Huaxian Paper” exporting to Korea

⊙1

2. "紫光宣纸书画社"的得名来历

2000年，吴报景在妹婿朱水兵开的北京琉璃厂的直销店做销售兼体验了解市场，该年末，自觉已历练有成的吴报景萌生回家正式办纸厂的念头。著名书法家谢冰岩听说有志的造纸友人吴报景即将回乡办造纸企业，便为其起名为"紫光宣纸书画社"，寓意为朝气蓬勃、紫气东来、光照行业。

## 七 泾县紫光宣纸书画社的传承现状与发展思考

## 7 Current Status of Business Inheritance and Thoughts on Development of Ziguang Calligraphy and Painting Xuan Paper Agency in Jingxian County

泾县紫光宣纸书画社以生产"绿杨宝"书画纸为主，"绿杨宣"系列占据整个产品的比例较大，因添加较高比例的青檀皮材料，其书画性能较优，属于特色高端书画纸。宣纸品种也有生产，种类包括点墨、韵墨等，但份额相对有限，并不是紫光宣纸书画社的发展重点。

据梳理调查中张世坤与吴报景等人介绍的信息后总结，"绿杨宣"的生产模式较有特点：一是加入青檀皮料比例高，有达到40%的，从而提高了书画纸的强度和亲墨性，既有宣纸宜书画的一般性能，又明显比宣纸价格低，性价比高；二是全部采用喷浆工艺，劳动强度与效率都比传统手工有所改善，因而产品的市场竞争力一直较大，工厂的生产也一直呈现良好状态；三是主要原料之一的青檀皮料直接购买泾县当地已加工好

⊙1 紫光宣纸书画社厂牌
Plaque of Ziguang Calligraphy and Painting Xuan Paper Agency

⊙2

的浆料，而主要原料之二的龙须草料本身也为外购浆板，从而完全避开了制浆环节的污染和建环保设施问题。

较强的市场拓展力和竞争力为紫光宣纸书画社在行业整体下行的背景下保持10%左右的净利润率提供了支撑，同时，利润空间也成为生产性传承得以坚守的核心要件。

访谈中，吴报景与调查组成员探讨了如何在满足现代市场需求和传承传统技艺之间取得平衡的问题。吴报景表示：紫光宣纸书画社在2010年开始挖掘和研究特色新品生产，在产品定位上，未来紫光宣纸书画社打算以中档色纸、中档白纸、特色复古纸为主营方向，既不打算往低端书画纸市场发展，也没有挑战“高处不胜寒”的纯宣纸消费市场的想法。

吴报景认为：一方面，由于生产成本、技工资源、劳动强度、大众消费定位等多重影响，中国书画用纸由非木浆原料的手工制作走向半机械和机械化工艺造纸是当代市场的较大趋势。但是，另一方面，“好字画”和“好书画家”需要匹配好的纸，需要保持中国书画纯正的纸墨基因。面对当代方向相逆的矛盾市场，紫光宣纸书画社只能设法找到适合自己的发展路线。

⊙2
吴报景描绘未来发展前景
Wu Baojing forecasting the future development

# 泾县紫光宣纸书画社

# 书画纸

Calligraphy and Painting Paper of Ziguang Calligraphy and Painting Xuan Paper Agency in Jingxian County

『绿杨宝』书画纸透光摄影图

A photo of "Lvyangbao" calligraphy and painting paper seen through the light

# 泾县紫光宣纸书画社 仿古汉纸

Vintage Paper
of Ziguang Calligraphy and Painting Xuan Paper
Agency in Jingxian County

紫光仿古汉纸透光摄影图
A photo of Ziguang vintage paper seen
through the light

# 第五节
# 泾县小岭西山宣纸工艺厂

安徽省
Anhui Province

宣城市
Xuancheng City

泾县
Jingxian County

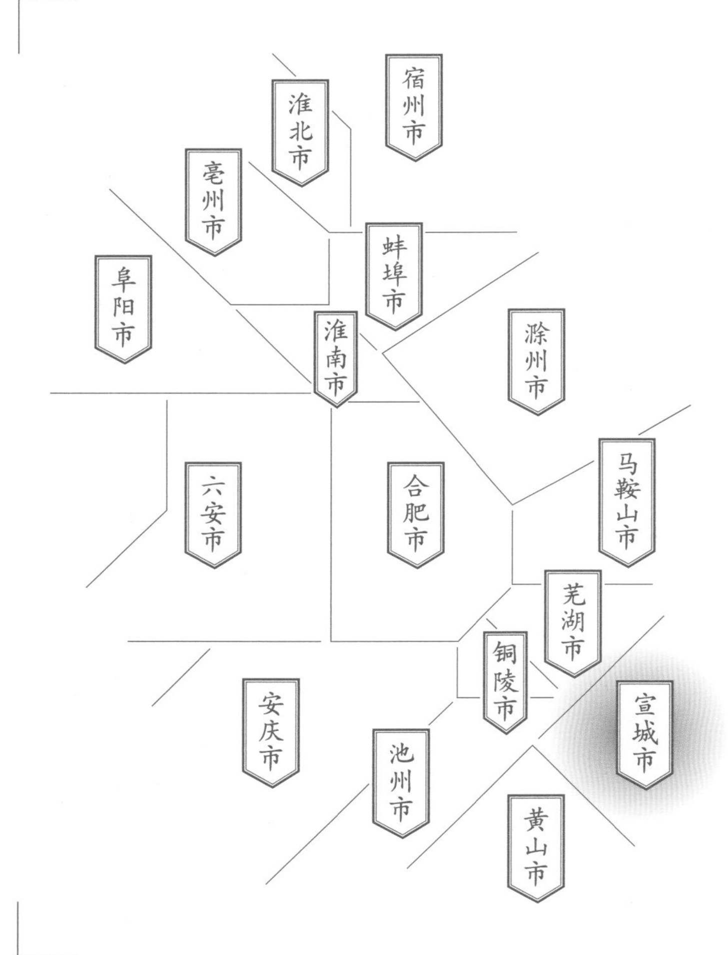


**调查对象**

丁家桥镇
泾县小岭西山宣纸工艺厂
书画纸

Section 5
## Xiaoling Xishan Xuan Paper Craft Factory in Jingxian County

**Subject**

Calligraphy and Painting Paper
of Xiaoling Xishan Xuan Paper Craft Factory
in Jingxian County
in Dingjiagiao Town

# 一 泾县小岭西山宣纸工艺厂的基础信息与生产环境

# 1 Basic Information and Production Environment of Xiaoling Xishan Xuan Paper Craft Factory in Jingxian County

泾县小岭西山宣纸工艺厂，工商登记原名为泾县百顺宣纸工艺品有限公司，坐落于泾县丁家桥镇小岭行政村西山村民组原泾县小岭宣纸厂劳动服务公司旧址，地理坐标为：东经118° 18′ 36″，北纬30° 41′ 21″。小岭西山宣纸工艺厂创办于2001年，主要以书画纸原纸和加工纸生产为主，厂长为曹柏胜。

调查组先后于2015年8月6日和2016年4月25日两次前往小岭西山宣纸工艺厂进行田野调查。调查组通过实地访谈和工艺考察获得的基础生产信息如下：截至2015年8月，共有员工14人、2个手工捞纸槽。按照常规四尺计算，一天可以生产书画纸40刀左右，年销售额约160万元。

⊙1

⊙2

⊙1
百顺宣纸工艺品有限公司大门
Gate of Baishun Xuan Paper Artwork Co., Ltd.

⊙2
小岭西山宣纸工艺厂边的林荫道
Rood by Xiaoling Xishan Xuan Paper Craft Factory

# 泾县小岭西山宣纸工艺厂位置示意图

# Location map of Xiaoling Xishan Xuan Paper Craft Factory in Jingxian County

路线图
泾县县城
↓
泾县小岭西山宣纸工艺厂
Road map from Jingxian County centre to Xiaoling Xishan Xuan Paper Craft Factory in Jingxian County

考察时间
2015年8月 / 2016年4月

Investigation Date
Aug. 2015 / Apr. 2016

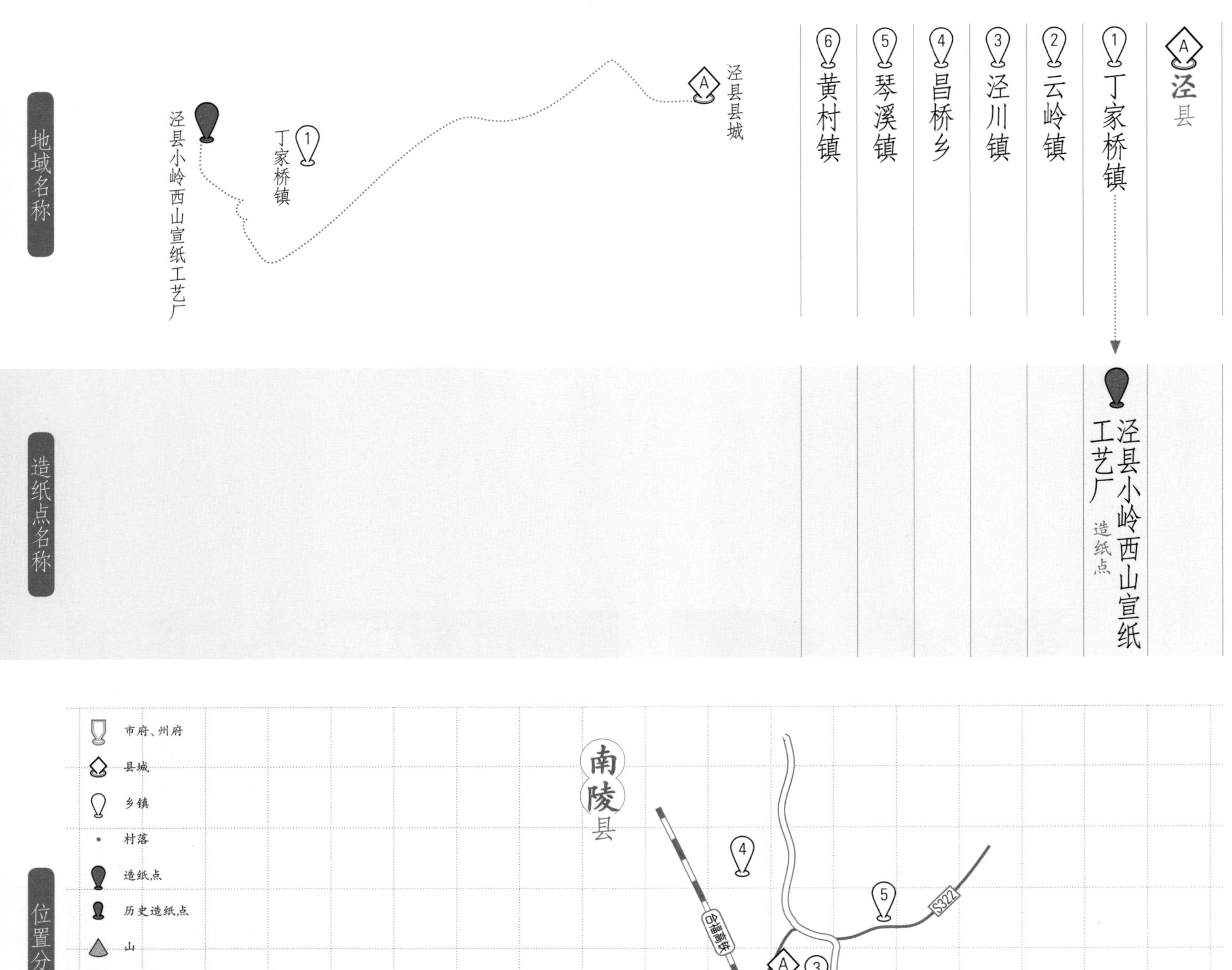

# 二 泾县小岭西山宣纸工艺厂的历史与传承情况

## 2 History and Inheritance of Xiaoling Xishan Xuan Paper Craft Factory in Jingxian County

据调查中创始人兼厂长曹柏胜介绍，泾县小岭西山宣纸工艺厂创办于2001年，办厂时共设2个手工纸槽，有14名左右的工人，十几年来生产槽位与工人人数一直保持相对稳定，技术工人多系小岭宣纸厂手艺在身的老工人，造纸团队工艺水平整齐。工厂初建时，投入约20万元，其中生产启动资金5万多元，购买改制后的小岭宣纸厂劳动服务公司厂房3万多元，设备投入10余万元。2010年，工商登记信息为百顺宣纸工艺品有限公司。2011年，申请注册"曹柏胜"商标。2012年，申请注册"徽家纸号""漫澍"等商标。小岭西山宣纸工艺厂严格分级分类管理产品与商标，"曹柏胜"商标主要用于厂里生产的宣纸[1]、书画纸，"徽家纸号"商标主要用于该厂的加工纸，"漫澍"商标主要用于自产的描红系列纸等宣纸、书画纸的衍生品。

曹柏胜，1961年出生于泾县，小岭曹氏宣纸世家传人。17～18岁时跟随造纸老技工父亲曹康水开始学习制作宣纸原料——燎草，19岁在集体所有制的小岭宣纸厂晒纸车间做学徒，6个月后正式顶班晒纸，在此岗位上工作了六七年后，开始担任晒纸车间负责人。1999年调到小岭宣纸厂劳动服务公司从事宣纸加工纸生产工作，先后从事过洒金等传统熟宣的加工。2000年小岭宣纸厂改制为私营后，曹柏胜筹措资金于2001年创办了小岭西山宣纸工艺厂。

第三代传承人中，儿子曹昊和儿媳妇刘曼曼负责"曹柏胜"古法檀皮宣和"徽家纸号"仿古打印纸等产品的销售，同时经营"小岭西山宣纸工艺厂"网站和淘宝网上的"曹柏胜宣纸直营店""徽家纸号文房馆""漫澍写经室"三个针对不同纸品类别的网店。

⊙1 曹柏胜 Cao Bosheng

[1] 泾县书画纸体系里将采购外省制作的稻草浆板经过浸泡后掺入青檀皮浆料制作的纸称为"宣纸"。

## 三
## 泾县小岭西山宣纸工艺厂的代表纸品及其用途与技术分析

## 3
## Representative Paper and Its Uses and Technical Analysis of Xiaoling Xishan Xuan Paper Craft Factory in Jingxian County

### （一）
### 代表纸品及其用途

调查组成员2015年8月6日调查的信息如下：小岭西山宣纸工艺厂主要生产书画纸和加工纸。曹柏胜认为：目前小岭西山宣纸工艺厂的代表纸品为"曹柏胜"古法檀皮宣（书画纸）和"徽家纸号"仿古打印纸。古法檀皮宣主要用于人物画、山水画、写意画和书法创作练习使用；而仿古打印纸可以使用打印机，实现图案文字打印功能，多用于家谱等需要保存时间较久的文字印制用途。

⊙1

### （二）
### 代表纸品的技术分析

测试小组对采样自小岭西山宣纸工艺厂"曹柏胜"古法檀皮宣所做的性能分析，主要包括厚度、定量、紧度、抗张力、抗张强度、撕裂度、湿强度、白度、耐老化度下降、尘埃度、吸水性、伸缩性、纤维长度和纤维宽度等。按相应要求，每一指标都重复测量若干次后求平均值，其中厚度抽取10个样本进行测试，定量抽取5个样本进行测试，抗张力（强度）抽取20个样本进行测试，撕裂度抽取10个样本进行测试，湿强度抽取20个样本进行测试，白度抽取10个样本进行测试，耐老化度下降抽取10个样本进行测试，尘埃度抽取4个样本进行测试，吸水性抽取10个样本进行测试，伸缩性抽取4个样本进行测试，纤维长度测试了200根纤维，纤维宽度测试了300根纤维。对"曹柏胜"古法檀皮宣进行测试分析所得到的相关性能参数见表3.5。表中列出了各参数的最大值、最小值及测量若干次所得到的平均值或者计算结果。

⊙2

⊙3

⊙1
『曹柏胜』古法檀皮宣
"Cao Bosheng" Xuan Paper made of *Pteroceltis tatarinowii* Maxim. bark in ancient methods

⊙2
『徽家纸号』仿古打印纸
Antique printing paper with "Hui Family Seal"

⊙3
『徽家纸号』不干胶书画加工纸标签
Self-adhesive paper label with "Hui Family Seal"

表3.5 "曹柏胜"古法檀皮宣相关性能参数
Table 3.5 Performance parameters of "Cao Bosheng" Xuan paper made of *Pteroceltis tatarinowii* bark in ancient methods

| 指标 | | | 单位 | 最大值 | 最小值 | 平均值 | 结果 |
|---|---|---|---|---|---|---|---|
| 厚度 | | | mm | 0.125 | 0.085 | 0.098 | 0.098 |
| 定量 | | | g/m² | — | — | — | 33.0 |
| 紧度 | | | g/cm³ | — | — | — | 0.337 |
| 抗张力 | 纵向 | | N | 15.4 | 12.5 | 14.2 | 14.2 |
| | 横向 | | N | 10.9 | 8.0 | 8.3 | 8.3 |
| 抗张强度 | | | kN/m | — | — | — | 0.750 |
| 撕裂度 | 纵向 | | mN | 430 | 360 | 394 | 394 |
| | 横向 | | mN | 470 | 410 | 442 | 442 |
| 撕裂指数 | | | mN·m²/g | — | — | — | 11.7 |
| 湿强度 | 纵向 | | mN | 910 | 820 | 853 | 853 |
| | 横向 | | mN | 520 | 430 | 492 | 492 |
| 白度 | | | % | 67.1 | 66.9 | 67.0 | 67.0 |
| 耐老化度下降 | | | % | — | — | — | 1.2 |
| 尘埃度 | 黑点 | | 个/m² | — | — | — | 48 |
| | 黄茎 | | 个/m² | — | — | — | 28 |
| | 双浆团 | | 个/m² | — | — | — | 0 |
| 吸水性 | | | mm | — | — | — | 15 |
| 伸缩性 | 浸湿 | | % | — | — | — | 0.48 |
| | 风干 | | % | — | — | — | 0.73 |
| 纤维 | 皮 | 长度 | mm | 4.28 | 0.68 | 1.58 | 1.58 |
| | | 宽度 | μm | 23.0 | 1.0 | 10.0 | 10.0 |
| | 草 | 长度 | mm | 2.43 | 0.42 | 1.19 | 1.19 |
| | | 宽度 | μm | 13.0 | 1.0 | 5.0 | 5.0 |

由表3.5中的数据可知，"曹柏胜"古法檀皮宣最厚约是最薄的1.471倍，经计算，其相对标准偏差为0.011，纸张厚薄较为一致。所测"曹柏胜"古法檀皮宣的平均定量为33.0 g/m²。通过计算可知，"曹柏胜"古法檀皮宣紧度为0.337 g/cm³，抗张强度为0.750 kN/m，抗张强度值较小。所测"曹柏胜"古法檀皮宣撕裂指数为11.7 mN·m²/g，撕裂度较大；湿强度纵横平均值为673 mN，湿强度较大。

所测"曹柏胜"古法檀皮宣平均白度为67.0%，白度较高，这是由于其加工过程中有漂白工序。白度最大值是最小值的1.003倍，相对标准偏差为0.067，白度差异相对较小。经过耐老化测试后，耐老化度下降1.2%。

所测"曹柏胜"古法檀皮宣尘埃度指标中黑点为48个/m²，黄茎为28个/m²，双浆团为0个/m²。吸水性纵横平均值为15 mm，纵横差为2.4 mm。伸缩性指标中浸湿后伸缩差为0.48%，风干后伸缩差为0.73%，说明"曹柏胜"古法檀皮宣伸缩性差异不大。

“曹柏胜”古法檀皮宣在10倍、20倍物镜下观测到的纤维形态分别见图★1、图★2。所测“曹柏胜”古法檀皮宣皮纤维长度为：最长4.28 mm，最短0.68 mm，平均长度为1.58 mm；纤维宽度为：最宽23.0 μm，最窄1.0 μm，平均宽度为10.0 μm。草纤维长度为：最长2.43 mm，最短0.42 mm，平均长度为1.19 mm；纤维宽度为：最宽13.0 μm，最窄1.0 μm，平均宽度为5.0 μm。“曹柏胜”古法檀皮宣润墨效果见图⊙1。

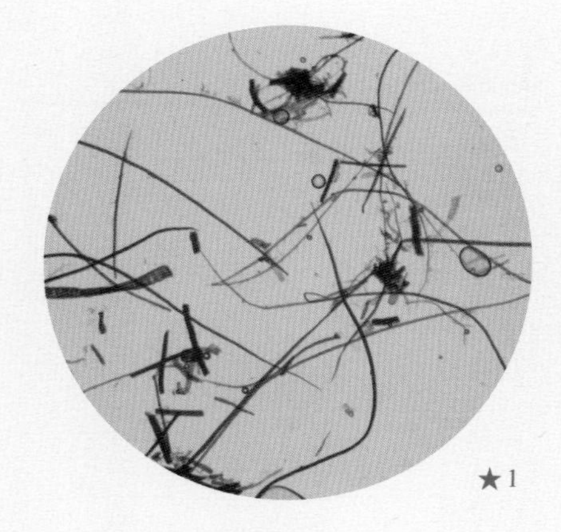

★1

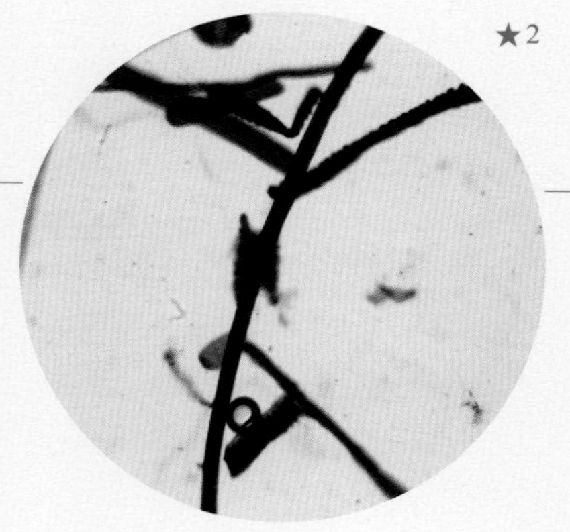

★2

⊙1

## 四 泾县小岭西山宣纸工艺厂代表纸品生产的原料、工艺流程与工具设备

## 4 Raw Materials, Papermaking Techniques and Tools of Representative Paper in Xiaoling Xishan Xuan Paper Craft Factory in Jingxian County

### （一）“曹柏胜”古法檀皮宣和“徽家纸号”仿古打印纸的原料

#### 1. 主料

“曹柏胜”古法檀皮宣主要使用龙须草浆板和檀皮浆混合制成，“徽家纸号”仿古打印纸则使用龙须草浆板制成。2015年8月调查时曹柏胜介绍：龙须草浆板从河南郑州运过来，价格约为11 000元/t；檀皮则是以850元/担（1担为50 kg）的价格从小岭农户手中收购毛皮，再在

★1 『曹柏胜』古法檀皮宣纤维形态图（10×）
Fibers of "Cao Bosheng" Xuan paper made of *Pteroceltis tatarinowii* Maxim. bark in ancient methods (10× objective)

★2 『曹柏胜』古法檀皮宣纤维形态图（20×）
Fibers of "Cao Bosheng" Xuan paper made of *Pteroceltis tatarinowii* Maxim. bark in ancient methods (20× objective)

⊙1 『曹柏胜』古法檀皮宣润墨效果
Writing performance of "Cao Bosheng" Xuan paper made of *Pteroceltis tatarinowii* Maxim. bark in ancient methods

⊙2

⊙3

⊙4

泾县园林牧场一厂家处以6元/kg的加工价格进行蒸煮、漂白，通常一锅375 kg干皮可以加工出400 kg湿白皮。

2. 辅料

（1）分张剂。“曹柏胜”古法檀皮宣和“徽家纸号”仿古打印纸使用日本进口分张剂，替代传统的纸药功能。据曹柏胜介绍，分张剂由泾县当地代理商代为购买，价格为300元/kg。

（2）水源。“曹柏胜”古法檀皮宣和“徽家纸号”仿古打印纸生产用水均来自厂房周边的山泉水。枯水期会使用水泵抽水。实测造纸用水的pH为6.87，呈弱酸性。

## （二）“曹柏胜”古法檀皮宣和“徽家纸号”仿古打印纸生产的工艺流程

据曹柏胜介绍，综合调查组2015年8月6日在小岭西山宣纸工艺厂的实地调查，“曹柏胜”古法檀皮宣和“徽家纸号”仿古打印纸的生产工艺流程为：

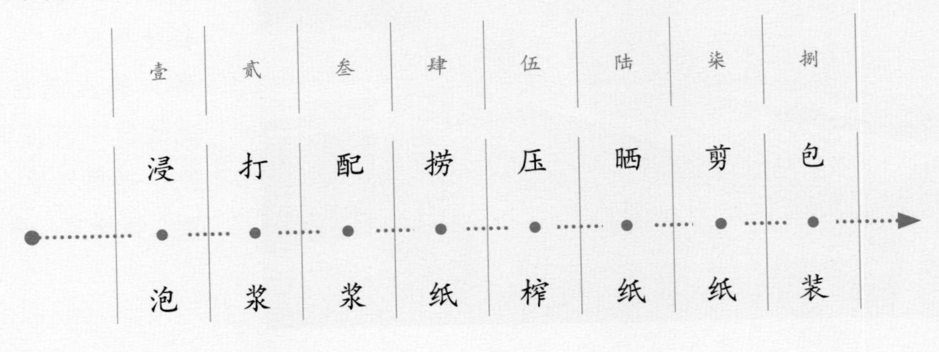


⊙2 古法制作檀皮宣的纸浆 Paper pulp of *Pteroceltis tatarinowii* Maxim. bark for making Xuan paper in ancient methods

⊙3 进口分张剂标签 Label of imported mucilage for separating the paper layers

⊙4 流经厂区旁的山泉水 Mountain spring flowing by the factory

## 壹 浸泡

### 1 ⊙1

制作“曹柏胜”古法檀皮宣和“徽家纸号”仿古打印纸时，均要将龙须草浆板浸泡一天，使浆板充分吸收水分软化，在浸泡的同时将浆板中的杂质去除。

⊙1

## 贰 打浆

### 2 ⊙2⊙3

制作“曹柏胜”古法檀皮宣时，需将浸泡好的龙须草浆板与漂白好的檀皮放在一起打浆。制作“徽家纸号”仿古打印纸时，只需将龙须草浆板打浆即可。

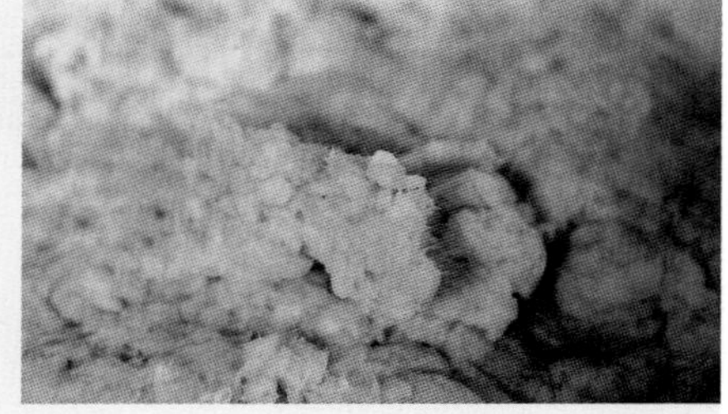

⊙2

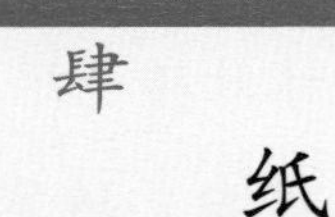

⊙3

## 叁 配浆

### 3

制作“曹柏胜”古法檀皮宣的常规配比是一次20 kg檀皮、55 kg龙须草浆板，但在实际生产中，皮、草的具体配比是根据市场的需要动态调整的，客户对纸的韧性要求越高，皮料的比例越高。制作“徽家纸号”仿古打印纸则是在龙须草浆板中配适量仿古色水性颜料。

## 肆 捞纸

### 4 ⊙4

“曹柏胜”古法檀皮宣和“徽家纸号”仿古打印纸捞纸工艺与其他手工捞书画纸工艺相同。两个捞纸工人站在捞纸槽两侧，一人为掌帘工，一人为抬帘工。捞纸时，两人同时将纸帘放入捞纸槽的水中，上下晃动两次，立即将纸帘端平抬出水面，待多余浆料从纸帘旁边漏出，留在纸帘上的浆料就形成了一张湿纸。采用两遍水形成湿纸后，由掌帘师傅把纸帘从帘床上平稳地取下，倒扣放置在纸槽旁边的湿纸板上，轻轻地揭起纸帘，使纸帘与湿纸分离，将湿纸留在湿纸板上。在此过程中，抬帘师傅用计数器计数。据曹柏胜介绍，一天一个一改二纸槽可生产24刀左右的纸。

⊙4

⊙1 浸泡龙须草浆板
Soaking pulp board of *Eulaliopsis binata*

⊙2 『曹柏胜』古法檀皮宣浆团
Pulp balls of "Cao Bosheng" Xuan paper made of *Pteroceltis tatarinowii* Maxim. bark in ancient methods

⊙3 『徽家纸号』仿古打印纸浆团
Pulp balls of antique printing paper with "Hui Family Seal"

⊙4 捞纸
Papermaking

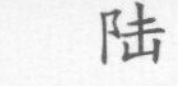



## 伍 压榨

### 5

⊙5

捞纸师傅在每天捞完纸后，将自己捞好的纸放在木榨上使用千斤顶进行压榨。压榨时力度由小变大，并且要缓慢，否则会压坏纸张。湿纸不再出水时，压榨结束。压榨时间约1小时。

⊙5

## 陆 晒纸

### 6

⊙6

（1）将压榨好的纸帖靠在钢板制作的焙墙上烤，让水分和纸药蒸发。

（2）将半干纸帖的额头切除一条小边（称为“杀额”），使额头规整以便于分张。

（3）把杀额后的纸帖放在纸架上，用额枪将额头、顺边、反边三边打松，以便更好地分张。

（4）晒纸工人用手指将纸帖的掐角揭下纸角，然后先额头后中间，从左往右将整张湿纸揭下来，用刷把将湿纸张贴在纸焙上。整个纸焙晒满后，将先晒的纸揭下来。如此往复。

“徽家纸号”仿古打印纸最后上墙工序稍有不同，是将两张纸放在一起晒，贴满整个晒纸墙后，就从开始晒纸处将已经蒸发干燥的纸取下来。

⊙6

## 柒 剪纸

### 7

⊙7⊙8

剪纸工序包含检纸和剪纸两步。检纸是对晒好的纸进行人工检验，将有洞、有杂质的不合格纸挑出。再完成对纸边的剪切。“曹柏胜”古法檀皮宣以50张为一个刀口，压上石头，剪纸工人持箭步姿态用特制的大剪刀一气呵成地完成剪纸。“徽家纸号”仿古打印纸则用裁纸机根据客户需求裁切成不同尺寸，一般为A3、A4等常用规格。

⊙7

⊙8

## 捌 包装

### 8

⊙9

剪好的纸按照100张/刀加盖“曹柏胜”或“徽家纸号”商标、厂名、纸张品种以及尺寸等印章。包装完毕后运入贮纸仓库。

⊙9

⊙5 榨纸 Pressing the paper

⊙6 晒纸 Drying the paper

⊙7 检纸 Checking the paper

⊙8 裁纸机 Machine for cutting the paper

⊙9 盖章 Stamping the paper

## （三）“曹柏胜”古法檀皮宣和“徽家纸号”仿古打印纸生产的工具设备

### 壹 打浆机 1

用来制作浆料，机械自动搅拌。

⊙1

### 贰 捞纸槽 2

盛浆设施，水泥浇筑。实测小岭西山宣纸工艺厂所用的捞纸槽尺寸为：长345 cm，宽185 cm，高80 cm。可捞一改二（即一帘为两张四尺）的纸。

⊙2

### 叁 帘床 3

用于捞纸时放置纸帘，木头外框，里面用塑料杆铺成。实测小岭西山宣纸工艺厂所用的帘床尺寸为：长309 cm，宽88 cm。

⊙3

### 肆 切帖刀 4

晒纸前用来切帖、杀额，钢制。实测小岭西山宣纸工艺厂所用切帖刀尺寸为：长38 cm，宽7 cm。其中刀长22 cm。

⊙4

⊙1 打浆机 Beating machine

⊙2 捞纸槽 Papermaking trough

⊙3 帘床 Frame for supporting the papermaking screen

⊙4 切帖刀 Knife for cutting the paper pile

## 伍 晒纸刷 5

晒纸时将纸刷上晒纸墙，刷柄为木制，刷毛为松针。实测小岭西山宣纸工艺厂所用晒纸刷尺寸为：长48 cm，宽13 cm。

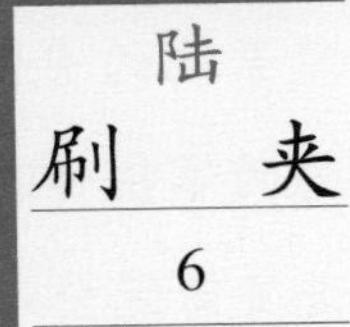

## 陆 刷夹 6

用来夹晒纸刷的工具。实测小岭西山宣纸工艺厂所用刷夹尺寸约为：长55 cm，宽5 cm。

⊙6

⊙5

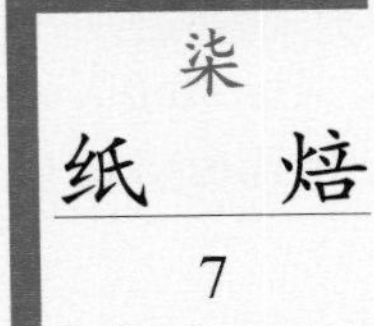

## 柒 纸焙 7

用来晒纸的设施，用两块长方形钢板焊接而成，中间贮水，烧柴加热，双面墙，可以两边晒纸。

⊙7

## 捌 压纸石 8

检纸和剪纸时压纸的石头，一般为河滩里没有棱角的普通石头。

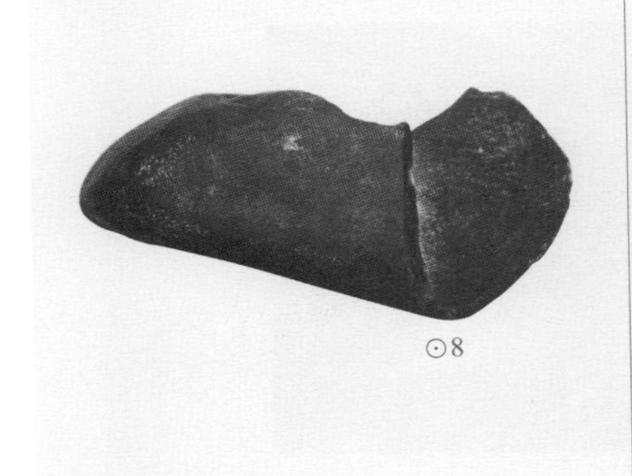

⊙8

## 玖 剪刀 9

用于剪纸，剪刀口为钢制，其余部分为铁制，泾县为造纸业专门打造的工具。

⊙9

⊙5 晒纸刷 Brush for drying the paper

⊙6 刷夹 Brush holder

⊙7 工人正在往焙墙上刷米汤，使焙面洁净 Workers brushing rice paste to clean the drying wall

⊙8 压纸石 Stone for pressing the paper

⊙9 剪刀 Shears

# 五 泾县小岭西山宣纸工艺厂的市场经营状况

## 5 Marketing Status of Xiaoling Xishan Xuan Paper Craft Factory in Jingxian County



"曹柏胜"古法檀皮宣和"徽家纸号"仿古打印纸等产品的销售由曹柏胜的儿子曹旻负责，主要通过实体经销、网络销售和微信销售。

实体经销主要是向在上海、合肥、南京、保定等地的实体店以配货方式营销，其中河北保定的两家实体店店主与曹柏胜是亲戚关系，其余的进货商完全是多年建立的业务联系。

网络销售除经营有"小岭西山宣纸工艺厂"网站，在网站上实现直销外，还在淘宝网分别开设了"曹柏胜宣纸直营店""徽家纸号文房馆""漫澍写经室"三个针对不同纸品类别的网店，同时通过微信微店开展微营销。

据曹柏胜介绍，儿子曹旻和儿媳妇刘曼曼一直以市场为导向，以订单定生产，不断开发新产品，如不干胶书画加工纸标签、热熔胶贴、植物云龙皮纸文化用品等。

截至调查组2016年4月25日第二次入厂调查时，四尺"曹柏胜"古法檀皮宣根据配比不同销售价为200～500元/刀，普通四尺书画纸销售价为100～200元/刀；"徽家纸号"仿古打印纸销售价格为150元/包（300张/包）。

小岭西山宣纸工艺厂年销售额约为160万元，利润率在5%～10%范围波动。

⊙1

⊙1 仓库 Warehouse

## 六
## 泾县小岭西山宣纸工艺厂的业态传承现状与发展思考

## 6
## Current Status of Business Inheritance and Thoughts on Development of Xiaoling Xishan Xuan Paper Craft Factory in Jingxian County

泾县丁家桥镇特别是小岭村多家民营造纸厂家或经销商都有从原“公家大厂”小岭宣纸厂改制解体获取生存发展资源的血缘背景。小岭西山宣纸工艺厂也不例外，不仅创始人曹柏胜有在小岭宣纸厂约22年的从业经历，而且厂里的技术工人大多数来自小岭宣纸厂，属于成熟的活态工艺链整体进入西山宣纸工艺厂。这些小岭宣纸厂的老工人群体较原生态地传承了小岭造纸源远流长的精湛技艺及工匠精神，成为小岭西山宣纸工艺厂良好传承生态得以接续的关键要素。

在整体手工造纸行业传统消费空间萎缩的背景下，小岭西山宣纸工艺厂与其他手工纸厂家一样，在激烈的竞争中努力放大传统优质基因，但值得关注的是：曹柏胜创新产品与渠道，寻找和发掘传统纸的新用途，突破性地服务市场新需求，体现出传统工艺行业难能可贵的发展路径探索。

（1）“徽家纸号”仿古打印纸颠覆性地拓宽了泾县宣纸文化体系里手工纸的规模化使用范围。虽然目前只是初步试水打印用纸大市场，在适用性、特色提炼、性价比优化、商务推广方案等诸多方面尚需持续提升，但这一产品化、品牌化的尝试形成了对地域传统的积极突破。

（2）不仅整体快速地由实体经销向网上商务营销的“互联网+手工纸”的渠道新体系发展，而且一开始就按照纸品类别实施网络营销的品牌区隔规划，如在淘宝网分别开设了“曹柏胜宣纸直营店”“徽家纸号文房馆”“漫澍写经室”三个针对不同纸品类别的网店，这种一家小手工纸厂分纸品开网上商店的做法在手工纸行业中尚不多见。同时，通过微信微店开展社交微营销的工作也在三代传承人的运营下展开。

⊙2

⊙2 小岭西山宣纸工艺厂的常用标签章 Common label seal of Xiaoling Xishan Xuan Paper Craft Factory

（3）植物云龙皮纸文化用品（如笔记本）、不干胶书画加工纸标签、热熔胶贴等系列手工纸新用品使手工纸与创意文化跨界交融。这些创新产品开拓了手工纸的新市场，强化了面向当代生活新需求的手工纸发展。

# 泾县小岭西山宣纸工艺厂 书画纸

## Calligraphy and Painting Paper of Xiaoling Xishan Xuan Paper Craft Factory in Jingxian County

「曹柏胜」古法檀皮宣透光摄影图
A photo of "Cao Bosheng" Xuan paper made of *Pteroceltis tatarinowii* Maxim. bark in ancient methods seen through the light

# 泾县小岭西山宣纸工艺厂

# 仿古打印纸

Vintage Printing Paper of Xiaoling Xishan Xuan Paper Craft Factory in Jingxian County

仿古打印纸透光摄影图

A photo of vintage printing paper seen through the light

# 第六节

# 安徽澄文堂宣纸艺术品有限公司

安徽省
Anhui Province

宣城市
Xuancheng City

泾县
Jingxian County

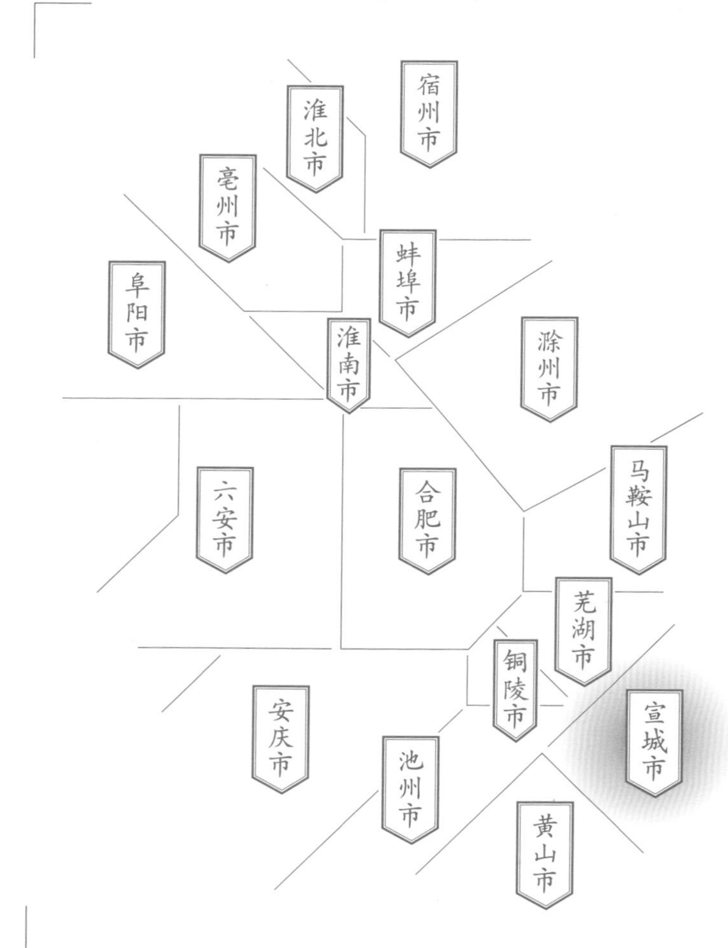


**调查对象**

黄村镇
安徽澄文堂宣纸艺术品有限公司
书画纸

Section 6

## Chengwentang Xuan Paper Artwork Co., Ltd. in Anhui Province

**Subject**

Calligraphy and Painting Paper
of Chengwentang Xuan Paper Artwork Co., Ltd.
in Anhui Province
in Huangcun Town

## 一
## 安徽澄文堂宣纸艺术品有限公司的基础信息与生产环境

### 1
### Basic Information and Production Environment of Chengwentang Xuan Paper Artwork Co., Ltd. in Anhui Province

安徽澄文堂宣纸艺术品有限公司坐落于泾县黄村镇九峰行政村，厂区所在地海拔120 m，地理坐标为：东经118° 19′ 7″，北纬30° 33′ 11″。安徽澄文堂宣纸艺术品有限公司（简称“澄文堂”）创办于2004年，主要生产手工捞制的书画纸。

2016年5月15日，调查组前往九峰村澄文堂生产厂区进行田野调查。通过实地访谈和现场考察获得的基础生产信息如下：2016年春季，澄文堂有员工18人，共设16帘纸槽，其中一改二纸槽3帘、单槽13帘。调查时，因手工书画纸行情较差，只有2帘一改二纸槽在正常生产。据企业负责人王四海的说法，照常规四尺计算，2帘槽一天可以生产书画纸40刀左右。澄文堂年销售额为200余万元，按照满负荷产能来算，澄文堂是泾县最大的手工书画纸生产企业。

⊙1

黄村镇九峰村位于泾县第二高峰——承流峰山下，是王姓、李姓集聚地，素有大康（地名）王府与李村的说法，人口最多时，常年有3000多人居住。九峰村村头百岁坊，建造于清代嘉庆年间，是泾县现存的唯一一座清代牌坊。此外，九峰村还有进士第、李氏宗祠、凤池门等明清时期建筑，显示出该村古风幽远、人文景观厚重的特点。

⊙2

承流峰以“承流积翠”景观位列泾县古代八大美景之一，因传说古代有承流府君住在山中而得名。《大清一统志》载：“山有九峰，皆极耸秀。”这应为九峰村村名的来源。山顶古有握月庵，半山中有元虚洞，山下有仙坛，唐代有游方僧人在坛基上建屋居住，有诗云：“远望疏密树，云外有无山。倚松看子落，隔竹听泉流。”据调查中村人回忆，大康王府和李村曾制作过竹纸，旧日曾是泾县竹纸制作的重要区域之一，20世纪60年代停产，至今遗迹已无存。

⊙1 百岁坊 Baisui (hundred years) Memorial Gateway

⊙2 承流峰 Chengliu Peak

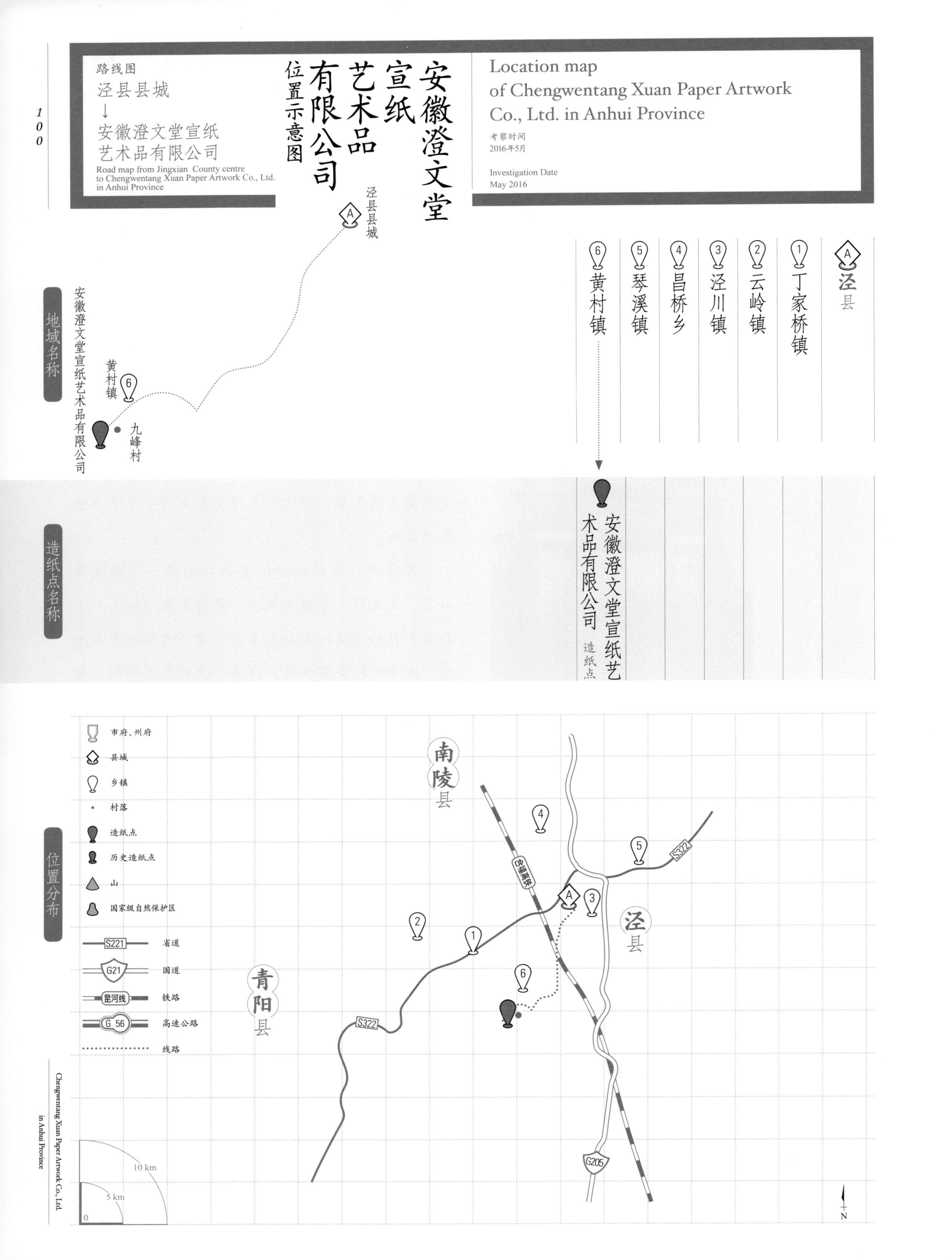
路线图
泾县县城
↓
安徽澄文堂宣纸艺术品有限公司
Road map from Jingxian County centre to Chengwentang Xuan Paper Artwork Co., Ltd. in Anhui Province
安徽澄文堂宣纸艺术品有限公司位置示意图
Location map of Chengwentang Xuan Paper Artwork Co., Ltd. in Anhui Province
考察时间
2016年5月
Investigation Date
May 2016
泾县县城
地域名称
安徽澄文堂宣纸艺术品有限公司
黄村镇
九峰村
泾县
丁家桥镇
云岭镇
泾川镇
昌桥乡
琴溪镇
黄村镇
造纸点名称
安徽澄文堂宣纸艺术品有限公司
造纸点
位置分布
市府、州府
县城
乡镇
村落
造纸点
历史造纸点
山
国家级自然保护区
S221 省道
G21 国道
昆河线 铁路
G 56 高速公路
线路
南陵县
青阳县
泾县
合福高铁
S322
G205
10 km
5 km
0
N

## 二
## 安徽澄文堂宣纸艺术品有限公司的历史与传承情况

## 2
## History and Inheritance of Chengwentang Xuan Paper Artwork Co., Ltd. in Anhui Province

安徽澄文堂宣纸艺术品有限公司注册地为泾县黄村镇九峰村。据调查中澄文堂法人代表王四海介绍，澄文堂宣纸艺术品有限公司创办于2004年，办厂时的最初名称是泾县黄村镇九峰宣纸厂，使用“九峰”为商标，有4个槽的生产规模，就近在九峰村一带招收熟练工。这些熟练工原为丁家桥一带宣纸厂、书画纸厂的技术工人，见村里办厂招人，遂选择辞去原工作就近务工。

2005年，澄文堂宣纸艺术品有限公司与中国宣纸集团公司（现中国宣纸股份有限公司）签订合同，按照约定标准专门为中国宣纸集团公司生产书画纸，中国宣纸集团公司定期派人现场验收，在产品上打上“红星”书画纸印章并包装、装箱后运回中国宣纸集团公司。因代工生产产能扩大，2013年加建3帘槽。2014年，收购原九峰小学（因为生源少，所以学校停办，学龄儿童转到黄村镇中心学校读书，原校址拍卖）和九峰粮食收购点后进行厂房扩建，扩建后的厂房占地10 000 m$^2$，设16帘槽生产书画纸，多数产品为替中国宣纸集团公司代工，少量书画纸打上“澄文堂”品牌自行销售。

2015年，因企业属于个体工商户，在与业务发包大户中国宣纸股份有限公司财务往来时，开具票据比较麻烦，加上年纳税达100多万元，已具备一般纳税人资质，便注销了泾县黄村镇九峰宣纸厂，易名为安徽澄文堂宣纸艺术品有限公司，同时提交了“澄文堂”商标的注册申请，截至调查时的2016年5月，“澄文堂”商标已被国家商标局正式受理并公告。

王四海，1970年生于泾县黄村镇九峰村，父亲原为九峰村村主任。初中毕业后，他先到李园宣纸厂食堂当厨师，两年后辞职单独做生意，贩卖过毛竹、木材，也专门从事过古建筑的拆建生意。2004年，他见到李园宣纸厂的原来同事先后开办了书画纸厂，也萌生了开厂的念头，就与当

澄文堂
商品/服务 绘画用纸; 墨汁; 毛笔; 宣
具; 查看详细信息
类似群 1601;1602;1611;1612
申请/注册号 18695156 申请日期 2015年12
申请人名称（中文） 安徽澄文堂宣纸艺术品有限公司
申请人名称（英文）
申请人地址（中文） 安徽省泾县黄村镇九峰村
申请人地址（英文）

⊙1

⊙2

⊙1 中国商标网公告信息截图 Screen shot of bulletin information on official website of National Trademark Office

⊙2 『澄文堂』厂区大门 Gate of Chengwentang Xuan Paper Artwork Co., Ltd.

⊙1

⊙2



时的九峰村村主任郑礼红商量，各筹资4万元投资租赁村里空房合资办厂。初建时，王四海既要在厂里管理纸的生产，又要在外联系纸的销售。

郑礼红，1970年生于九峰村，1995年高中毕业后到杭州打工，1999年底回家务农，2000年当选为九峰村村主任。2004年与王四海合资的企业开业后，在村里工作不忙时，郑礼红便到厂里帮王四海打下手。2005年，根据地方政府提高村干部综合素质的要求，他参加了安徽农业大学经济管理专业大专班函授学习，2007年毕业。2011年，随着安徽澄文堂宣纸艺术品有限公司规模扩大，郑礼红便辞去村主任一职，专门在厂里管理生产和内勤。



## 三 安徽澄文堂宣纸艺术品有限公司的代表纸品及其用途与技术分析

## 3 Representative Paper and Its Uses and Technical Analysis of Chengwentang Xuan Paper Artwork Co., Ltd. in Anhui Province

### (一) 代表纸品及其用途

澄文堂生产的书画纸主要使用两个商标，除了经过中国宣纸集团公司现场验收后，打上“红星”商标外，其余由澄文堂接单后生产的产品均打上“澄文堂”商标。“红星”书画纸分一般书画纸和一星、二星、三星书画纸，主要规格为四尺。据王四海介绍，澄文堂自有品牌的书画纸品种较多，主要有四尺、五尺、六尺、尺八屏等规格，根据配料除生产一般书画纸和一星、二星、



⊙1 访谈王四海（左）
Interviewing Wang Sihai (left)

⊙2 郑礼红协助扳榨
Zheng Lihong assisting with the pressing step

三星书画纸外，还生产麻纸、槽底纸等。主要代表产品是三星书画纸，出厂售价为600元/刀。尽管在市场上以书画纸营销，但因产品货真价实，比很多打上宣纸名头的产品更好用，故不仅在市场上普遍受到欢迎，而且还成为中国宣纸集团公司出口到日本的主要产品之一。

## (二)
## 代表纸品的性能分析

测试小组对采样的澄文堂所产“红星”书画纸所做的性能分析，主要包括厚度、定量、紧度、抗张力、抗张强度、撕裂度、湿强度、白度、耐老化度下降、尘埃度、吸水性、伸缩性、纤维长度和纤维宽度等。按相应要求，每一指标都重复测量若干次后求平均值，其中厚度抽取10个样本进行测试，定量抽取5个样本进行测试，抗张力（强度）抽取20个样本进行测试，撕裂度抽取6个样本进行测试，湿强度抽取5个样本进行测试，白度抽取10个样本进行测试，耐老化度下降抽取10个样本进行测试，尘埃度抽取1个样本进行测试，吸水性抽取5个样本进行测试，伸缩性抽取4个样本进行测试，纤维长度测试了200根纤维，纤维宽度测试了300根纤维。对澄文堂所产“红星”书画纸进行测试分析所得到的相关性能参数见表3.6。表中列出了各参数的最大值、最小值及测量若干次所得到的平均值或者计算结果。

表3.6 “红星”书画纸相关性能参数
Table 3.6 Performance parameters of “Red Star” calligraphy and painting paper

| 指标 | | 单位 | 最大值 | 最小值 | 平均值 | 结果 |
|---|---|---|---|---|---|---|
| 厚度 | | mm | 0.097 | 0.079 | 0.089 | 0.089 |
| 定量 | | g/m² | — | — | — | 21.9 |
| 紧度 | | g/cm³ | — | — | — | 0.246 |
| 抗张力 | 纵向 | N | 17.1 | 10.5 | 14.8 | 14.8 |
| | 横向 | N | 11.1 | 6.2 | 9.4 | 9.4 |
| 抗张强度 | | kN/m | | | | 0.807 |
| 撕裂度 | 纵向 | mN | 235.1 | 203.1 | 221.1 | 221.1 |
| | 横向 | mN | 291.7 | 256.0 | 274.5 | 274.5 |
| 撕裂指数 | | mN · m²/g | — | — | — | 22.7 |
| 湿强度 | 纵向 | mN | 2 527 | 1 982 | 1 188 | 1 019 |
| | 横向 | mN | 1 936 | 1 537 | 850 | |
| 白度 | | % | 71.6 | 70.6 | 71.1 | 71.1 |
| 耐老化度下降 | | % | — | — | — | 2.3 |
| 尘埃度 | 黑点 | 个/m² | — | — | — | 48 |
| | 黄茎 | 个/m² | — | — | — | 52 |
| | 双浆团 | 个/m² | — | — | — | 0 |

续表

| 指标 | | 单位 | 最大值 | 最小值 | 平均值 | 结果 |
|---|---|---|---|---|---|---|
| 吸水性 | | mm | — | — | — | 15 |
| 伸缩性 | 浸湿 | % | — | — | — | 0.13 |
| | 风干 | % | — | — | — | 0.18 |
| 纤维 | 长度 | mm | 1.34 | 0.10 | 0.56 | 0.56 |
| | 宽度 | μm | 19.5 | 0.4 | 5.3 | 5.3 |





由表3.6中的数据可知，“红星”书画纸最厚约是最薄的1.23倍，经计算，其相对标准偏差为0.006 5，纸张厚薄较为一致。所测“红星”书画纸的平均定量为21.9 g/m²。通过计算可知，“红星”书画纸抗张强度为0.807 kN/m。所测“红星”书画纸撕裂指数为22.7 mN · m²/g；湿强度纵横平均值为1 019 mN，湿强度较大。

所测“红星”书画纸平均白度为71.1%，白度较高。白度最大值是最小值的1.014倍，相对标准偏差为0.37，白度差异相对较小。经过耐老化测试后，耐老化度下降2.3%。

所测“红星”书画纸尘埃度指标中黑点为48个/m²，黄茎为52个/m²，双浆团为0个/m²。吸水性纵横平均值为15 mm，纵横差为3 mm。伸缩性指标中浸湿后伸缩差为0.13%，风干后伸缩差为0.18%。

“红星”书画纸在10倍和20倍物镜下观测到的纤维形态分别见图★1、图★2。所测“红星”书画纸纤维长度：最长1.34 mm，最短0.10 mm，平均长度为0.56 mm；纤维宽度：最宽19.5 μm，最窄0.4 μm，平均宽度为5.3 μm。“红星”书画纸润墨效果见图⊙1。

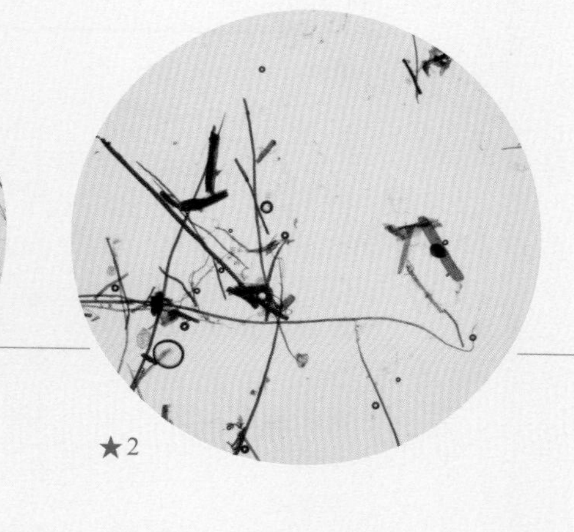

★1　★2

⊙1

★1
『红星』书画纸纤维形态图(10×)
Fibers of "Red Star" calligraphy and painting paper (10× objective)

★2
『红星』书画纸纤维形态图(20×)
Fibers of "Red Star" calligraphy and painting paper (20× objective)

⊙1
『红星』书画纸润墨效果
Writing performance of "Red Star" calligraphy and painting paper

## 四
## "澄文堂"书画纸生产的原料、工艺流程与工具设备

## 4
## Raw Materials, Papermaking Techniques and Tools of Chengwentang Xuan Paper Artwork Co., Ltd. in Anhui Province

### (一)
### "澄文堂"书画纸生产的原料

#### 1. 主料

"澄文堂"书画纸主要使用龙须草浆板和青檀皮浆混合制成。据访谈时王四海介绍，安徽澄文堂宣纸艺术品有限公司使用的龙须草浆板从泾县浆板商处购买，浆板商则从河南安阳等地运来，2015年精品浆板价格约为11 000元/t。青檀皮浆由中国宣纸股份有限公司制作，按绝干浆240元/kg供给。

⊙2

⊙2 浆料 Pulp materials

安徽澄文堂宣纸艺术品有限公司书画纸的原料配比如下：三星级书画纸（以贴"红星"牌为主）皮、草比例为3∶7，二星级书画纸（以贴"红星"牌为主）的皮、草比例为2∶8，一星级书画纸（以贴"红星"牌为主）的皮、草比例为1∶9，普通书画纸的皮、草比例为1∶19。除此之外，"澄文堂"系列书画纸还有罗纹、龟纹等品种。

#### 2. 辅料

（1）纸药。"澄文堂"书画纸使用化学纸药聚丙烯酰胺。

（2）水源。生产用水主要取自九峰村的九峰水库，该水库中的水来自九峰附近的山泉，属于徽水河源头水之一。调查组实测安徽澄文堂宣纸艺术品有限公司造纸用水的pH为6.0，呈微酸性。

⊙1

⊙1 眺望九峰水库 Overlooking Jiufeng Reservoir

## （二）安徽澄文堂宣纸艺术品有限公司书画纸生产的工艺流程

综合调查组2016年5月15日在安徽澄文堂宣纸艺术品有限公司的实地调查，以及王四海、郑礼红的介绍，总结“澄文堂”书画纸（包括代工贴牌的“红星”产品）生产工艺流程为：

### 壹 浸泡

1 ⊙2

制作“澄文堂”书画纸要将龙须草浆板中的杂质通过人工选拣方式去除，然后将纯净浆板浸泡一天，使浆板充分吸收水分软化。

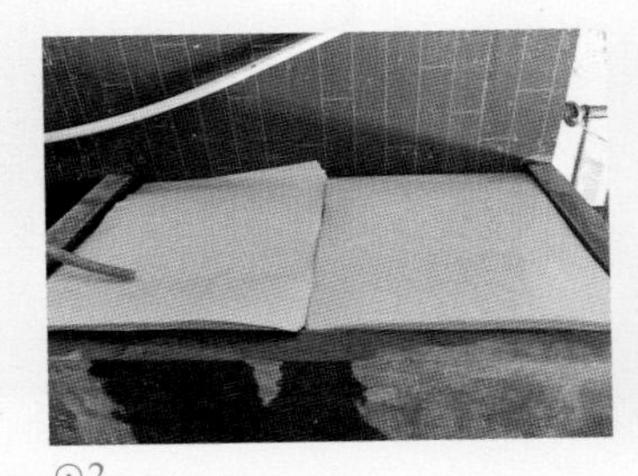
⊙2

### 贰 打浆

2 ⊙3

将浸泡好的龙须草浆板直接放入打浆机打成成熟龙须草浆料。

⊙3

### 叁 配浆

3 ⊙4

按照企业内定等级书画纸配浆标准，将中国宣纸股份有限公司供应的青檀皮浆料掺入龙须草成熟浆料中，通过打浆机将浆料混合，形成成熟浆料。

⊙4

### 肆 捞纸

4 ⊙5⊙6

“澄文堂”书画纸的捞纸工艺与宣纸工艺相同。掌帘、抬帘两名工人分别站在纸槽两头，其动作要领与捞制宣纸一样。不同的是捞制一改二书画纸在提帘上档时，需将纸帘在帘床上拖动，便于提帘上档时能取纸帘的中间部位，以掌握平衡。安徽澄文堂宣纸艺术品有限公司制作普通书画纸和一星书画纸时，均采用一改二纸帘捞纸，制作二星、三星书画纸时，则以正常的四尺单帘捞纸。

⊙6

⊙5

⊙2 浸泡 Soaking the pulp board

⊙3 打浆 Beating the pulp

⊙4 配浆 Mixing the pulp

⊙5/6 捞纸 Papermaking

伍

## 压榨

5 ⊙7

捞纸工下班后，由帮槽工将盖纸帘盖在湿纸帖上，再盖上纸板，架上榨杆，使用千斤顶进行压榨。压榨时力度逐步由小变大，直至压到纸帖不再出水时压榨结束。

⊙7

陆

## 晒纸

6 ⊙8⊙9

“澄文堂”普通书画纸和一星书画纸在晒纸之前，要用切帖刀将纸帖的额头切掉一小边，一般以1～2 cm为宜；在晒制二星、三星书画纸之前，需将纸帖放在纸焙上炕干，然后浇水润帖，再将润好帖的额头切去一条边，称为“切额”。将切好额的纸帖放在纸架上，用鞭帖板打纸帖整面，让纸帖发松后，用手将纸帖的四边往外翻，再用额枪将纸边打松便可晒纸了。晒纸时，先从掐角开始起头，将单张湿纸从纸帖上揭下后，用刷把将纸张贴在纸焙上，晒完整焙后逐张揭下，平放在纸桌上。

⊙9

⊙8

柒

## 检验、剪纸

7 ⊙10

对晒好的纸进行人工检验，将有洞、有杂质的不合格纸挑出来。将检查好的纸用特制大剪刀剪裁整齐，要求纸的四边每边一刀成功。纸长时，边剪边向前移动脚步，一气呵成，否则纸边就不会成一条直线。

⊙10

捌

## 包装

8

剪好的纸按照100张/刀的规格分装，加盖“澄文堂”商标、厂名、纸张品种以及尺寸等的印章，包装完毕后运入贮纸仓库。如果是中国宣纸股份有限公司贴牌的产品，则定期现场由发包单位的验收人将验收合格的产品加盖“红星”书画纸印章后，运回中国宣纸股份有限公司仓库。

⊙7 用以压榨的设备 Tools for pressing the paper
⊙8 揭纸 Peeling the paper
⊙9 晒纸 Drying the paper
⊙10 检纸 Checking the paper

## (三)

## “澄文堂”书画纸（包括代工贴牌的“红星”品牌）生产的工具设备

### 壹 打浆机 1

用来制作浆料，机械自动搅拌。

⊙11

### 贰 捞纸槽 2

水泥浇筑的盛浆设施。实测安徽澄文堂宣纸艺术品有限公司所用的一改二捞纸槽尺寸为：长345 cm，宽185 cm，高80 cm。

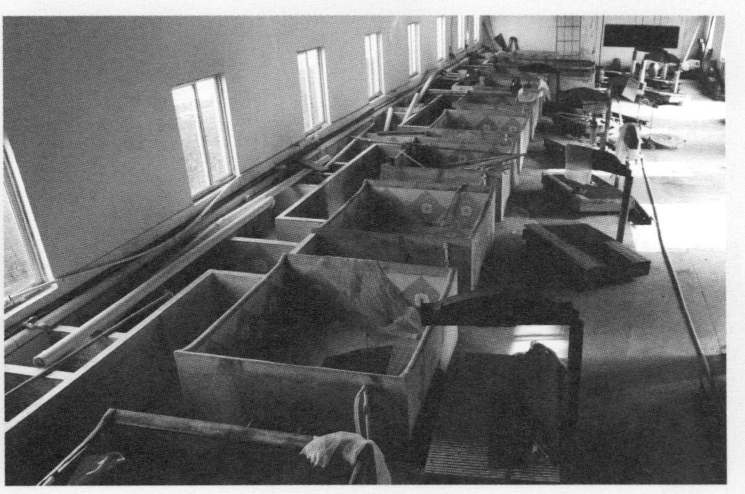

⊙12

### 叁 帘床 3

捞纸工具，木头外框，中间装芒秆。实测安徽澄文堂宣纸艺术品有限公司所用的一改二纸帘尺寸为：长309 cm，宽88 cm。

⊙13

### 肆 切帖刀 4

晒纸前用来切帖额，刀刃为钢制。实测安徽澄文堂宣纸艺术品有限公司所用切帖刀尺寸为：长38 cm，宽7 cm。其中刀刃长22 cm。

⊙14

⊙11 打浆机 Beating machine

⊙12 捞纸槽生产线 Production line of papermaking trough

⊙13 帘床 Frame for supporting the papermaking screen

⊙14 切帖刀 Knife for cutting the paper

伍

## 刷把 5

晒纸工具，用刷把将纸刷上纸焙，刷柄为木制，刷毛为松针。实测安徽澄文堂宣纸艺术品有限公司所用刷把尺寸为：长48 cm，宽13 cm。

⊙15

⊙16

陆

## 纸焙 6

晒纸设备，用两块长方形钢板焊接而成，中间贮水，通过烧柴、烧煤加热后保温，可以两边晒纸。

柒

## 剪刀 7

用来剪纸，剪刀口为钢制，其余部分为铁制，泾县本地特色工具。

⊙17

⊙15 刷把 Brush

⊙16 纸焙 Drying wall

⊙17 剪刀 Shears

# 五 安徽澄文堂宣纸艺术品有限公司的市场经营状况

## 5 Marketing Status of Chengwentang Xuan Paper Artwork Co., Ltd. in Anhui Province

安徽澄文堂宣纸艺术品有限公司自2005年以来，一直是中国宣纸股份有限公司外协生产单位，所产书画纸主体由中国宣纸股份有限公司定时派人到安徽澄文堂宣纸艺术品有限公司验货，现场将验收合格的产品加盖"红星"书画纸印章后运回公司，根据验收合格数量按合同付款，安徽澄文堂宣纸艺术品有限公司开具正式票据到中国宣纸股份有限公司平账。依托这一委托贴牌生产模式，在泾县宣纸、手工书画纸行业中，安徽澄文堂宣纸艺术品有限公司的实际纳税额居全县手工造纸企业前三名，也由此成为泾县最具影响力的专业生产手工书画纸企业之一。

中国宣纸股份有限公司将澄文堂制作的书画纸验收后，纳入该公司销售体系，作为"红星"宣纸附属产品在国内外销售，最高年份可委托代生产近2 000万元的书画纸。2014年前，安徽澄文堂宣纸艺术品有限公司根据中国宣纸股份有限公司下达的生产任务组织生产。2014年开始，随着行业形势发生变化，中国宣纸股份有限公司的销售也在逐步下行，年订单任务迅速下降，安徽澄文堂宣纸艺术品有限公司的库存量快速增加。为缓解库存压力，王四海、郑礼红将产量一再压缩，曾一度处于半停产状态。

安徽澄文堂宣纸艺术品有限公司是黄村镇最大的书画纸生产企业，在镇政府的要求下，王四海、郑礼红全力拓展各自的渠道，加大"澄文堂"纸品的营销，2015年以来勉强维持2帘槽的生产。截至调查时的2016年5月，"澄文堂"四尺规格三星书画纸和四尺罗纹纸的售价约为300元/刀，掺入稻草浆的四尺三星书画纸为460元/刀，麻纸为200元/刀，由于大行情下行和企业经营低谷，产品的平均利润率较低。

⊙18

⊙19

⊙18/19 安徽澄文堂宣纸艺术品有限公司贮纸仓库
Paper storing warehouse of Chengwentang Xuan Paper Artwork Co., Ltd in Anhui Province

# 六 “澄文堂”书画纸的品牌文化与民俗故事

## 6 Brand Culture and Stories of “Chengwentang” Calligraphy and Painting Paper

“澄文堂”书画纸本来用的是“九峰”商标。据王四海介绍，用“九峰”做品牌名称的主要原因是该厂落户在九峰村和九峰山，但准备提交申请注册时，没想到已经有人注册过该商标了。由于该厂的产品主要由中国宣纸股份有限公司代销，自己本身对品牌的诉求不大，加上一时也找不到好名字，就将这件事搁下来了。

2013年前后，经人介绍，王四海认识了北京雨辰科创文化发展有限公司董事长张应。张应原供职于新闻界，后挂靠华夏文化遗产保护中心，注册成立北京雨辰科创文化发展有限公司后，开创了“宣纸中国”网上销售平台。该平台有独立服务器、域名、线上线下交易中心，也开发了“宣纸中国”系列产品，与一般挂靠在“淘宝”“京东”等平台上的网络销售模式有较大区别。在两人的交往中，“宣纸中国”网也不间断地向九峰宣纸厂定制书画纸产品，但总体销量不大。王四海通过张应认识了一些北京书画家。2015年，王四海认识了画家蔡小汀，蔡小汀是已故著名画家蔡鹤汀（1909～1976）的儿子，他得知王四海、郑礼红陷于品牌命名的尴尬局面后，联想到南唐后主李煜曾创制过文化名纸“澄心堂”，思绪灵动，提议取名“澄文堂”。王四海闻之非常喜欢，当即委托北京的一家代理机构向国家商标局提交了商标注册申请。回到泾县，他干脆将企业名称也改成了“安徽澄文堂宣纸艺术品有限公司”。

⊙1

⊙1 蔡小汀与王四海 Cai Xiaoting and Wang Sihai

# 七
# 安徽澄文堂宣纸艺术品有限公司的传承现状与发展思考

## 7
## Current Status of Business Inheritance and Thoughts on Development of Chengwentang Xuan Paper Artwork Co., Ltd. in Anhui Province

澄文堂的创建人王四海、郑礼红均非宣纸行业出身，澄文堂也不具有造纸工艺传承的常规传统，与泾县其他手工造纸厂家有着不同的发展路径。

澄文堂之所以能在10年内发展成较大规模，与中国宣纸股份有限公司授权生产“红星”书画纸有很大关系，在较大批量贴牌生产书画纸的情况下，资金、产品销售基本不用担忧。

从2014年开始，中国宣纸股份有限公司受到艺术用纸行业断崖式下行的冲击，澄文堂也受到了极大的关联性影响，贴牌订单大幅萎缩，从16帘槽的满负荷生产跳水般地下降至2帘槽，产品迅速积压，资金链危机顿起，王四海、郑礼红被迫重新寻找企业出路。

调查中了解到他们的做法包括：与泾县加工纸企业协作，以更低的价格将低档书画纸供应给加工户制成加工纸拓展销售；与全国各地的文房四宝经营户联系，力求迅速扩大澄文堂自有品牌书画纸的市场，等等。

访谈中王四海深有感触地表示，这些年依托中国宣纸股份有限公司协作生产的发展模式，尽管挣了一点钱，但基本上都投入到厂房建设之中了，工厂现在几乎没有外债了，但行业这么不景气，澄文堂在完全没有来得及准备的情况下陷入了困境。目前只希望整个行业尽快好起来，否则能否维持企业生存确实是个很大的问题。

⊙2

⊙3

⊙2
积压在库的『红星』书画纸
Excess inventory of "Red Star" calligraphy and painting paper

⊙3
在库房中若有所思的王四海
Wang Sihai thinking in the warehouse

# 安徽澄文堂宣纸艺术品有限公司书画纸

Calligraphy and Painting Paper of Chengwentang Xuan Paper Artwork Co., Ltd. in Anhui Province

『红星』书画纸透光摄影图
A photo of "Red Star" calligraphy and painting paper seen through the light

# 第四章 皮纸

# Chapter IV Bast Paper

# 第一节

# 泾县守金皮纸厂

安徽省
Anhui Province

宣城市
Xuancheng City

泾县
Jingxian County

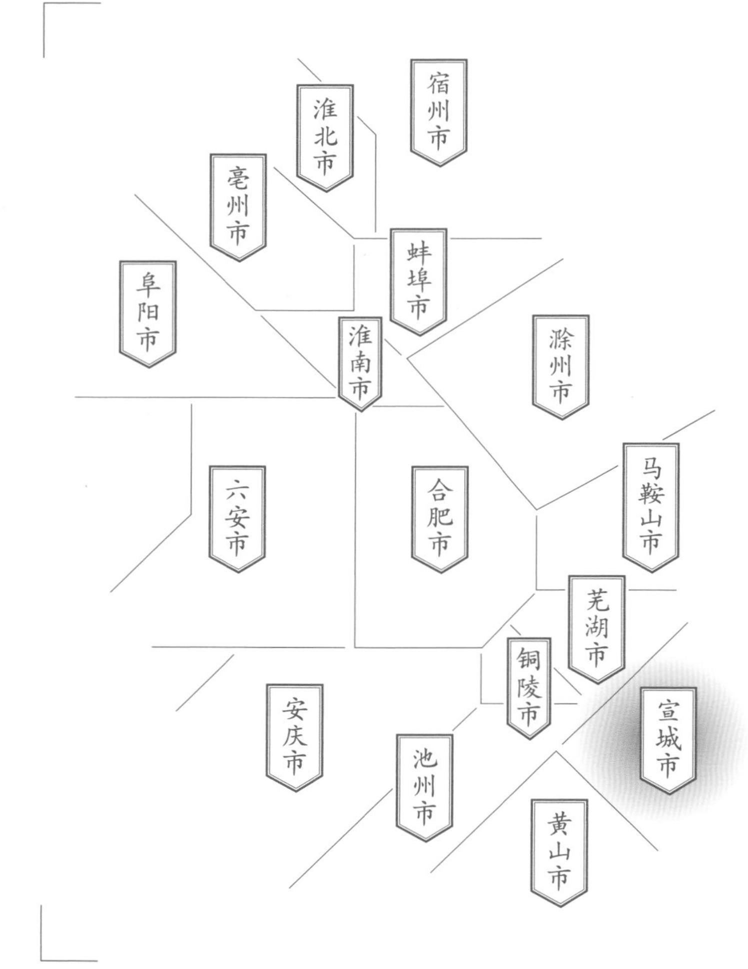


**调查对象**

泾川镇
泾县守金皮纸厂
皮纸

Section 1

## Shoujin Bast Paper Factory in Jingxian County

**Subject**

Bast Paper
of Shoujin Bast Paper Factory
in Jingxian County
in Jingchuan Town

## 一
## 泾县守金皮纸厂的基础信息与生产环境

## 1
## Basic Information and Production Environment of Shoujin Bast Paper Factory in Jingxian County

泾县守金皮纸厂坐落于泾县泾川镇城西社区园林行政村程家村民组，当地人也习惯性地称为程家皮纸厂。地理坐标为：东经118° 21′ 54″，北纬30° 42′ 13″。纸厂依山而建，厂区背倚山脉起伏叠嶂环境错落而建。守金皮纸厂的前身始创于1990年，早期生产书画纸，因经营困难中间被迫停产数年，2003年开始转而生产皮纸。产品包括构皮纸、楮皮纸和雁皮纸三大类。调查组于2015年7月16日和2016年4月25日两次前往守金皮纸厂进行调查，获得的基础生产信息如下：截至调查时的2016年初，纸厂有员工20多人，共有11帘皮纸槽位，其中包括1帘丈二纸槽、1帘丈八纸槽、1帘八尺纸槽，其余均为四尺和六尺纸槽。守金皮纸厂皮纸生产工艺已全部采用手工吊帘捞纸方式。

园林行政村位于泾川镇，当地地名为百岭坑的区域，百岭坑是著名的小岭“九岭十三坑”之一，境内群山环绕，山间优质溪水潺流不息。在泾县，“九岭”指环绕在宣纸历史著名产区小岭村周围的9座山峰，而“十三坑”以前是山凹中的13个传统造纸聚集的自然村落，如今被“改编”成村民组。程家村民组是园林村的一个村民组。

⊙1

⊙2

⊙1 程家皮纸厂路边指示牌 Road sign of Chengjia Bast Paper Factory

⊙2 通往程家皮纸厂的乡道 Country road to Chengjia Bast Paper Factory

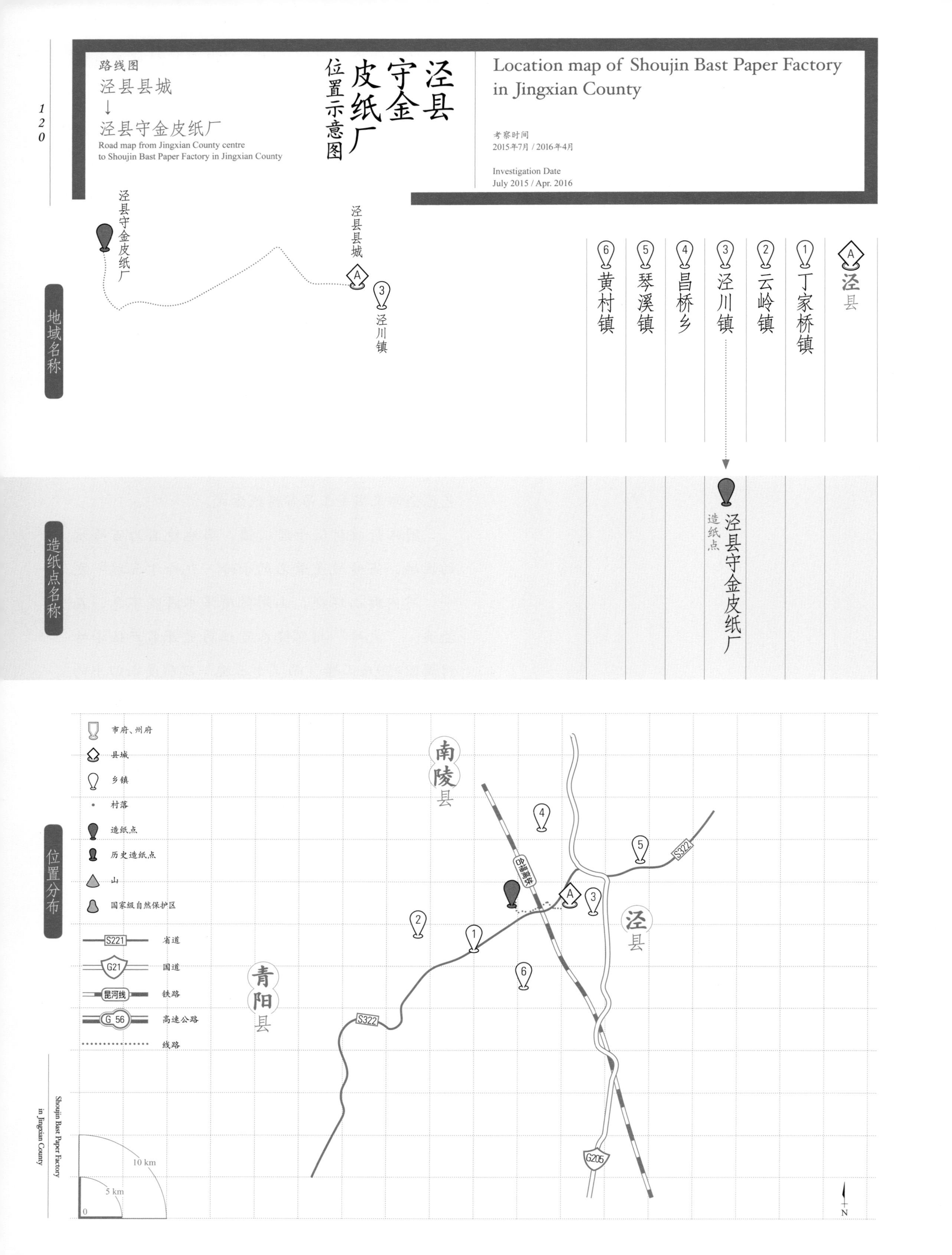

路线图
泾县县城
↓
泾县守金皮纸厂
Road map from Jingxian County centre
to Shoujin Bast Paper Factory in Jingxian County
泾县守金皮纸厂位置示意图
Location map of Shoujin Bast Paper Factory in Jingxian County
考察时间
2015年7月 / 2016年4月
Investigation Date
July 2015 / Apr. 2016
泾县守金皮纸厂
泾县县城
A
3
泾川镇
地域名称
A 泾县
1 丁家桥镇
2 云岭镇
3 泾川镇
4 昌桥乡
5 琴溪镇
6 黄村镇
造纸点名称
造纸点
泾县守金皮纸厂
位置分布
市府、州府
县城
乡镇
村落
造纸点
历史造纸点
山
国家级自然保护区
S221 省道
G21 国道
昆河线 铁路
G 56 高速公路
线路
南陵县
青阳县
泾县
S322
G205
合福高铁
10 km
5 km
0
N

## 二 泾县守金皮纸厂的历史与传承情况

## 2 History and Inheritance of Shoujin Bast Paper Factory in Jingxian County

泾县守金皮纸厂自1990年创立原书画纸厂至2016年调查时已有26年的历史，公司法人代表为程守海。

调查组于2015年7月16日对程守海进行了初次访谈，2016年4月25日调查组对纸厂进行补充调查时，对程守海进行了第二次访谈。程守海叙述的守金皮纸厂发展历程为：20多年前（1990年左右）纸厂就开始生产书画纸，是泾县生产书画纸较早的企业之一。那时启动资金约2万元，用家里的老房子当厂房，开设了1帘槽生产。当时，整个书画纸市场行情不错，而泾县本地少有人做，原料也没有用龙须草浆板，使用的是各个加工纸厂的纸边、染色废纸等。因20世纪90年代这些属于边角料的原纸，除了少量的册页加工会使用温州皮纸、富阳竹纸外，基本是纯正的宣纸，经过浸泡、打浆加上后期的捞、晒、剪工艺后，使用效果比现在流行的浆板书画纸要好，市场行情也特别旺。

值得注意的是，交流时程守海一再强调，如今的书画纸原料多是龙须草浆板，其中采用喷浆工艺捞出来的纸，短期来看效果蛮好，但从长远来看喷浆纸会被逐步淘汰。程守海认为原因有两点：一是喷浆捞出来的纸书写绘画时间长了容易变色；二是喷浆书画纸比较脆，拉力不够。

书画纸生产的好日子维系5～6年后，由于同类厂家逐渐增多，供给过剩，市场行情变得不景气，纸厂资金周转也因此出现问题，1996年生产

⊙1 守金皮纸厂周边环境 Surrounding environment of Shoujin Bast Paper Factory

⊙2 程守海 Cheng Shouhai

书画纸的原厂倒闭了，程守海被迫卖掉生产设备支付了工人的工资，自己则去了李园村的三星宣纸厂打工。他在三星宣纸厂共上了三年班，其间当了一年的捞纸师傅。2002年，程守海重新起步，在原先老厂房的基础上成立了“守金皮纸厂”，开始走上制作皮纸之路。

说起当年第二次办厂选择造皮纸的原因，程守海回忆：实际上，在皮纸厂正式筹建之前，他先做了一年多的皮料加工活，那时候资金很有限，程守海就把自家山上的竹子砍掉卖了，辛辛苦苦筹集了8000多元，做皮料也慢慢筹集了几万元。但皮料加工只做了一年多，一是因为当时皮料价格被压得太低，二是因为泾县皮纸在国外市场（韩国）卖得相当好。考虑到生存压力和受到国外皮纸市场发展潜力的启发，程守海终于痛下决心用几万元买了一台二手打浆机，自己开始造起了皮纸。

守金皮纸厂当初只有2帘槽，规模很小。随着市场的开拓，效益逐步转好，开始投资扩建厂房。2011年前后，程守海在老厂房旁边盖起了新的房子，生产车间也随之搬至新的厂房，截至2015年7月16日调查时，纸厂已有生产槽位11个，生产能力迅速提高。到2016年4月25日二次入厂调查时，守金皮纸厂又新建了办公区域和仓库、车间，规模也在持续扩大中。

程守海，1965年出生于泾川镇园林村。1979年，年仅14岁的他开始在程家生产队做农工，16岁开始在泾县百岭坑宣纸厂[1]做学徒，跟师傅张彪飞学习捞纸，共学了两年多的时间。1984年，19岁的程守海学成出师，离开百岭坑宣纸厂后在李园宣纸厂从事捞纸工作。1987年，程守海又到汪同和宣纸厂捞纸。创办纸厂之前从事捞纸多年，程守海堪称资深的宣纸技术工人出身。但访谈时发现，程守海家中父祖辈没有从事手工造纸行业的，在创办纸厂之前他也几乎没有客户和外界联系，是比较典型的白手起家造纸的类型。

调查时，程守海主要管理生产，儿子程玮则全权负责销售。程玮，1989年出生，2011年大学毕业后就回到厂里开起了网店，厂里产品的销售（含实体店销售、网络营销）都由他管理。

⊙1

⊙2

⊙1 调查组成员再次访谈程守海 A researcher reinterviewing Cheng Shouhai

⊙2 程玮 Cheng Wei

[1] 据程守海介绍，20世纪70年代百岭坑有个“百岭坑宣纸厂”，是当时泾县三家宣纸厂之一，另外两家是“红星宣纸厂”和“小岭宣纸厂”。百岭坑宣纸厂当初为村办企业，后来企业改制后，纸厂倒闭并进行了厂址搬迁，其旧址即现在的“安徽泾县双鹿宣纸有限公司”。

# 三 泾县守金皮纸厂的代表纸品及其用途与技术分析

# 3 Representative Paper and Its Uses and Technical Analysis of Shoujin Bast Paper Factory in Jingxian County

## (一) 代表纸品及其用途

守金皮纸厂的产品主要有构皮纸、楮皮纸和雁皮纸三种。根据客户的需求和产品用途的不同，产品包括白色皮纸、仿古本色皮纸和其他色纸，其中色纸是在浆料中加入诸如红花、黄花、茶叶、稻草、玉米须等物质生产而成的，多为工艺装饰用纸。

据程守海介绍，守金皮纸厂的三类皮纸中，雁皮纸售价最高，原料也最贵；其次是楮皮纸，再次是构皮纸。雁皮纸的原料是100%的雁皮，很轻很薄，拉力强，主要用于画工笔画、写书法、印刷古书和拓碑刻字。楮皮纸的原料是楮皮加木浆，配比根据需求来定，楮皮比例从50%到80%不等，可用于书法、绘画，做环保袋、纸扇、一次性纸衣等。构皮纸的原料也是根据需求来定的，构皮一般占60%～80%，木浆占20%～40%，构皮纸可用于书法、绘画和包装等。访谈中当调查人员比较疑惑地问起楮皮与构皮的区别时，程守海给出的说法是：楮皮就是大构皮，制浆时可以打成短纤维，生产中可使用平筛、除砂器，主要产于泰国、通过江西中间商转购的白皮，价格为12.66元/kg左右。构皮是指小构皮，直接从陕西调运生皮，价格为4.6元/kg，交由专事蒸煮皮的人员加工，按5元/kg收取加工费，出浆率为58%～60%。

守金皮纸厂的三类皮纸中，以构皮本色（茶叶）云龙皮纸和仿古本色皮纸最具代表性。构皮本色云龙皮纸可用于包装和装饰，如包装普洱茶、茅台酒，贴墙纸等；仿古本色皮纸主要用于书法、绘画和古书修复。

⊙3 构皮本色云龙皮纸 Yunlong bast paper made of mulberry bark and tea-leaves

⊙4 成品纸仓库 Warehouse of paper products

(二)

## 代表纸品的性能分析

测试小组对采样自守金皮纸厂的构皮本色云龙皮纸所做的性能分析，主要包括厚度、定量、紧度、抗张力、抗张强度、撕裂度、湿强度、色度、耐老化度下降、吸水性、伸缩性、纤维长度和纤维宽度等。按相应要求，每一指标都需重复测量若干次后求平均值，其中厚度抽取10个样本进行测试，定量抽取5个样本进行测试，抗张力（强度）抽取20个样本进行测试，撕裂度抽取10个样本进行测试，湿强度抽取20个样本进行测试，色度抽取10个样本进行测试，耐老化度下降抽取10个样本进行测试，吸水性抽取10个样本进行测试，伸缩性抽取4个样本进行测试，纤维长度测试了200根纤维，纤维宽度测试了300根纤维。对构皮本色云龙皮纸进行测试分析所得到的相关性能参数见表4.1。表中列出了各参数的最大值、最小值及测量若干次所得到的平均值或者计算结果。

表4.1　构皮本色云龙皮纸相关性能参数
Table 4.1　Performance parameters of unbleached Yunlong bast paper made of mulberry bark

| 指标 | | 单位 | 最大值 | 最小值 | 平均值 | 结果 |
|---|---|---|---|---|---|---|
| 厚度 | | mm | 0.122 | 0.101 | 0.112 | 0.112 |
| 定量 | | g/m² | — | — | — | 32.4 |
| 紧度 | | g/cm³ | — | — | — | 0.289 |
| 抗张力 | 纵向 | N | 13.6 | 9.3 | 12.8 | 12.8 |
| | 横向 | N | 12.4 | 4.6 | 6.8 | 6.8 |
| 抗张强度 | | kN/m | — | — | — | 0.653 |
| 撕裂度 | 纵向 | mN | 940 | 740 | 804 | 804 |
| | 横向 | mN | 860 | 440 | 668 | 668 |
| 撕裂指数 | | mN · m²/g | — | — | — | 22.7 |
| 湿强度 | 纵向 | mN | 350 | 250 | 300 | 300 |
| | 横向 | mN | 200 | 100 | 150 | 150 |
| 色度 | | % | 43.8 | 41.7 | 42.86 | 42.9 |
| 耐老化度下降 | | % | — | — | — | 0.7 |
| 吸水性 | | mm | — | — | — | 6 |
| 伸缩性 | 浸湿 | % | — | — | — | 0.33 |
| | 风干 | % | — | — | — | 0.93 |
| 纤维 | 皮 长度 | mm | 8.18 | 0.78 | 3.36 | 3.36 |
| | 皮 宽度 | μm | 39.0 | 2.0 | 9.0 | 9.0 |
| | 草 长度 | mm | 2.05 | 0.29 | 0.79 | 0.79 |
| | 草 宽度 | μm | 46.0 | 4.0 | 19.0 | 19.0 |

由表4.1中的数据可知，构皮本色云龙纸最厚约是最薄的1.21倍，经计算，其相对标准偏差为0.008，纸张厚薄较为一致。所测构皮本色云龙皮纸的平均定量为32.4 g/m$^2$。通过计算可知，构皮本色云龙皮纸紧度为0.289 g/cm$^3$，抗张强度为0.653 kN/m，抗张强度值较大。所测构皮本色云龙皮纸撕裂指数为22.7 mN · m$^2$/g，撕裂度较大；湿强度纵横平均值为225 mN，湿强度较大。

所测构皮本色云龙皮纸平均色度为42.9%，色度较高。色度最大值是最小值的1.050倍，相对标准偏差为0.790，色度差异相对较小。经过耐老化测试后，耐老化度下降0.7%。

所测构皮本色云龙皮纸吸水性纵横平均值为6 mm，纵横差为2.6 mm。伸缩性指标中浸湿后伸缩差为0.33%，风干后伸缩差为0.93%，说明构皮本色云龙皮纸伸缩性差异不大。

构皮本色云龙皮纸在10倍、20倍物镜下观测的纤维形态分别见图★1、图★2。所测构皮本色云龙皮纸构皮纤维长度：最长8.18 mm，最短0.78 mm，平均长度为3.36 mm；纤维宽度：最宽39.0 μm，最窄2.0 μm，平均宽度为9.0 μm。草纤维长度：最长2.05 mm，最短0.29 mm，平均长度为0.79 mm；纤维宽度：最宽46.0 μm，最窄4.0 μm，平均宽度为19.0 μm。构皮本色云龙皮纸润墨效果见图⊙1。

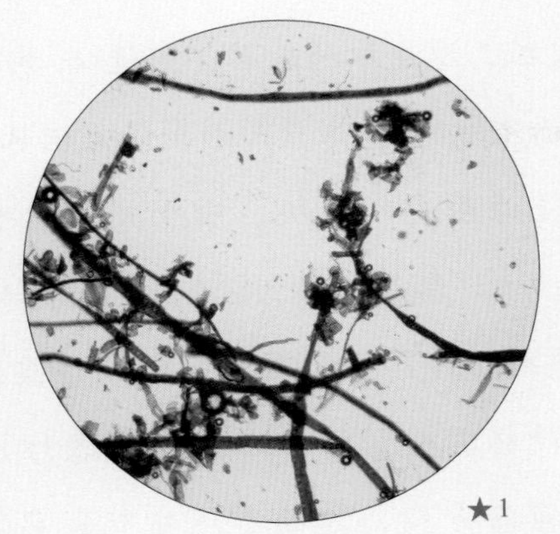

★1 构皮本色云龙皮纸纤维形态图（10×）
Fibers of unbleached Yunlong bast paper made of mulberry bark (10× objective)

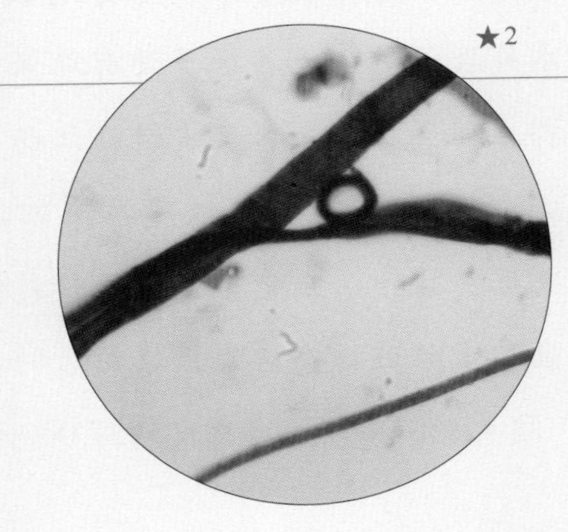

★2 构皮本色云龙皮纸纤维形态图（20×）
Fibers of unbleached Yunlong bast paper made of mulberry bark (20× objective)

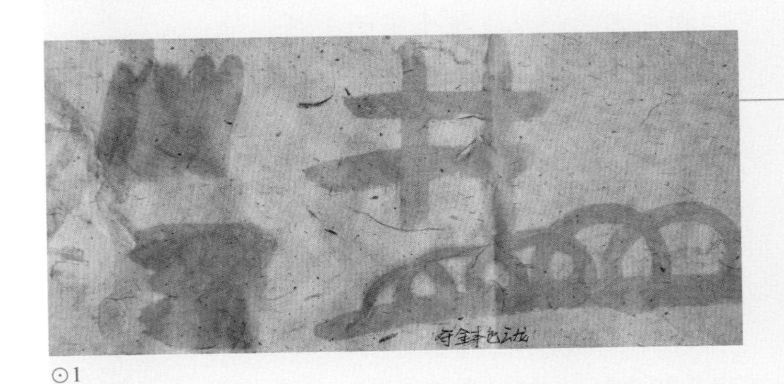

⊙1 构皮本色云龙皮纸润墨效果
Writing performance of unbleached Yunlong bast paper made of mulberry bark

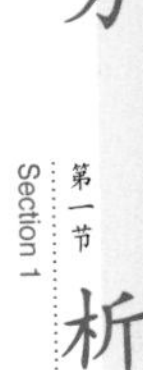

## 四
## 构皮本色云龙皮纸与仿古本色皮纸生产的原料、工艺流程与工具设备

## 4
## Raw Materials, Papermaking Techniques and Tools of Yunlong Bast Paper Made of Mulberry Bark and Antique Unbleached Bast Paper

泾县当地基本不产构皮、楮皮、雁皮、三桠皮和木浆等原料，守金皮纸厂所需原料均从外地购买。守金皮纸厂雁皮原料从菲律宾进口，构皮和楮皮原料从陕西等地购买，木浆原料则是直接从泾县当地经销商和上海购买的浆板。由于皮料原料加工会产生大量污水，环境污染大，泾县当地环保管控很严格，考虑到环境污染和治污成本，守金皮纸厂将购买到的皮料都运输到江西省萍乡市进行代加工。

据程守海介绍，江西省萍乡市有完备的皮料加工和污水处理设施，泾县很多没有环保设备的造纸企业都在那里加工皮料。皮料在江西省萍乡市经过蒸煮处理后，守金皮纸厂再买回漂白后的皮料，根据实际产品生产需要再进行补漂白。2015年调查时了解到的价格为：漂白后的构皮约6 000元/t，楮皮约8 000元/t，雁皮约37 000元/t。

### (一)
### 构皮本色云龙皮纸生产的原料和工艺流程

#### 1. 构皮本色云龙皮纸生产的原料

(1) 主料

构皮本色云龙皮纸的主要原料包括构皮、木浆和茶叶片。其中，构皮、木浆均从外地购买，茶叶片是从邻近的芜湖市购买的工艺茶叶。一般情况下，木浆价格为5 600元/t，茶叶价格为70～80元/kg。据程守海介绍，构皮是长纤维，木浆是短纤维，两种原料混合可增加纸张纤维的密度。

(2) 辅料

① 纸药。作为辅料的纸药，守金皮纸厂使用的是化学纸药聚丙烯酰胺，化学纸药在泾县当地即可购买到。

② 水源。守金皮纸厂造纸使用的是地下井水。据程守海的说法，水质和造纸关联性很大，水质好坏会决定纸的质量高低，百岭坑的水质在小岭“十三坑”中算是最好的，造出的纸不会发黄。调查组通过实测，守金皮纸厂的生产用水pH约为6.8，呈弱酸性。

### 2. 构皮本色云龙皮纸生产的工艺流程

通过调查组于2015年7月16日和2016年4月25日对守金皮纸厂构皮本色云龙皮纸生产工艺进行的实地调查和访谈，归纳其主要工艺流程如下：

| 壹 | 贰 | 叁 | 肆 | 伍 | 陆 | 柒 | 捌 | 玖 | 拾 |
|---|---|---|---|---|---|---|---|---|---|
| 拣皮 | 浸泡、洗漂 | 榨皮 | 第二次拣皮 | 打浆 | 捞纸 | 压榨 | 晒纸 | 检验 | 入库 |

壹
#### 拣皮
1

购买的漂白构皮料需要进行人工拣选，挑出木棍等杂质。

贰
#### 浸泡、洗漂
2 ⊙1

拣选后的构皮料需要进行浸泡和洗漂，浸泡使用清水在浸泡池浸泡，漂白使用漂白液。构皮半成品需要漂白和清洗以得到漂白构皮料，一般来说，洗漂需要耗时7～8小时。

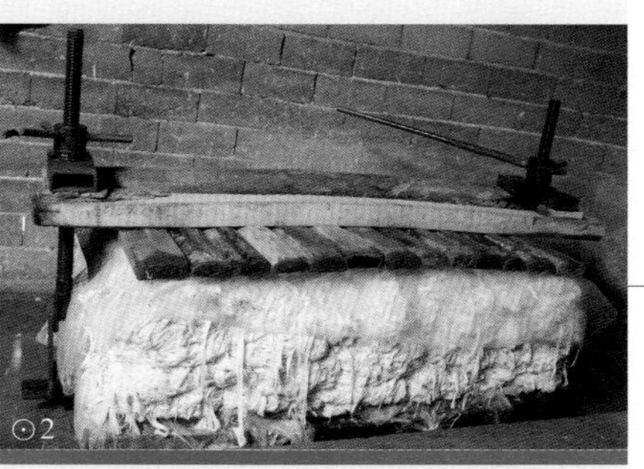

叁
#### 榨皮
3 ⊙2

洗漂后的构皮料需要用千斤顶榨干。从洗漂池捞出来的皮料一般需要榨3小时左右。

⊙1 洗漂构皮料 Cleaning and bleaching mulberry bark materials

⊙2 压榨洗漂后的构皮料 Pressing the cleaned and bleached mulberry bark materials

## 肆 第二次拣皮 4

⊙3

榨干后的构皮料需要再次进行人工拣选，挑选出细棍等杂质和未经完全漂白的皮料。

⊙3

## 伍 打浆 5

⊙4

（1）将木浆浆板放入打浆机内打浆。

（2）将精选后的构皮料放入打浆机内与木浆混合打浆，至打碎时一般需要10多分钟。

（3）在混合的纸浆中再次放入构皮料，粗打1～2分钟后形成云龙长纤维。

⊙4

## 陆 捞纸 6

⊙5⊙6

将混合浆放入纸槽后，需要按照一定比例放入茶叶片，采用手工吊帘方式捞纸。吊帘捞纸方式早年从台湾地区引进，与传统宣纸两人捞纸方式不同，这里只需要一个人捞纸。

（1）捞纸工将纸帘放在帘床上，双手按住帘床，平放在浆料水面之上。

（2）将帘床在浆池中前后摆动两次，使纸帘上沾上浆料。

（3）捞纸工从帘床上取下纸帘，将湿纸轻轻从纸帘放至纸槽旁边的纸架上，这样捞出的湿纸带有茶叶，如此循环往复。

一般而言，工人一天工作12小时，每个周日休息，捞纸工一天多则能捞10刀左右，少则能捞7刀左右。

⊙5

⊙6

⊙7

## 柒 压榨 7

⊙7

捞好的湿纸先盖上盖纸帘，帘上压上压纸板，压到纸帖不再出水为止。据程守海介绍，因化学纸药浓度高，当天捞的纸张要第二天压榨，如果当天压榨，那么纸帖很容易被压坏。

⊙3 工人正在拣皮 Workers picking and choosing the bark

⊙4 工人在打浆 A worker beating the pulp

⊙5 捞纸车间 Papermaking workshop

⊙6 捞纸 Papermaking

⊙7 压榨 Pressing the paper

捌

## 晒纸

8 ⊙8⊙9

压榨后的纸帖晒前需要在四个边角浇水以软化边缘，浇水后的纸帖直接进行晒纸。晒纸时，晒纸工将纸帖上的纸一张一张揭离下来刷上铁焙，用刷子来回刷几下将纸贴平整。铁焙温度很高，几十秒后，刷上去的湿纸即可烘干取下。

⊙9

⊙8

玖

## 检验

9

晒干后的本色云龙皮纸不需要进行剪裁，可直接进行人工检验，将有水洞或缺角的纸张挑选出来进行回笼打浆。

拾

## 入库

10 ⊙10

检验好的皮纸可直接包装入库。

⊙10

(二)

## 仿古本色皮纸生产的原料和工艺流程

### 1. 仿古本色皮纸生产的原料

(1) 主料

仿古本色皮纸的主要原料包括构皮和木浆。其中构皮是未经漂白的皮料，木浆是直接购买的木浆浆板。

(2) 辅料

① 纸药。调查中发现，作为辅料的纸药，守金皮纸厂仿古本色皮纸使用的化学纸药是通过经销商从日本购进的，品名上只显示出分张剂字样。

② 水源。守金皮纸厂造纸使用的是地下井水。调查组通过实测，所用水pH约为6.8，呈弱酸性。

⊙8 纸帖浇水 Watering the paper pile

⊙9 晒纸上墙 Drying the paper on a wall

⊙10 检纸 Checking the paper

## 2. 仿古本色皮纸生产的工艺流程

调查组于2015年7月16日对守金皮纸厂仿古本色皮纸生产工艺进行了实地调查，通过观察和访谈得知其主要工艺流程如下：

### 壹 浸泡、清洗

1

针对仿古本色皮纸的原料本色木浆，需要浸泡，使其软化易于打浆；针对未漂白的构皮原料，需要清洗除去杂质。

### 贰 拣皮

2 ⊙1

未经漂白的构皮皮料需要进行人工选皮、撕皮，挑选出木棍等杂质。

⊙1

### 叁 打浆

3 ⊙2

（1）将木浆放入打浆机内打浆。
（2）将未经漂白的构皮料放入木浆内打浆直至皮料打碎。
（3）在混合浆料中再次放入构皮料，粗打约2分钟后即可，能形成云龙效果的长纤维。

⊙2

⊙1 拣皮料 Picking and choosing the bark

⊙2 调查组成员在拍摄工人打浆 A researcher taking photos of the beating step

肆

## 捞纸

4 ⊙3

将混合浆放入纸槽后，需要按照一定比例放入化学纸药，采用手工吊帘方式捞纸。一般而言，工人一天工作12小时，每个周日休息，捞纸工一天多则能捞10刀左右，少则能捞7刀左右。

伍

## 压榨

5 ⊙4

捞好的湿纸当天需要用千斤顶压榨，压榨力度要适中，用力从小到大，压到纸帖不再出水为止。据程守海介绍，压榨最多数量为35刀，一般压榨20刀。

陆

## 晒纸

6

晒纸时，晒纸工将纸帖上的纸一张一张揭离下来刷上铁焙，铁焙温度很高，几十秒后，刷上去的湿纸即可烘干取下。

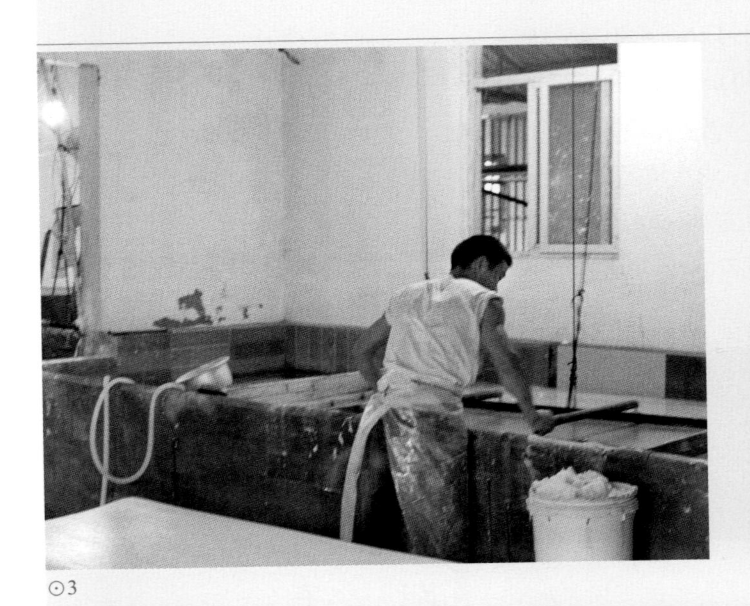

⊙3

⊙4

柒

## 检验

7

晒干后的仿古本色云龙皮纸不需要进行剪裁，可直接进行人工检验，将破损或有瑕疵的纸张挑选出来。

捌

## 入库

8

检验好的皮纸可直接包装入库存放。

⊙3 捞纸 Papermaking

⊙4 压榨 Pressing the paper

## (三)
## 构皮本色云龙皮纸与仿古本色皮纸生产的工具设备

### 壹 打浆机 1

用来打碎构皮料和木浆浆板，生产出用于捞纸的混合浆。

⊙1

⊙2

### 贰 纸槽 2

捞纸时用来盛浆，水泥浇筑而成。槽位尺寸根据产品的规格不同而异，实测守金皮纸厂的四尺槽尺寸为：长195 cm，宽174 cm，高80 cm；六尺槽尺寸为：长250 cm，宽180 cm，高80 cm；八尺槽尺寸为：长200 cm，宽205 cm，高80 cm；丈八槽尺寸为：长600 cm，宽300 cm，高80 cm；丈二槽尺寸为：长410 cm，宽215 cm，高 80 cm。

⊙3

### 叁 纸帘 3

抄纸的重要工具，吊帘纸帘由竹子编织而成，表面光滑平整，帘纹细密。

⊙4

### 肆 帘床 4

捞纸时用来放置纸帘，帘床大小一般比纸帘大一点。

⊙5

### 伍 铁焙 5

铁焙由光滑的两块钢板制成，墙体内装有大量的热水，通过加热来升温。守金皮纸厂共有2个晒纸车间、2条铁焙。1个晒纸车间一天能晒40刀纸左右。

### 陆 刷子 6

晒纸时使用的工具。实测守金皮纸厂正在使用的刷子尺寸为：长约50 cm，宽约14 cm。刷毛由松针制成。

⊙6

⊙1 打浆机 Beating machine

⊙2 待打浆的木浆浆板原料 Raw material of wood pulp board to be beaten

⊙3 纸槽 Papermaking trough

⊙4 纸帘 Papermaking screen

⊙5 帘床 Frame for supporting the papermaking screen

⊙6 刷子 Brush

## 柒 枪棍 7

晒纸时用来划动纸张边缘，使纸边软化便于揭纸上铁焙。

⊙7

## 捌 钉耙 8

洗漂浆料时用来捞起皮料。

⊙8

## 玖 压纸石 9

晒好的纸张从铁焙上揭下后需要整齐堆放在纸板上，此时压纸石用来固定纸张。

⊙9





## 五 泾县守金皮纸厂的市场销售情况

## 5 Sales and Marketing Status of Shoujin Bast Paper Factory in Jingxian County

调查中，与2015～2016年对泾县宣纸和书画纸企业的普遍感受不同，程守海对守金皮纸厂的现状和未来发展空间均持乐观态度，并且多少有些庆幸当年选择了泾县手工纸行业的冷门 — 造皮纸。程守海表示：市场对于皮纸的需求量逐渐增加起来，皮纸的使用范围越来越广，纸品销售状况良好，市场前景较为广阔，因而厂里的生产也保持扩张的态势。2015年至2016年初，守金皮纸厂为扩大生产规模，新建了办公区域和纸品库房；程守海表示还准备在2016年下半年至2017年再新建一个厂房。

守金皮纸厂生产的皮纸销售渠道既有内销，也有外贸。构皮纸类全部用于国内销售，楮皮纸类和雁皮纸类以出口至韩国、日本为主。从2016年初了解的售价看，构皮本色云龙

⊙7 枪棍 Tool for separating the paper layers

⊙8 钉耙 Rake for picking up the bark materials

⊙9 压纸石 Stone for pressing the paper

皮纸市场价为260元/刀，仿古本色皮纸市场价为200元/刀，雁皮纸为500～600元/刀。

从销售模式看，国内产品销售以经销商为主，由于纸厂皮纸产量较大，经销商遍布全国各大城市，主要以北京、天津及各省会城市为主，共有100多家。2011年公司正式开通了网络销售渠道（淘宝店），开始了网店销售模式，2016年初调查的数据显示，网上产品销售量已占总销量的1/3左右。守金皮纸厂年产量为33 000多刀，产销量基本持平。

⊙2

⊙1



## 六 泾县守金皮纸厂的品牌文化与民俗故事

## 6 Brand Culture and Stories of Shoujin Bast Paper Factory in Jingxian County

### 1. 厂名“守金”的由来

守金皮纸厂已走过近30年的风雨，从起初的“小打小闹”成长为泾县规模最大的皮纸厂，发展成果的取得与程守海多年的执着坚守有很大关系。访谈中说起“守金皮纸厂”的厂名，程守海会心一笑地说：“厂名‘守金’两字是我的妻子李金梅起初想出来的，各取两人姓名中间的一个字。‘守金’既寓意夫妻感情和睦、同心同德，也表达了对财源兴旺的期盼。”程守海表示：十分感激妻子多年的辛劳和支持，没有两人共同努力地“守金”，纸厂也难以发展到今天。

### 2.“在被打压中开始了造纸人生”

程守海在访谈中还提到当年的一段经历：在开办皮纸厂之前，他干了一年多的皮料加工，当初的想法是专心致志地把原料加工做好也是不错

⊙1 新建办公兼展示区域
New office and display area

⊙2 坐落在山脚下的守金皮纸厂外部环境
External view of Shoujin Bast Paper Factory

⊙3

⊙4

的选择，加工好的皮料主要卖给泾县皮纸生产厂家。但不久后，当地的厂家开始有意过度压低他的皮料价格，导致他在外销渠道未能建立的背景下皮料销路不畅，利润空间被大幅挤压。"我是个好强的人，自尊心强，看到别人这样打压我，我心里很不是滋味，想着一定要另谋出路干出一番事业……"凭借自己多年一线造纸的经验，加上皮料加工自己也熟，程守海索性破釜沉舟开始自己造起了皮纸，从此踏上造皮纸的路。





⊙5

⊙6

## 七 泾县守金皮纸厂的业态传承现状与发展思考

## 7 Current Status and Development of Shoujin Bast Paper Factory in Jingxian County

调查时泾县守金皮纸厂的生产、经营和销售状况良好，而且产能、设施和技工都处于良性扩张状态；同时，皮纸在国内外市场的前景短期内均被看好，手工纸的这一业态类型并未面临行业低谷。

调查中程守海表示：发展中遇到的难题，最主要的是环保问题。守金皮纸厂每年都要将购买的皮料运送到江西萍乡市加工再运输回来，加工和运输成本使生产成本明显上升。泾县当地环保要求很高、管控很严格，对乱排乱放的污染企业惩处力度很大，而企业现有规模不够大，难以上环保设备，因此，环保是制约企业发展的一大障碍。如果泾县政府能够对中小型企业加大扶持力度，提供公共性的环保设备供企业使用，或者针对中小型手工造纸企业集中建设原料加工的公共服务中心（加工厂），那对守金皮纸厂来说就是一大福利了。

⊙3 程守海和李金梅共同守护的家园 Cheng Shouhai and Li Jinmei's house

⊙4 守金皮纸厂外景 External view of Shoujin Bast Paper Factory

⊙5/6 守金皮纸厂周围环境 Surroundings of Shoujin Bast Paper Factory

# 构皮纸

Paper Mulberry Bark Paper
of Shoujin Bast Paper Factory
in Jingxian County

构皮本色云龙纸透光摄影图
A photo of Yunlong paper made of paper mulberry bark (original color) seen through the light

# 第二节

# 泾县小岭驰星纸厂





安徽省
Anhui Province

宣城市
Xuancheng City

泾县
Jingxian County

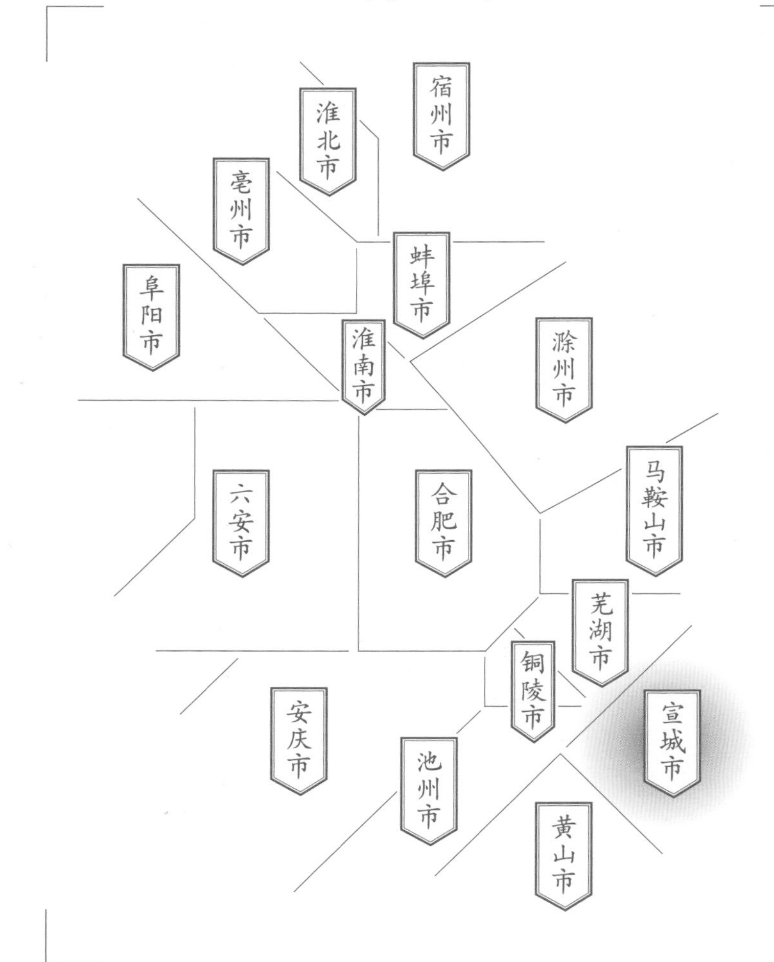


**调查对象**

丁家桥镇
泾县小岭驰星纸厂
皮纸

## Section 2
## Xiaoling Chixing Paper Factory in Jingxian County

Subject

Bast Paper
of Xiaoling Chixing Paper Factory
in Jingxian County
in Dingjiaqiao Town

# 一
# 泾县小岭驰星纸厂的基础信息与生产环境

## 1
## Basic Information and Production Environment of Xiaoling Chixing Paper Factory in Jingxian County

泾县小岭驰星纸厂坐落于泾县丁家桥镇小岭行政村占云村民组，地理坐标为：东经118° 19′ 27″，北纬30° 39′ 17″。驰星纸厂创办于2003年，生产厂区占地6 667 m$^2$左右，主产雁皮、三桠皮、构皮等各类皮纸，共有13帘槽的产能，注册商标为“忆宣”。2015年7月16日和2016年4月20日，调查组对驰星纸厂进行了两次田野调查，获知的基础信息如下：截至2015年底，小岭驰星纸厂有20多名工人，保持了7帘槽的皮纸生产，2014～2015年，平均年产量14 000余刀，销售额300多万元。

⊙1

⊙2

⊙1 驰星纸厂背后的山峰 Mountain peak behind Chixing Paper Factory

⊙2 驰星纸厂大门 Entrance of Chixing Paper Factory

路线图
泾县县城
↓
泾县小岭驰星纸厂
Road map from Jingxian County centre to Xiaoling Chixing Paper Factory in Jingxian County

# 泾县小岭驰星纸厂位置示意图

# Location map of Xiaoling Chixing Paper Factory in Jingxian County

考察时间
2015年7月 / 2016年4月

Investigation Date
July 2015 / Apr. 2016

地域名称

泾县县城
丁家桥镇
泾县小岭驰星纸厂

A 泾县
1 丁家桥镇
2 云岭镇
3 泾川镇
4 昌桥乡
5 琴溪镇
6 黄村镇

造纸点名称

造纸点 泾县小岭驰星纸厂

位置分布

市府、州府
县城
乡镇
村落
造纸点
历史造纸点
山
国家级自然保护区
S221 省道
G21 国道
昆河线 铁路
G 56 高速公路
线路

南陵县
青阳县
泾县
合福高铁
S322
G205
10 km
5 km
0
N

## 二
## 泾县小岭驰星纸厂的历史与传承情况

## 2
## History and Inheritance of Xiaoling Chixing Paper Factory in Jingxian County

泾县小岭驰星纸厂由曹志平、许富贵和沈树杭于2003年合资创办，法人代表由曹志平担任，中国工商总局全国企业信用信息显示，驰星纸厂于2005年由曹志平注册。

曹志平，1959年3月出生，曹氏宣纸世家传人，世代居住在小岭村占云村民组。据调查时曹志平的自述，1980年，他在泾县小岭宣纸厂扩建西山分厂时被招入捞纸车间从事捞纸工作，师从小岭人曹康贵，当时学徒一个月工资17元。一般学徒需要三年才出师，因西山分厂刚刚扩建投产，需要大量的技术工人，曹志平便提前站槽顶岗，负责掌帘。他因个子高、体力好，专门捞六尺纸。按照当时集体所有制工人定级制度，对于曹志平等人提前站槽，且又不能拿学徒工资的情况，小岭宣纸厂采取了按月计件工资制度。四尺纸掌帘工，每刀纸为0.2元，每天任务8刀，超额部分按0.3元/刀计酬；六尺掌帘工，每刀纸为0.28元，每天任务6刀，超额部分按0.42元/刀计酬。

⊙1

曹志平1986年被小岭宣纸厂委派采购宣纸原材料，1988年担任厂长助理兼任新工艺车间（小岭宣纸厂氧–碱法）主任，1989～2001年（这期间，小岭宣纸厂改为安徽红旗宣纸有限公司）担任小岭宣纸厂副厂长、副总经理，分管厂里新工艺、原材料（燎草生产）、保卫、生产、供应科等。2001年，红旗宣纸有限公司改制，曹志平以每年400元由企业买断工龄。同年6月，曹志平在小岭宣纸厂西山车间从事檀皮加工工作，将加工后的青檀化学皮卖给当地的书画纸厂。

2002年，常年在韩国经商的沈阳籍商人叶明委托蔡先生（调查中曹志平已经记不起全名）寻找合适出口韩国的纸，蔡先生认识在南京开文房四宝店的沈树杭（原小岭宣纸厂销售科科长），通过沈树杭的介绍，曹志平认识了叶明，双方达成协议，共同开办纸厂，专做出口到韩国的皮纸。

交流中曹志平表示，当年之所以选择专做皮

⊙1
在厂区咨询曹志平纸的原料加工
Consulting Cao Zhiping about raw materials processing step of paper

纸，是因为考虑到制作宣纸需要净化设备，加上泾县竞争厂家较多，投资风险较大，而当时的泾县皮纸生产企业少，又有出口韩国的订单保底，自己投资皮纸厂一开始信心很足。但是，筹办初期没想到的是，泾县当地的技工资源储备基本以宣纸工人为主，找会造皮纸的工人一时不容易。不过天无绝人之路，凑巧的是，当时韩国人在陕西宝鸡开办的皮纸厂因为搬迁到上海，宝鸡厂的一部分员工仍留在宝鸡。而小岭人曹康造在丁家桥镇开办了一个皮纸厂，宝鸡厂的一部分技工便来到了泾县的皮纸厂。曹康造的厂由于效益不太好而难以维持，这些工人便顺利转入曹志平所办的新企业。

2003年曹志平与红旗宣纸厂同事许富贵、沈树杭合作，每人投入2万元，贷款4万多元，借款4万元，在占元的空地（公私合营泾县宣纸厂占元生产车间的原址）上建设厂房，开设了6帘槽的皮纸生产区。2004年沈树杭退股，驰星纸厂的股东变为曹志平和许富贵两人，2005年完成了安徽省泾县小岭驰星纸厂的工商登记。

许富贵，1957年生于泾县小岭村方家门，初中毕业后随家人专门为小岭宣纸原料合作社加工宣纸原料。小岭宣纸原料合作社从20世纪60年代就开始在许湾生产宣纸，使用“红旗”作为商标。1981年西山车间投产后，小岭宣纸原料合作社正式改名为“安徽省泾县小岭宣纸厂”，许富贵被招入新成立的小岭宣纸厂制浆车间担任蒸煮工，专事檀皮蒸煮工作。

1984年泾县金竹坑宣纸厂（现为安徽泾县金星宣纸有限公司）创办时，通过金竹坑宣纸厂首任厂长曹金修的协调，借调了吴召玉、曹国富、许富贵等技术工人，许富贵主要负责蒸煮。1985年金竹坑宣纸厂的蒸煮学徒工顶岗后，许富贵回到小岭宣纸厂继续从事原岗位工作，直至小岭宣纸厂停产。

据曹志平介绍，其父曹康贵（与其师傅同名，但不是一人）一直在村里务农，大妹妹曹银平在驰星纸厂从事检验工作，五妹妹曹五平在博古堂工艺品厂做册页加工工作。

⊙1

⊙1 曹志平 Cao Zhiping

# 三

# 泾县小岭驰星纸厂的代表纸品及其用途与技术分析

3

Representative Paper and Its Uses and Technical Analysis of Xiaoling Chixing Paper Factory in Jingxian County

## （一）

## 代表纸品及其用途

调查组2015年7月16日和2016年4月20日调查获知的信息如下：驰星纸厂生产多种皮纸，如雁皮纸、三桠皮纸、楮皮纸、构皮纸。产品通常按照四尺、六尺、八尺、尺八屏生产。据曹志平介绍，目前驰星纸厂以订单定生产，按照客户要求生产。其中雁皮纸主要供应给美术学院进行写意人物工笔画创作，还可以用于写小楷；偏白色的构皮纸加了四川凉山州产小叶苦丁茶作为云南普洱茶的包装纸，本色黑构皮纸加中药材用来做出口国外的酒包装；三桠皮纸在沈阳、北京、广州等一些地方的博物馆做修复用纸。

⊙2

⊙3

## （二）

## 代表纸品的性能分析

测试小组对采样自驰星纸厂的雁皮罗纹纸所做的性能分析，主要包括厚度、定量、紧度、抗张力、抗张强度、撕裂度、湿强度、白度、耐老化度下降、尘埃度、吸水性、伸缩性、纤维长度和纤维宽度等。按相应要求，每一指标都需重复测量若干次后求平均值，其中厚度抽取10个样本进行测试，定量抽取5个样本进行测试，抗张力（强度）抽取20个样本进行测试，撕裂度抽取10

⊙2 普洱茶包装纸 Packaging paper for Pu'er Tea

⊙3 正在蒸煮的中药材 Boiling the Chinese medicinal materials

个样本进行测试，湿强度抽取20个样本进行测试，白度抽取10个样本进行测试，耐老化度下降抽取10个样本进行测试，尘埃度抽取4个样本进行测试，吸水性抽取10个样本进行测试，伸缩性抽取4个样本进行测试，纤维长度测试了200根纤维，纤维宽度测试了300根纤维。对驰星纸厂雁皮罗纹纸进行测试分析所得到的相关性能参数见表4.2。表中列出了各参数的最大值、最小值及测量若干次所得到的平均值或者计算结果。

表4.2　泾县小岭驰星纸厂雁皮罗纹纸相关性能参数
Table 4.2　Performance parameters of *Wikstroemia pilosa* Cheng Luowen paper of Xiaoling Chixing Paper Factory in Jingxian County

| 指标 | | 单位 | 最大值 | 最小值 | 平均值 | 结果 |
|---|---|---|---|---|---|---|
| 厚度 | | mm | 0.035 | 0.031 | 0.033 | 0.033 |
| 定量 | | g/m² | — | — | — | 21.3 |
| 紧度 | | g/cm³ | — | — | — | 0.645 |
| 抗张力 | 纵向 | N | 10.3 | 6.3 | 8.7 | 8.7 |
| | 横向 | N | 6.3 | 5.0 | 5.4 | 5.4 |
| 抗张强度 | | kN/m | — | — | — | 0.470 |
| 撕裂度 | 纵向 | mN | 150 | 130 | 142 | 142 |
| | 横向 | mN | 230 | 190 | 210 | 210 |
| 撕裂指数 | | mN · m²/g | — | — | — | 16.1 |
| 湿强度 | 纵向 | mN | 1 100 | 1 000 | 210 | 210 |
| | 横向 | mN | 800 | 600 | 140 | 140 |
| 白度 | | % | 69.0 | 67.5 | 68.1 | 68.1 |
| 耐老化度下降 | | % | — | — | — | 3.0 |
| 尘埃度 | 黑点 | 个/m² | — | — | — | 48 |
| | 黄茎 | 个/m² | — | — | — | 16 |
| | 双浆团 | 个/m² | — | — | — | 0 |
| 吸水性 | | mm | — | — | — | 7.0 |
| 伸缩性 | 浸湿 | % | — | — | — | 0.75 |
| | 风干 | % | — | — | — | 0.18 |
| 纤维　皮 | 长度 | mm | 8.30 | 0.99 | 3.19 | 3.19 |
| | 宽度 | μm | 11.0 | 1.0 | 5.0 | 5.0 |

由表4.2中的数据可知，驰星纸厂雁皮罗纹纸最厚约是最薄的1.13倍，经计算，其相对标准偏差为0.001，纸张厚薄较为一致。所测驰星纸厂雁皮罗纹纸的平均定量为21.3 g/m²。通过计算可知，驰星纸厂雁皮罗纹纸紧度为0.645 g/cm³。抗张强度为0.470 kN/m，抗张强度值较小。所测驰星纸厂雁皮罗纹纸撕裂指数为16.1 mN · m²/g，撕裂度较大；湿强度纵横平均值为175 mN，湿强度较大。

所测驰星纸厂雁皮罗纹纸平均白度为68.1%，白度较高。白度最大值是最小值的1.022倍，相对标准偏差为0.595，白度差异相对较小。经过耐老化测试后，耐老化度下降3.0%。

★1 ★2

所测驰星纸厂雁皮罗纹纸尘埃度指标中黑点为48个/m²，黄茎为16个/m²，双浆团为0个/m²。吸水性纵横平均值为7.0 mm，纵横差为2.2 mm。伸缩性指标中浸湿后伸缩差为0.75 %，风干后伸缩差为0.18 %，说明驰星纸厂雁皮罗纹纸伸缩性差异不大。

驰星纸厂雁皮罗纹纸在10倍、20倍物镜下观测的纤维形态分别见图★1、图★2。所测驰星纸厂雁皮罗纹纸皮纤维长度：最长8.30 mm，最短0.99 mm，平均长度为3.19 mm；纤维宽度：最宽11.0 μm，最窄1.0 μm，平均宽度为5.0 μm。驰星纸厂雁皮罗纹纸润墨效果见图⊙1。

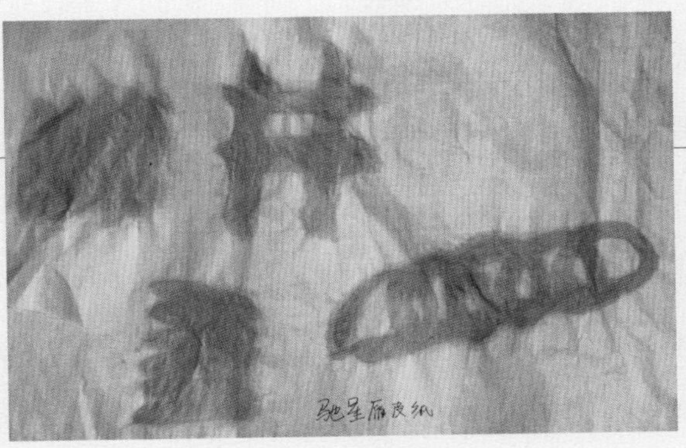


⊙1

## 四 驰星纸厂“忆宣”皮纸生产的原料、工艺流程与工具设备

## 4 Raw Materials, Papermaking Techniques and Tools of “Yixuan” Bast Paper in Chixing Paper Factory

### （一）“忆宣”皮纸生产的原料

#### 1. 主料

(1) 雁皮。雁皮又称山棉皮、荛花皮，植株高约2 m，皮质纤维纤细柔软，主要用于生产伞纸、扇纸、皮纸、云母纸、蜡纸等，也可用于做绳索和编织袋。据曹志平介绍，“忆宣”皮纸所用的雁皮是从江西省萍乡市经销商处购买的蒸煮好的菲律宾雁皮，2015年的购买价为30 000多元/t。

★1 驰星纸厂雁皮罗纹纸纤维形态图（10×）
Fibers of *Wikstroemia pilosa* Cheng Luowen paper in Chixing Paper Factory (10× objective)

★2 驰星纸厂雁皮罗纹纸纤维形态图（20×）
Fibers of *Wikstroemia pilosa* Cheng Luowen paper in Chixing Paper Factory (20× objective)

⊙1 驰星纸厂雁皮罗纹纸润墨效果
Writing performance of *Wikstroemia pilosa* Cheng Luowen paper in Chixing Paper Factory

（2）三桠皮。三桠皮又称结香皮、金腰带、萝冬花。三桠枝条韧皮部的纤维层叫三桠皮。三桠是落叶灌木类植物，高2 m以上，枝呈三叉状，所以叫三桠树。三桠皮纤维坚韧、细柔，中部较宽，两端较细。据曹志平介绍，“忆宣”皮纸所用的三桠皮是从江西省萍乡市经销商处购买的蒸煮好的绝干浆，购买价为50 000多元/t。

（3）构皮。构皮又称楮皮，桑科构属，为多年生落叶乔木。构叶表背两面都为深绿色，两面都有毛刺，是最古老的优良造纸原料。构皮纤维较长，平均长度介于亚麻与红麻之间，纤维壁上有明显的横节纹，两端尖细，常呈分枝状。据曹志平介绍，“忆宣”皮纸所用的构皮分为黑构皮和白构皮两种：黑构皮一般出产于陕西等地，纤维较粗，拉力与韧性强，通常用来做中低档纸；白构皮一般产自泰国、缅甸和我国的广西、云南等地，纤维较细，捞出来的纸较为光滑。驰星纸厂的构皮是从江西省萍乡市经销商处购买的蒸煮好的绝干浆，2015年黑构皮购买价为30 000元/t，白构皮购买价为40 000多元/t。

（4）木浆。据曹志平介绍，制作“忆宣”中低档皮纸时还会加入少量的木浆，从上海经销商处购买加拿大木浆，2015年的购买价为5 000元/t。

### 2. 辅料

（1）纸药。据曹志平介绍，“忆宣”皮纸使用的是化学纸药，一般以22.5元/kg的价格从泾县代理商处购买，由代理商整体从日本进口。

（2）水源。“忆宣”皮纸的生产用水主要是从小岭周坑村山上流下的山涧水。据调查组在现场的测试，“忆宣”皮纸制作所用的山涧水pH为6.5，呈弱酸性。

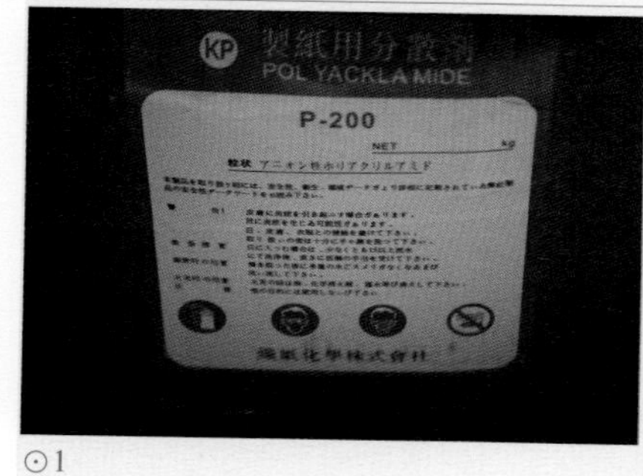

⊙1

⊙2

⊙1
日本进口的化学纸药
Imported Japanese chemical papermaking mucilage

⊙2
周坑村的山涧水
Stream of Zhoukeng Village

(二)

## "忆宣"皮纸生产的工艺流程

调查组于2015年7月16日和2016年4月20日在驰星纸厂实地调查时，根据曹志平的说法，因原料不同，各种纸的加工工艺自然也有所不同，不同点主要体现在制浆工艺上，成纸制作工艺几乎都一样。因驰星纸厂所用的原料均为外购的成熟浆料，成纸的加工工艺差异性很小。调查组根据曹志平的口述和现场观察，"忆宣"皮纸生产的工艺流程为：

| 壹 | 贰 | 叁 | 肆 | 伍 | 陆 | 柒 | 捌 | 玖 | 拾 |
|---|---|---|---|---|---|---|---|---|---|
| 皮料拣选 | 清洗 | 打浆 | 捞纸 | 压榨 | 烘帖 | 醒帖 | 晒纸 | 检验、剪纸 | 包装 |

### 壹 皮料拣选

1

将收购来的漂白皮料先通过人工拣选，拣选出杂质和不合格皮料，合格的皮料放在一边备用。

### 贰 清洗

2 ⊙3

将拣选出的合格皮料过清水进行清洗。

### 叁 打浆

3 ⊙4⊙5

将皮料通过打浆机打碎。在打浆时，根据各种纸的需要，掺入木浆形成混合浆，通常的配比为60%的木浆加40%的构皮。如果遇到一些拉力不强的浆料，适量放入一些在山东泰安购买的拉力剂，增加纸的拉力和强度，提高纸张纤维的结合力。



⊙4

⊙5

⊙3

⊙3 清洗皮料 Cleaning the bark materials

⊙4 打浆 Beating the pulp

⊙5 拉力剂外包装 Package of paper adhesives

## 肆 捞纸 4

⊙6~⊙12

将打好的浆料通过传输管道输入到纸槽中，放入一定比例的纸药（0.5 kg纸药一般可制作约500张纸），搅拌后，采用吊帘方式捞纸。在捞纸时，捞纸工人站在纸槽一边，将铺好帘子的帘架前端往上竖，由外向身边下水，纸浆布满整个帘面后，前后晃动帘架

⊙6

⊙7

⊙8

⊙9



荡水后将浆料倒掉。架上帘架后，先揭开固定帘子的木框，右手拿起湿纸帘，左手托住纸帘下端，将纸帘移送到纸板时，先将纸帘下端抵住纸板上固定的木桩，将纸帘形成筒状由身边向外平稳倒扣在纸板上，而后再将纸帘由身边向外揭走，移送到纸槽的帘架上，捞下一张纸。

⊙10

⊙6 捞纸车间 Papermaking workshop

⊙7/10 捞纸 Papermaking

## 压 榨

5 ⊙13⊙14

捞纸工人下班后，先在湿纸帖上盖上盖纸帘，帘上再加木板，先后逐渐扳上由工字钢制作的榨杆，榨杆上方垫上木枕、千斤顶，连到钢铁浇铸的将军柱上，将前一天所捞的湿纸放在木榨上，不断使用千斤顶，缓缓挤压榨出水分。

⊙11

⊙12

⊙13

⊙14

⊙11 捞纸结束做记号 Making the paper after papermaking

⊙12 做好记号等待压榨的纸帖 Paper with marks to be pressed

⊙13 压榨工具 Tool for pressing the paper

⊙14 压榨 Pressing the paper

## 陆 烘帖 6

将压榨后的纸帖放在焙墙上烤，将纸帖中的水分进一步蒸发。

## 柒 醒帖 7

将蒸发过的纸帖用清水慢慢淋湿，水量多少完全凭工人自己掌握。淋湿后的纸帖不能直接晒纸，需摆放几小时，使水分充分润湿纸帖。这道工序称为“醒帖”。

## 捌 晒纸 8

⊙15⊙16

将待晒的纸帖架上纸架，用左手手指在纸左上角先起一纸角，逐步将一张湿纸揭离下来，用刷把由上到下刷贴在铁制的晒纸焙上，再晒下一张纸。纸焙全部贴满后，再依次将干燥的纸揭取下来。

⊙15

⊙16



## 玖 检验、剪纸 9

将晒好的纸进行逐张检验，遇到不合格的纸立即抽出作废，合格的纸整理好，每100张放上一张套皮纸，整块纸帖压上压低石，按照50张1个刀口，持平剪刀裁剪四边，规整纸张。

## 拾 包装 10

⊙17⊙18

剪好后的纸按100张/刀分好，再加盖“忆宣”各类章。包装完毕后放入仓库。

⊙17

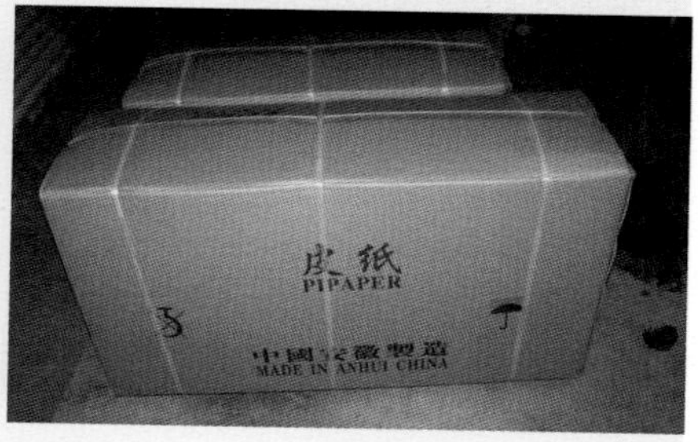


⊙18

⊙15 晒纸上墙 Dring the paper on a wall

⊙16 揭纸 Peeling the paper down

⊙17/18 仓库中的成品纸 Paper products in the warehouse

(三)

"忆宣"皮纸制作的工具设备



壹

## 捞纸槽

1

盛浆工具，由水泥浇筑而成。实测驰星纸厂所用的四尺捞纸槽尺寸为：长200 cm，宽180 cm，高80 cm；六尺捞纸槽尺寸为：长250 cm，宽220 cm，高80 cm；八尺捞纸槽尺寸为：长320 cm，宽225 cm，高80 cm；尺八屏槽尺寸为：长300 cm，宽180 cm，高80 cm。

贰

## 纸 帘

2

用于捞纸，由慈竹编织而成，表面光滑平整，帘纹细而密集。据曹志平介绍，驰星纸厂所用纸帘从福建省的厦门与龙岩等地购买，近几年的价格均为800多元/张。实测驰星纸厂所用纸帘尺寸为：长162 cm，宽90 cm。

⊙19

叁

## 松毛刷

3

晒纸时将纸刷上晒纸墙，刷柄为木制，刷毛为松毛。实测驰星纸厂所用刷子长49 cm，带刷毛一共宽10 cm。

陆

## 木 头

6

在晒纸时充当镇纸，将晒干的纸用木头压住。晒的时候用木头压一下干得快，只适用于云龙纸。实测驰星纸厂所用木头尺寸为：长13.5 cm，宽5.5 cm。

⊙22

肆

## 铲 子

4

晒纸时遇到晒纸墙上不好牵扯的纸头时，用于牵起纸头。实测驰星纸厂所用铲子长21 cm。

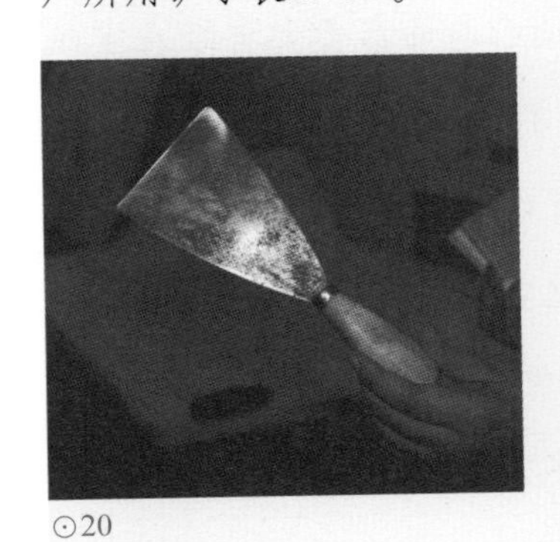

⊙20

伍

## 压纸石

5

用于压晒好的纸。实测驰星纸厂所用压纸石长19 cm。

⊙21

⊙19 纸帘 Papermaking screen

⊙20 铲子 Shovel for separating the paper layers

⊙21 压纸石 Stone for pressing the paper

⊙22 木头 Wood for pressing the paper

## 五 泾县小岭驰星纸厂的市场经营状况

## 5 Marketing Status of Xiaoling Chixing Paper Factory in Jingxian County

据访谈时曹志平的叙述：初创期间，驰星纸厂的皮纸主要以出口韩国及日本为主，出口占总销售的80%左右。自2006年以来，国内市场逐步扩大。截至调查时的2015年底，驰星纸厂一直以订单来定产量和品种，主要客户分布在北京、杭州、成都、合肥、郑州、西安等地。其中北京主要销售渠道的经销商为曹义虎、张志联、曹亮和曹友泉，杭州主要销售渠道的经销商为蔡小坤，成都主要销售渠道的经销商为周新华，合肥主要销售渠道的经销商为余一付，西安主要销售渠道的经销商为曹国富。

⊙1 营业执照 Business license

## 六 泾县小岭驰星纸厂的品牌文化与民俗故事

## 6 Brand Culture and Stories of Xiaoling Chixing Paper Factory in Jingxian County

### 1. “忆宣”品牌的来历

访谈中谈及为什么会选择“忆宣”这个品牌名时，曹志平表示：当年准备注册时，才发现驰星商标已被他人注册，转而一想，自己与许富贵都曾经在宣纸行业工作20多年，是不折不扣的老宣纸工人，对宣纸这个行业及小岭宣纸厂有难以忘怀的情感，于是同许富贵商量，一致认同取“忆宣”这个品牌名称，其寓意就是纪念自己青春年华里曾经奉献过的“宣纸岁月”。

### 2. 割舍不下的“小岭宣纸厂”情怀

曹志平与许富贵都曾经在泾县小岭宣纸厂从事造纸工作20多年，可以说从青春到中年的生涯都贡献给了小岭宣纸厂。许富贵20多年一直是皮料蒸煮技工，曹志平则先是捞纸工，以后又担任过车间主任、副厂长、副总经理。访谈中，两人都对当年宣纸发展历程中占有重要地位的小岭宣纸厂的解体深感痛惜，谈到动情处，曹志平拿出2014年他作为主要动议人，发起恢复小岭宣纸厂并有大批旧日员工摁手印向政府申请的原件，细述了事件的来龙去脉和发起之心，令调查人员也深受他们“小岭宣纸厂情怀”的感染。

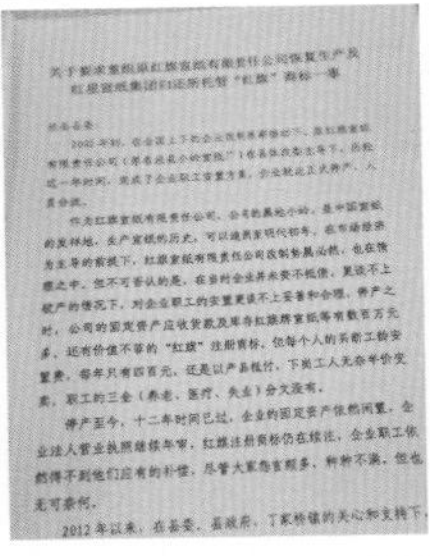
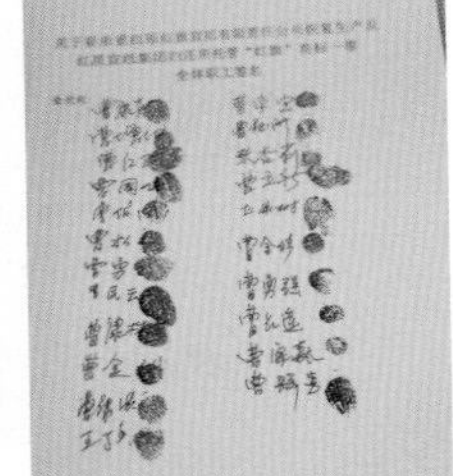
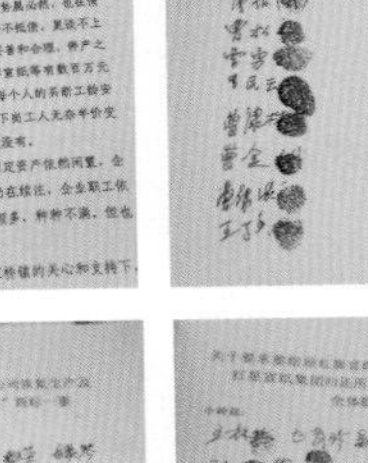
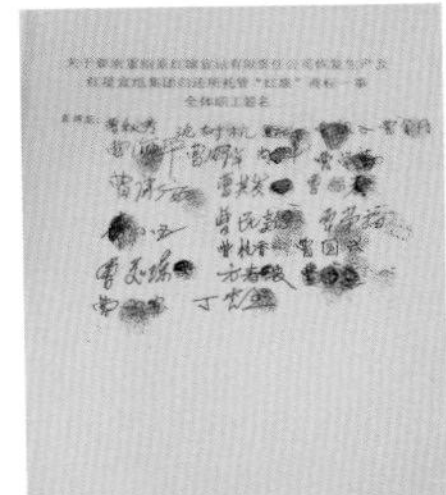
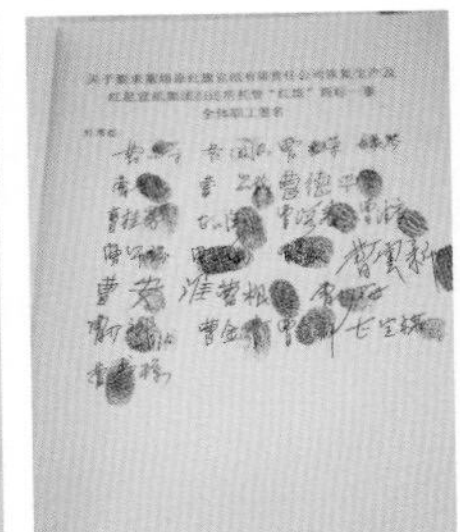
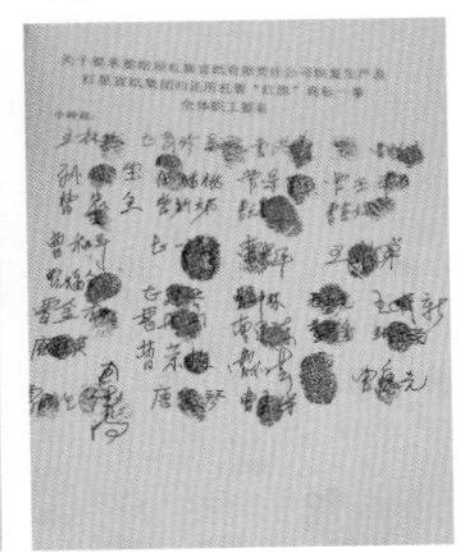

⊙2

⊙3

⊙4

## 七 泾县小岭驰星纸厂的业态传承现状与发展思考

## 7 Current Status of Business Inheritance and Thoughts on Development of Xiaoling Chixing Paper Factory in Jingxian County

驰星纸厂作为泾县为数不多的皮纸厂，从建厂开始，产品就主要外销供应给韩国，所以生产标准基本上参照了韩国的手工纸体系标准，引进的陕西宝鸡原韩国人开办皮纸厂的技术工人，在调查时仍有三人在驰星纸厂工作。曹志平介绍说，自己与许富贵一直在努力创造一个稳定的收入环境来稳定技术人员。2015年驰星纸厂捞纸工人每月5 000元，晒纸工人每月10 000元。工人较高的收入加上稳定的发展目标，让驰星纸厂在良性的经营中不断发展，所产皮纸不断拓展和培育新的用途和领域，使其生产和销售一直处于稳定状态。

⊙2 『恢复小岭宣纸厂上书』原件照 Original photos of "Asking for Resuming Xiaoling Paper Factory"

⊙3 陕西宝鸡的捞纸工 A papermaker from Baoji City of Shaanxi Province

⊙4 调查组成员和曹志平等在驰星纸厂大门口合影 A group photo of researchers and Cao Zhiping etc. at the gate of Chixing Paper Factory

# 泾县小岭驰星纸厂

# 雁皮纸

*Wikstroemia pilosa* Cheng Paper
of Xiaoling Chixing Paper Factory
in Jingxian County

雁皮罗纹纸透光摄影图
A photo of *Wikstroemia pilosa* Cheng Luowen paper seen through the light

# 第三节

# 潜山县星杰桑皮纸厂

安徽省
Anhui Province

宣城市
Xuancheng City

潜山县
Qianshan County

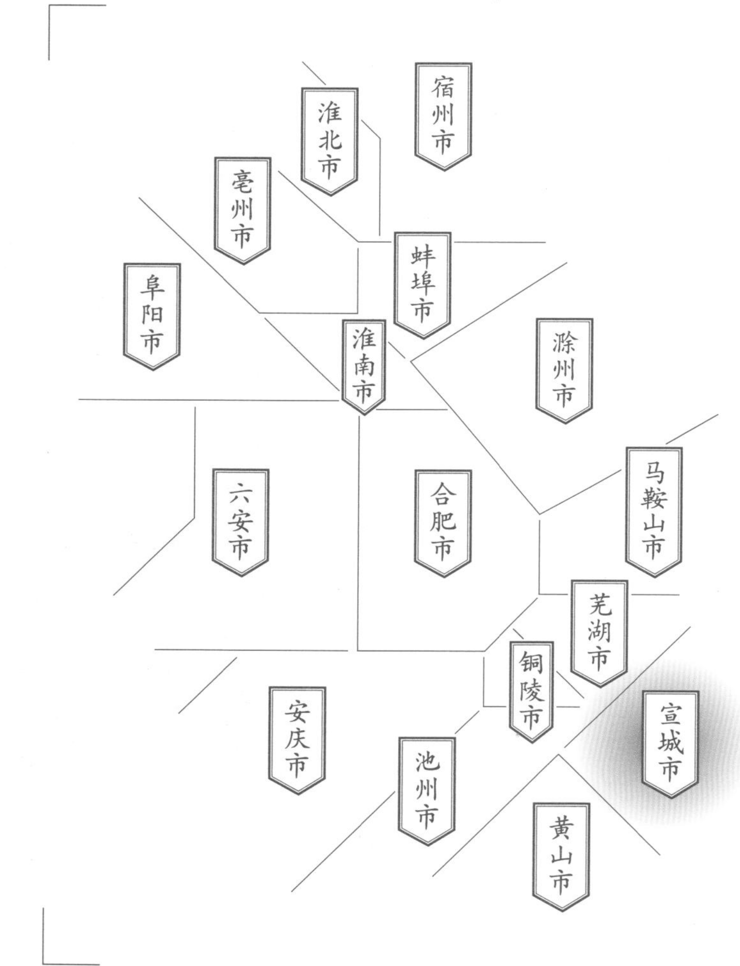


**调查对象**

官庄镇
潜山县星杰桑皮纸厂
皮纸

Section 3

## Xingjie Mulberry Bark Paper Factory in Qianshan County

**Subject**

Bast Paper
of Xingjie Mulberry Bark Paper Factory
in Qianshan County
in Guanzhuang Town

一

# 潜山县星杰桑皮纸厂的基础信息与生产环境

1

Basic Information and Production Environment of Xingjie Mulberry Bark Paper Factory in Qianshan County

潜山县星杰桑皮纸厂坐落于安庆市潜山县官庄镇坛畈行政村，地理坐标为：东经116° 36′ 11″，北纬31° 1′ 22″。桑皮纸为潜山县历史上著名的手工纸品，在当地乡土的口传记忆中，这种造桑皮纸的技艺始于东汉末年，因此民间一直又有汉皮纸的别称，按照尺寸分别有大汉、中汉和小汉的称谓。如果传说有据，那么潜山桑皮纸迄今已有1 700多年的历史，只可惜目前尚未在任何地方文献中发现有涉及早期起源的记述。

桑皮纸具有柔韧、拉力强、不断裂、不褪色、防虫、无毒性、吸水力强的特点，传统用途主要是做高级包装和制伞、糊篓、做炮引及制作某些工艺品。同时，作为书画用纸和出版修复用纸性能也颇佳。21世纪初，故宫博物院对乾隆皇帝晚年退隐后专享的宫殿倦勤斋大修，其殿内通景画抢修工程急需用当年进贡的优质桑皮褙纸，经过在国内外的“铲地皮”式寻访，2004～2005年潜山县与相邻岳西县的桑皮纸脱颖而出，成为闻名全国的“倦勤斋通景画修复专用纸”。2008年，潜山桑皮纸制作技艺入选第二批国家级非物质文化遗产保护名录；2012年，星杰桑皮纸厂厂长刘同烟被评为桑皮纸制作技艺国家级非物质文化遗产代表性项目代表性传承人。

2015年10月8日、12月3日，调查组两次入星杰桑皮纸厂调查，获得的基础生产信息如下：截至调查时，纸厂共有2个槽位，员工12人左右（动态）。因加工纯桑皮原料所需流动的

⊙1

⊙1 星杰桑皮纸厂外景 External view of Xingjie Mulberry Bark Paper Factory

路线图
潜山县城
↓
潜山县星杰桑皮纸厂
Road map from Qianshan County centre to Xingjie Mulberry Bark Paper Factory in Qianshan County

# 潜山县星杰桑皮纸厂位置示意图

# Location map of Xingjie Mulberry Bark Paper Factory in Qianshan County

考察时间
2015年10月 / 2015年12月

Investigation Date
Oct. 2015 / Dec. 2015

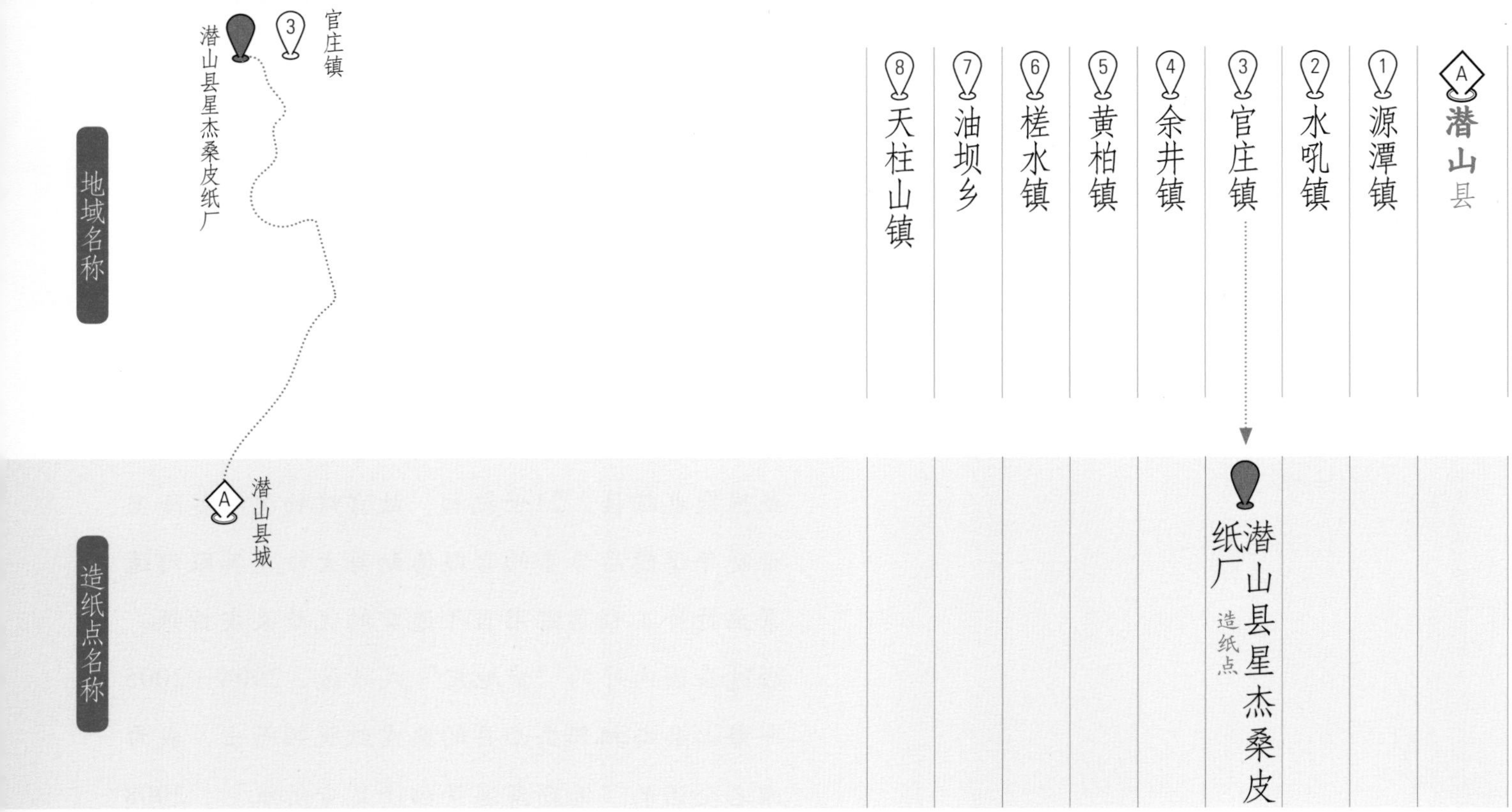

地域名称

A 潜山县
1 源潭镇
2 水吼镇
3 官庄镇
4 余井镇
5 黄柏镇
6 槎水镇
7 油坝乡
8 天柱山镇

造纸点名称

潜山县星杰桑皮纸厂 造纸点

官庄镇
潜山县星杰桑皮纸厂
潜山县城

位置分布

市府、州府
县城
乡镇
村落
造纸点
历史造纸点
山
国家级自然保护区
S221 省道
G21 国道
昆河线 铁路
G56 高速公路
线路

岳西县
潜山县
怀宁县
G105
10 km
5 km
0
N

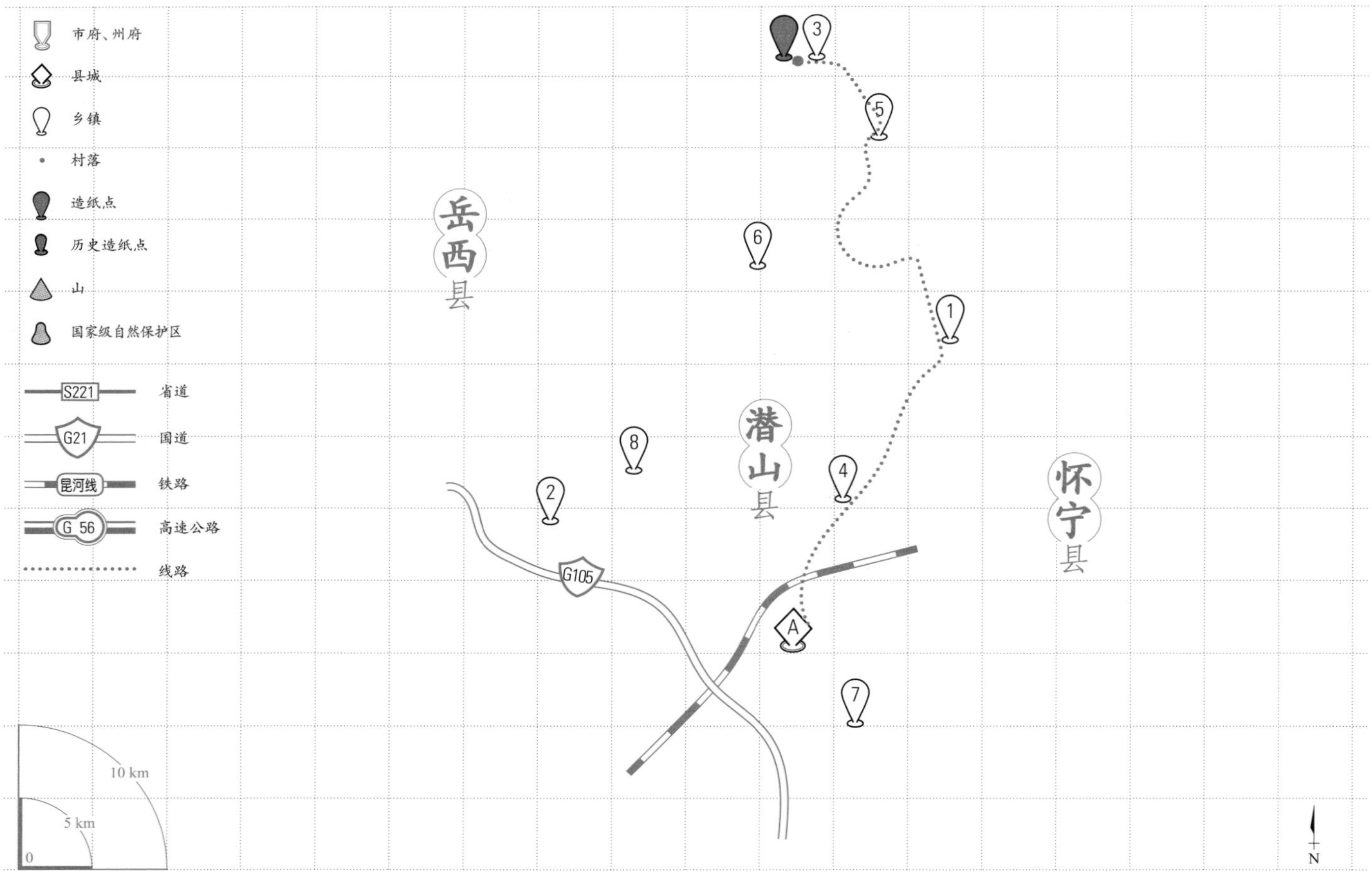

河水严重缺乏，10月8日第一次调查时纯桑皮纸未生产，纸厂正在生产书画纸；12月3日调查前接到刘同烟的电话，获知由于连日下雨，河水充沛，正在制作桑皮纸，因此记录与补拍了漂洗、袋料等工序。

据访谈中刘同烟的介绍，星杰桑皮纸厂年产量5 000多刀纸，年销售额约100万元，其中书画纸产量占80%左右，均使用“紫烟”商标。

⊙1

坛畈村地处潜山县官庄镇的西北角，金紫山森林公园脚下。这里三面环山，南有安徽省重点文物保护单位大香山寺，西面是金紫山森林公园与平洋河景点，风景秀丽，气候宜人。村内有平阳路和坛香路两条山区公路通过。2014年底的数据显示，全村有550户约2 100人，其中流动人口406人。农业总产值72.45万元，人均收入345元。坛畈村特色产业为石材加工，调查组两次调查进村时，均能看见规模很大的石材切割加工现场，时不时会听见响彻全村的切割噪声。



## 二 潜山县星杰桑皮纸厂的历史与传承情况

## 2 History and Inheritance of Xingjie Mulberry Bark Paper Factory in Qianshan County

潜山县星杰桑皮纸厂注册成立于2007年，2015年调查时新厂已走过8年的历程，而纸厂在老厂房至少能追忆到至今已有六代上百年生产桑皮纸的历史。据刘同烟访谈时回忆的排序，老厂房从祖辈流传至今，到他这一代能说得上来的至少已有五代的传承史，他侄子刘绍成则是在这幢厂房造纸的第六代了。传承谱系分别是：继高公—如耿公—孝虞公—友英公—刘同烟—刘绍成。

老厂房主人刘同烟2012年被评为桑皮纸制作技艺国家级非物质文化遗产代表性项目代表性传承人，妻子陈爱荣和侄子刘绍成2011年分别被评为桑皮纸制作技艺省级非物质文化遗产项目代表性传承人，两位兄长也从事桑皮纸制作，是典型的乡村造纸世家。

⊙1
调查组成员在村中与刘同烟交流
Researchers communicating with Liu Tongyan in the village

作为有家族记忆的第五代造纸传承人，刘同烟1964年出生于坛畈村，从事造纸已有30多年的时间。1981年刘同烟初中毕业，1982年开始跟着两位兄长学习制作桑皮纸，二哥主要负责捞纸，大哥则负责煮皮、袋料等原料加工环节。刘同烟回忆：最初家庭作坊主要生产大、中、小汉纸，三种纸的尺寸、质量、用途、厚薄各不相同，原料主要是桑皮和文化纸边。

1985年，家庭纸坊开始生产书画纸，原料主要有桑皮、龙须草、木浆和竹浆，当时厂里员工只有4～5人。1985年之前，纸坊每年产量维持在500～600刀；1985年后，纸厂开始生产六尺书画纸，年总产量为700～800刀，主要销售到当地文化用品部。1998年，受机械纸的冲击，书画纸市场萧条，当地很多纸厂倒闭，刘家纸坊也暂时停产。1999～2001年，由于家庭纸坊停工，刘同烟便去了位于黄山市黄山区的黄山白天鹅宣纸厂打工，主要从事抄纸和配料工作，两位兄长则在家务农。

2002年上半年，刘同烟回家开始重新造纸，为北京档案馆生产古书修复用纸。2005年，故宫倦勤斋通景画抢修工程急需褙纸，纸厂生产的纯桑皮纸成功中标，该订单一直持续至2008年。2007年，星杰桑皮纸厂成立，新厂区开建，刘同

⊙1

⊙2

⊙3

⊙1
坛畈村的农家在收稻谷
Farmers in Tanfan Village harvesting rice

⊙2
正在讲解工序的刘同烟（右二）
Liu Tongyan explaining the papermaking procedures

⊙3
刘同烟在观察野生桑树皮
Liu Tongyan observing wild mulberry bark

烟成为企业法人。新厂启动资金约25万元，主要生产书画纸；而原为家庭纸坊的老厂房则生产纯桑皮纸，形成两个产品体系并存的生产格局。

刘同烟的妻子陈爱荣1966年出生，娘家在官庄镇香山村，父母也是抄造大、小、中汉纸的造纸户，所以她对桑皮纸制作技艺十分熟悉。侄子刘绍成43岁，16岁初中毕业后就跟随刘同烟学造纸。由于纯桑皮纸工序繁杂、制作时间长、需要劳力多，每次造纸时，刘同烟都会带领妻子、侄子、侄媳等家人一起出工出力。

刘同烟夫妇共有三个子女，大女儿在大学教书，二女儿从事会计工作，小儿子经营服装生意，虽然三个孩子均未从事造纸行业，但刘同烟表示尊重他们的职业选择。除了同辈的妻子、兄长外，家中侄子、侄媳会在造纸繁忙期予以帮衬。桑皮纸制作技艺一直从祖辈流传至今，由于考虑自己年龄渐大，2015年访谈时刘同烟表示：目前正在将制作技艺逐步传承给侄子刘绍成。

⊙4

⊙4 坛畈村公路上的桑皮纸宣传牌
Billboard of mulberry bark paper by the road of Tanfan Village

## 三 潜山县星杰桑皮纸厂的代表纸品及其用途与技术分析

## 3 Representative Paper and Its Uses and Technical Analysis of Xingjie Mulberry Bark Paper Factory in Qianshan County

### (一) 代表纸品及其用途

星杰桑皮纸厂生产的纸品包括纯桑皮纸和书画纸两大类，其中以纯桑皮纸最具代表性。纯桑皮纸根据尺寸、用途的不同，可细分为大汉纸、中汉纸、小汉纸、针灸艾条纸、文物修复纸、拓印纸等。

书画纸主要用于书法、绘画。纯桑皮纸根据种类不同用途有所区别。大汉纸尺寸为80 cm × 60 cm，主要用来做爆竹的引火线、纸伞和包装。中汉纸尺寸为70 cm × 50 cm，小汉纸尺寸为60 cm × 40 cm，主要用来糊篓、贴窗户、制作小玩具纸品（灯笼）等。针灸艾条纸主要用于中医临床，文物修复纸和拓印纸用来修复古籍、文物等。它们的规格有大有小，大的约为138 cm × 68 cm（四尺），小的约为80 cm × 60 cm。

⊙1

⊙2

其中，书画纸是常年生产的，纯桑皮纸则根据客户订单和造纸坊的水源状况来生产，产品均在国内销售。

## （二）代表纸品的性能分析

测试小组对采样自潜山县星杰桑皮纸厂（本部分以下简称“潜山星杰”）生产的针灸艾条纸所做的性能分析，主要包括厚度、定量、紧度、抗张力、抗张强度、撕裂度、湿强度、色度、耐老化度下降、吸水性、伸缩性、纤维长度和纤维宽度等。按相应技术规范要求，每一指标都需重复测量若干次后求平均值。其中厚度抽取10个样本进行测试，定量抽取5个样本进行测试，抗张力（强度）抽取20个样本进行测试，撕裂度抽取10个样本进行测试，湿强度抽取20个样本进行测试，色度抽取10个样本进行测试，耐老化度下降抽取10个样本进行测试，吸水性抽取10个样本进行测试，伸缩性抽取4个样本进行测试，纤维长度测试200根纤维，纤维宽度测试300根纤维。对潜山星杰针灸艾条纸进行测试分析所得到的相关性能参数见表4.3。表中列出了各参数的最大值、最小值及测量若干次所得到的平均值或计算结果。

性能分析

表4.3 潜山星杰针灸艾条纸相关性能参数
Table 4.3 Performance parameters of Qianshan Xingjie Mulberry Bark Paper for acupuncture and moxibustion

| 指标 | | 单位 | 最大值 | 最小值 | 平均值 | 结果 |
|---|---|---|---|---|---|---|
| 厚度 | | mm | 0.096 | 0.054 | 0.080 | 0.080 |
| 定量 | | g/m$^2$ | — | — | — | 15.9 |
| 紧度 | | g/cm$^3$ | — | — | — | 0.199 |
| 抗张力 | 纵向 | N | 18.4 | 11.0 | 15.2 | 15.2 |
| | 横向 | N | 7.6 | 5.6 | 6.8 | 6.8 |
| 抗张强度 | | kN/m | — | — | — | 0.733 |
| 撕裂度 | 纵向 | mN | 250 | 230 | 242 | 242 |
| | 横向 | mN | 570 | 410 | 498 | 498 |
| 撕裂指数 | | mN · m$^2$/g | — | — | — | 18.6 |
| 湿强度 | 纵向 | mN | 850 | 750 | 810 | 810 |
| | 横向 | mN | 700 | 500 | 610 | 610 |

⊙1/2
刘同烟与侄子在打槽
Liu Tongyan and his nephew stirring the papermaking materials

续表

| 指标 | | 单位 | 最大值 | 最小值 | 平均值 | 结果 |
|---|---|---|---|---|---|---|
| 色度 | | % | 68.0 | 65.9 | 66.3 | 66.3 |
| 耐老化度下降 | | % | — | — | — | 3.2 |
| 吸水性 | | mm | — | — | — | 24 |
| 伸缩性 | 浸湿 | % | — | — | — | 0.25 |
| | 风干 | % | — | — | — | 0.60 |
| 纤维　　桑皮 | 长度 | mm | 15.44 | 0.13 | 7.29 | 7.29 |
| | 宽度 | μm | 35.0 | 1.0 | 12.0 | 12.0 |

由表4.3中的数据可知，潜山星杰针灸艾条纸最厚约是最薄的1.78倍，经计算，其相对标准偏差为0.014，纸张厚薄较为一致。所测潜山星杰针灸艾条纸的平均定量为15.9 g/m²。通过计算可知，潜山星杰针灸艾条纸紧度为0.199 g/cm³，抗张强度为0.733 kN/m，抗张强度值较大。所测潜山星杰针灸艾条纸撕裂指数为18.6 mN · m²/g，撕裂度较大；湿强度纵横平均值为710 mN，湿强度较小。

所测潜山星杰针灸艾条纸平均色度为66.3%，色度最大值是最小值的1.032倍，相对标准偏差为0.723，色度差异相对较大。经过耐老化测试后，耐老化度下降3.2%。所测潜山星杰针灸艾条纸吸水性纵横平均值为24 mm，纵横差为3.4 mm。伸缩性指标中浸湿后伸缩差为0.25%，风干后伸缩差为0.60%，说明所测桑皮纸伸缩性差异不大。

潜山星杰针灸艾条纸在10倍、20倍物镜下观测的纤维形态分别见图★1、图★2。所测桑皮纸纤维长度：最长15.44 mm，最短0.13 mm，平均长度为7.29 mm；纤维宽度：最宽35.0 μm，最窄1.0 μm，平均宽度为12.0 μm。潜山星杰针灸艾条纸润墨效果见图⊙3。

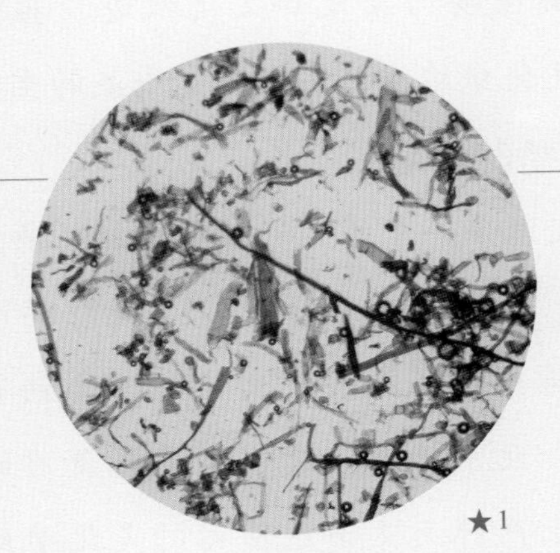

★1

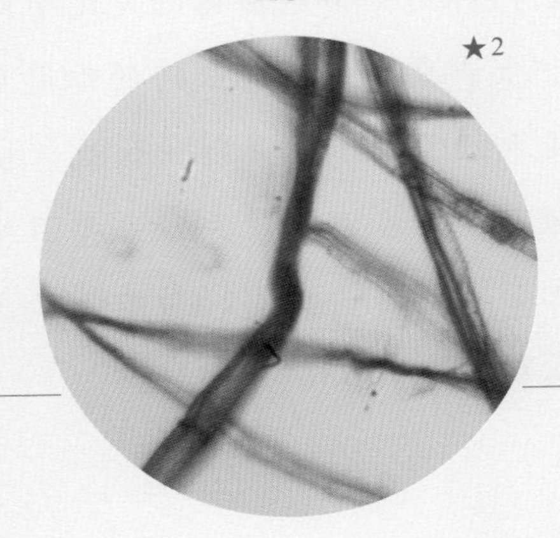

★2

⊙3

★1 针灸艾条纸纤维形态图（10×） Fibers of mulberry bark paper for acupuncture and moxibustion (10× objective)

★2 针灸艾条纸纤维形态图（20×） Fibers of mulberry bark paper for acupuncture and moxibustion (20× objective)

⊙3 针灸艾条纸润墨效果 Writing performance of mulberry bark paper for acupuncture and moxibustion

# 四 潜山县星杰桑皮纸厂代表纸品生产的原料、工艺流程与工具设备

# 4 Raw Materials, Papermaking Techniques and Tools of Xingjie Mulberry Bark Paper Factory in Qianshan County

据两次访谈中刘同烟的介绍，“紫烟”书画纸和纯桑皮纸均采用手工捞制方式，书画纸和纯桑皮纸的生产原料有所不同。书画纸的生产原料为桑皮、龙须草浆、竹浆和木浆，其中桑皮占10%，龙须草浆占60%，竹浆占20%，木浆占10%。龙须草浆从河南运过来，木浆是进口的，竹浆从四川运过来。大、中、小汉纸的生产原料主要为桑皮和文化纸边，其中桑皮占80%，文化纸边占20%。纯桑皮纸的生产原料为100%的野生桑皮。

⊙1

桑皮纸生产中用到的纸药包括植物纸药——杨桃藤、青丹皮、桐藤花根（当地的叫法）和化学纸药聚丙烯酰胺。生产故宫倦勤斋通景画用褙纸和北京档案馆古书修复纸使用的是植物纸药，生产针灸艾条纸使用的是化学纸药聚丙烯酰胺。生产桑皮纸添加纸药的作用主要有两个：一是利于分张；二是防虫防蛀。植物纸药中，杨桃藤多于秋冬季使用，主要因为其使用寿命较短，对保鲜要求高；而青丹皮、桐藤花根主要在夏季温度高时使用。

⊙2

## (一) “紫烟”纯桑皮纸生产的原料

### 1. 主料

野生桑树是生长在海拔500 m以上、1 000 m以下的可再生植物，根部再生能力强，今年将其杆枝砍去，来年又可从根部再生出新苗，于肥沃之地当年就可以长到2 m多高，有大拇指粗细。其叶的形状与枫叶相似，家蚕也可食用，成片的野生桑树春夏之季在潜山的山间乡野可以看见，野

⊙1 水池里浸泡的文化纸边料
Wenhua paper offcut soaking in the sink

⊙2 针灸艾条纸（右）与书画纸（白色）
Paper for acupuncture and moxibustion (right) and calligraphy and painting paper (white)

⊙3

⊙4

蚕食其叶，古称野桑。桑树最大的根杆部直径可达20 cm，其枝部伸展范围可达200 m²。每根枝头顶尖带藤系，能缠绕其他植物，如过大不砍伐，枝杆就会自然枯死，根部会再生。不过据刘同烟说，像上述这样老桑树的皮剥下来是不能用来做桑皮纸的。

调查中刘同烟特别表示：很多人会把桑皮纸的原料桑皮当作是家桑树剥皮制作的，这是一种误解。野桑与家桑除了生长的形状不同，最重要的区分点是家桑剥下来的皮薄、纤维短、粗糙，不论是用碓打还是机打都易成米粒状，拉力不好；而适龄的野桑剥下的皮厚、纤维长而细嫩，外表壳薄，碓打易分散，拉力强，所以古人可用其树皮造出优质的桑皮纸。

⊙5

2. 辅料

(1) 杨桃藤。又叫野生猕猴桃，是一种藤系植物，其叶呈圆形，枝杆直径可达2～3 cm，人们习称之为杨桃藤。它的枝杆中心部有一层汁液，取其枝杆经过浸泡、揉汁、过滤，放进纸浆内，起分解作用。杨桃藤跟桑树一样，太老的不能取汁，太嫩的汁液太少，都需要靠采伐人把握，采伐过的来年又可生出新芽，一般生长了3～4年的杨桃藤最为适宜。杨桃藤做纸药一般适用于气温偏低的秋冬季。

(2) 青丹皮。又称粑叶树、神丹皮，是一种阔叶植物，叶子最大的有扇子大小，呈圆形，杆细高，一般都是单挑枝杆，外形从杆部到枝叶都是青色。通常青丹树长到尺围大小就砍伐剥皮、浸泡取汁。表面一层青皮会自行脱落，中间厚厚的带纤维的皮用来制绳，叶子采下可以垫在蒸笼

⊙3 潜山山间的野桑树
Wild mulberry trees on the hills of Qianshan County

⊙4 坛畈村边的野桑树（樊嘉禄供图）
Wild mulberry trees near Tanfan Village (photo provided by Fan Jialu)

⊙5 浸湿的桑皮原料
Wetted mulberry bark

上蒸粑，故土语称之为粑叶树。青丹皮做纸药一般适用于炎热的夏天。在坛畈村，每逢夏天酷热时就会因缺少合适的纸药而无法造纸。为了寻找能在夏天分解纸浆的植物汁，还有一段神话传说，所以古人又称它为“神丹皮”。

（3）桐藤花根。这是一种很小的植物，其枝杆细小，最大的杆部也只能长到小拇指粗细，其叶细小，需经常砍伐，不然就会自然枯死。每逢夏末秋初，采其浸泡取汁，它的杆部、叶部、根部都有浓浓的汁液，砍伐时有一种黏黏的感觉。桐藤花根是老一辈造纸人用的纸药，刘同烟说现在造纸已不用了。

（4）水源。生产桑皮纸使用的是地下井水和山溪水。其中地下井水主要用来捞纸，而山溪水用来洗料。截至入村调查时，调查组成员实测地下井水pH约为6.8，呈偏弱酸性。

⊙1

⊙2

⊙1/2 山间洗料的山涧溪水
Stream for cleaning the materials

## (二)
## “紫烟”纯桑皮纸生产的工艺流程

制作纯手工桑皮纸工艺繁杂，工序繁多，从第一道工序到成品必须花费不少于3个月的时间。此外，还要具备一定的有利条件，其中水是最主要的，需要丰沛的山泉水，还需要地下水。刘同烟表示：潜山造纸人古代传下来的说法是造纯桑皮纸需要72道工序。

调查组于2015年10月8日和12月3日对“紫烟”纯桑皮纸生产工艺进行了实地调查和访谈，归纳“紫烟”纯桑皮纸制造的工艺流程如下：

| 壹 | 贰 | 叁 | 肆 | 伍 | 陆 | 柒 | 捌 | 玖 | 拾 | 拾壹 | 拾贰 | 拾叁 | 拾肆 |
|---|---|---|---|---|---|---|---|---|---|---|---|---|---|
| 剥皮 | 选拣皮 | 晒皮 | 打捆 | 扎皮 | 蒸皮、煮皮 | 揉皮、洗皮 | 浆皮 | 初洗 | 挤皮 | 初捡 | 中洗 | 中捡 | 水漂 |

| 拾伍 | 拾陆 | 拾柒 | 拾捌 | 拾玖 | 贰拾 | 贰拾壹 | 贰拾贰 | 贰拾叁 | 贰拾肆 | 贰拾伍 | 贰拾陆 | 贰拾柒 | 贰拾捌 |
|---|---|---|---|---|---|---|---|---|---|---|---|---|---|
| 再选 | 露漂 | 精选 | 打皮 | 搅拌 | 过滤 | 袋料 | 拌浆 | 滤浆 | 取浆 | 下槽 | 搅拌 | 划槽 | 锤汁 |

| 贰拾玖 | 叁拾 | 叁拾壹 | 叁拾贰 | 叁拾叁 | 叁拾肆 | 叁拾伍 | 叁拾陆 | 叁拾柒 | 叁拾捌 |
|---|---|---|---|---|---|---|---|---|---|
| 配汁 | 下帘 | 捞纸 | 榨帖 | 退塔 | 烧焙 | 晒纸 | 检纸 | 切纸 | 打包 |

其中，制浆工序可分为皮浆制作和草浆制作两个过程。

## 壹 剥皮

### 1

⊙1

农历二十四节气的“惊蛰”后，植物就开始行汁。这时的桑树从根部通过骨枝向上、外发汁。它的面皮和骨杆之间比较疏松。将其砍伐，去掉细枝，修成光杆；将皮剥离，脚踩杆部，手拉树皮即可剥下。

剥皮者必须把握以下要点：①选3～4龄的桑树，不能太老也不能太嫩；②树杆要滑直；③杆部不能有太多的枝丫、枝节。以上几点把握得好不好，会对所造纸的品质好坏有直接影响。小嫩桑树，皮薄，没有拉力，也制不出多少纸浆；太老的桑树皮虽厚，但纤维太粗无法打碎，制出的纸质粗糙，杂质多。杆部枝丫太多的桑树皮，每逢一个枝丫就有疤痕，皮上就有一定的老茧，难以除尽，也制不出多少纸浆。只有3～4龄的桑树，其杆直、有大拇指粗细，好剥，皮质较厚，纤维适中又没老茧，造出的纸质量最好。到“清明”后，桑树发出新芽，人们就停止采剥，因这时的树汁散至枝叶，树皮难剥。

⊙1

## 贰 选拣皮

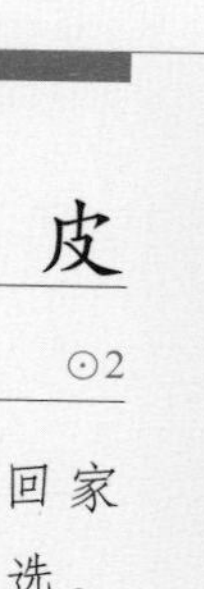

### 2

⊙2

也称刮青，剥好的桑皮挑回家后，将其打开，一把一把地选。将其根部老茧和枝杆部细嫩的地方用剪刀剪去，再用刮刀将皮杆中部的老茧刮掉，这道工序就完成了。

⊙2

## 叁 晒皮

### 3

将选剪好的皮摊在阳光下晾晒，并且要一字排开，一般在20 ℃左右的天气下晒5～6天，每天需翻晒两次，使桑皮卷成杆形，它的含湿量不能超过50%，皮收折时有响声。

## 肆 打捆

### 4

⊙3

⊙3

晒干后，因皮长短不一，将干皮按1.3～1.5 m长折成一把一把的，捆好叠放在仓库，库中必须架50 cm左右的架子，将皮放在上面，以防回潮。

⊙1
剥下的桑树皮（前为野桑皮，后为家桑皮）（樊嘉禄供图）
Mulberry bark (wild ones at the front, cultivated ones at the back) (photo provided by Fan Jialu)

⊙2
女工在拣皮
A female worker picking and choosing the bark

⊙3
入库存放的干皮料
Dried bark materials in the warehouse

## 伍 扎 皮 5

在煮或者蒸之前，从库房将皮取出，成捆打开后放在阳光下重新晒，晒干后把皮折成50 cm长，一把一把地进行蒸煮。

## 陆 蒸皮、煮皮 6

⊙4⊙5

（1）蒸皮。将蒸锅洗干净，放满清水，把扎好的皮料一层一层地放在皮甑内摆好，到甑满为止，封上盖子，一定要保证其密封性，再用干柴烧锅蒸皮。开始火苗要旺盛，如小火慢烧的话，水蒸气不能上升到顶部，皮就不能蒸熟。一直保持大火烧8～9小时，等到从远处能闻到皮的香味时，再用小火烧3～5小时。

⊙4

⊙5

（2）煮皮。同样将煮锅洗干净，放大半锅水，将水烧开后，打开锅，把晒干的皮一把一把地放入锅里，最大的锅能煮50 kg干皮。下锅后用厚塑料布盖严，用大火烧煮，使塑料布顶起，过3个半小时后翻动一次。这样煮6～7小时，观察皮面的壳是否脱落，如没有，还须再烧煮，直至脱落为止，具体时间得靠工人现场把握。

## 柒 揉皮、洗皮 7

⊙6

⊙6

（1）揉皮。将蒸好的皮用皮钩从皮甑里拉出来，放在揉皮板上，揉皮人脚穿草鞋，用脚一把一把地揉，揉一会拿起抖动一下再揉，直到揉好为止。注意：必须每揉一把就从甑里拉一把，如果拉出来凉了皮壳就不易揉掉。

（2）洗皮。就是把煮好的皮用皮钩一把一把地纳入流动的清水河内（河水尽量不要含有任何杂质），纳完后再用皮钩把河水内的皮一把把地摆散开，使细小的皮壳脱落，这样在河内存放并翻动三四次，放下洗皮篮按把清洗。

## 捌 浆 皮 8

将蒸好揉好的皮放入浆皮池内，浆皮池必须提前装好浓度为60%以上的石灰水，使皮拉起时整个皮面都呈石灰白的颜色。如颜色不白可加石灰，这样把皮像码柴火一样堆放在浆池边4～5天。

⊙4 蒸锅 Steaming pot

⊙5 观察炉膛的刘同烟 Liu Tongyan observing the stove heat

⊙6 刘同烟在洗皮 Liu Tongyan cleaning the bark

## 玖
## 初洗
9 ⊙7

将浆好存放4～5天的桑皮拉到无杂质的流动的清水河内，用手在水的上游一把把地排开并在水中摆动，使皮面上的石灰脱落，这时原先没洗掉的部分皮壳会跟着石灰一起脱落，皮的颜色也会变得白一些，不像原本那么黑。

⊙7

## 拾
## 挤皮
10

向河内放入洗皮篮，用手将篮内的桑皮料摆洗好，然后一把一把地挤压成圆柱形，存放在干净的地方，晾放3～4天，使没有挤干的水分自行排出。夏天则只能放1～2天。

## 拾壹
## 初捡
11 ⊙8

选一个阴凉的地方或单独的捡皮房，在不能有污染的地方放好皮料，用捡皮筛将挤压好的桑皮分开或抖散，一根根地从头到尾将没有脱落的皮壳和老茧用手捡出或用指甲将老茧刮掉，再装进捡皮袋内。

⊙8

## 拾贰
## 中洗
12 ⊙9

中洗的过程跟初洗基本一样，就是使初捡过程中刮掉的细小的皮壳或老茧在流动的清水河内摆洗流掉。

⊙9

## 拾叁
## 中捡
13

中捡的过程跟初捡基本一样，就是将中洗过后所留下的皮壳、皮茧再捡一次。

⊙7 初洗皮料（樊嘉禄供图）
Cleaning the bark materials for the first time (photo provided by Fan Jialu)

⊙8 初捡
Choosing the qualified bark for the first time

⊙9 中洗（樊嘉禄供图）
Cleaning the bark materials for the second time (photo provided by Fan Jialu)

⊙10

## 拾肆 水漂

### 14 ⊙10

古语又叫泱皮，就是在流动的清水河内找一个回水弯处，用河石砌制一个长3 m、宽2 m的流动的水池，把桑皮放进池内排好放匀，在进水口放好一个过滤网，不要让河内杂质流进，流进水池的水不要太少，这样5～7天皮质就能变白。

## 拾伍 再选

### 15

把漂好的桑皮洗好、挤干，放进挑皮篮内担回家，再用土筛挑选出桑皮在河内不小心沾染的杂质，把一些没有漂白的桑皮挑选出来放在一起，把漂好的放在一起。

## 拾陆 露漂

### 16

就是把水漂没有漂白的桑皮放在砌好的平台上，摆好散开，日晒、夜露7～9天，或根据天气、皮色来控制时间，直至漂白为止。

## 拾柒 精选

### 17 ⊙11

把原来水漂和露漂过的原料放在一起。水漂的原料，捡去在存放过程中或沾染杂质，或夏天少许变质、发霉的桑皮；露漂的原料，捡去漂不白的或在制作过程中沾染杂质的皮料。选下的桑皮就不能再用了。

## 拾捌 打皮

### 18 ⊙12

从这道工序起，以后每道都要做好准备工作，不能使桑皮沾染上杂质，如果沾染上杂质就会影响产品质量。先将皮碓擦干净，两人合作，一人踏碓，一人添皮，反复捶打。一般普通的桑皮纸要求捶打3～4遍就行了，精制桑皮纸则要捶打8～9遍。

⊙11

⊙12

⊙10 水漂
Bleaching mulberry bark

⊙11 精选
Choosing the bark materials carefully

⊙12 踏碓添皮（樊嘉禄供图）
Beating and adding in mulberry bark (photo provided by Fan Jialu)

拾玖

## 搅　　拌

19　⊙13

采用哪种搅拌方式要根据所要桑皮纸的拉力来定。一般机器搅拌出的桑皮浆拉力弱，而人工搅拌出的桑皮浆拉力强。人工搅拌，就是把打好的桑皮放在拌浆池内，加入地下水，用搅拌锤来回搅拌，一次不要拌得太多，以防不匀。而机器搅拌能多拌而匀，但它把桑皮的纤维大多切断、切散了，这样桑皮浆就失了拉力。

⊙13

贰拾

## 过　　滤

20

把拌好的桑皮浆放进滤池内，使水分滤干，在滤池口上盖好塑料布，以防杂质进入，这样存放24小时。

贰拾壹

## 袋　　料

21　⊙14⊙15

（1）在流动的清水河内，建一个长1.5 m、宽1 m、深1 m的潭子。再在潭的上游，将流水引进潭内，使潭里的水不停地流动，潭边设人站的地方和一块平台。

（2）将过滤干的桑皮浆从滤池取出，放进挑皮篮内，带上料袋、袋锤，放在平台上，然后放进料袋内，但不要太多，按每50 kg压6个饼的比例。

（3）灌气。装好料后，将袋放入潭内，打开袋口，将料锤放入料袋，将锤杆夹在两腿之间，双手使劲抖动袋口，使袋内灌满空气，快速扎死袋口。这道工序技巧较有难度，一般人不易完成。

（4）锤袋。灌好气后，料袋变成个大气球，双手捉杆一下一下地在水潭中锤袋。如果中间停下，则必须用手向袋上浇水，袋上水干了，里面的气就会跑掉，那又得再次灌气。这样一直锤到袋里皮料的滤水与河里的清水一样才行。

（5）压饼。将料袋拉起放到平台上，平扩袋口，双脚踏料，将袋内的水挤压干。这时还要观看挤压出来的水是清还是浑，如果不清还需再袋，挤干桑皮，料就成饼状了。

⊙14

⊙15

贰拾贰

## 拌　　浆

22

把饼状的桑皮浆，用手一块块分开放进拌浆池内，抽入地下水进行搅拌，直至拌匀。

⊙13 机器搅拌 Machine stirring the bark pulp

⊙14 灌气 Charge the bag with air

⊙15 锤袋 Beating the bag

## 贰拾叁
## 滤　　浆
### 23

跟前述过滤工序相似，但必须用清水冲洗滤池。

## 贰拾肆
## 取　　浆
### 24

用塑料桶从滤池里把浆取出。取浆时必须把握浆料的水分含量，一般水分多就难取，水分少又不容易散开。不同的皮浆有不同的滤水性能。

## 贰拾伍
## 下　　槽
### 25

将取出的浆料放入槽笼，这一环节很重要。按照槽笼的大小、所需纸的厚度和浆料的水分来判断怎么下槽。

⊙16

## 贰拾陆
## 搅　　拌
### 26

⊙16

将下槽后的料，先用搅拌锤从远向近、从下向上，在槽跟前搅拌匀；再由近向远，由上向下搅拌，使浆料散布在整个槽内。通常搅拌3～5分钟，槽内浆料会均匀分布。

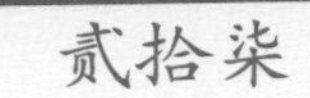

## 贰拾柒
## 划　　槽
### 27

⊙17

浆料搅拌均匀后，两人各站在槽的两头，手拿1.2 m长的竹竿，在槽内划动。两人配合要适当，一先一后，不能同时，这样一直划到用手捧起水浆，细看已经特别均匀、没有线条疙瘩为止。

⊙17

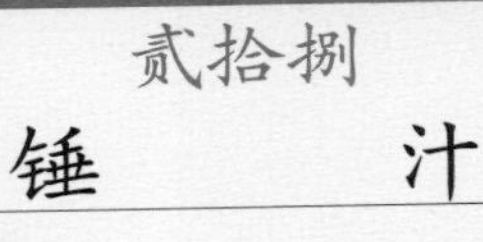

## 贰拾捌
## 锤　　汁
### 28

要看在什么季节捞纸，需要用什么植物纸药来做配料。如果是春冬，就用杨桃藤，炎热的夏天用青丹皮杆，秋天也可用少量青丹皮或桐藤花根。将杨桃藤每根从头到尾用小锤锤破，再用刀切成一尺来长，扎成一把一把的，放入水缸内浸泡。青丹皮杆也一样。而桐藤花根必须碓打，打烂后，放入滤篮中浸泡、取汁。

⊙16 搅拌
Stirring the materials

⊙17 刘同烟与侄子在划槽
Liu Tongyan and his nephew stirring the materials

## 贰拾玖 配汁

29

以上提到的锤汁一定要在捞纸前一天进行。浸泡、配汁时，双手揉杨桃藤，使杆部的汁流出来，再用手搅一搅，看汁能不能达到当时的天气热度和槽内纸料的浓度要求，配汁多少和浓度多高都要由操作人的经验来定。配制不好，纸不是薄就是厚。配制时不能有一点杂质和汁球流入槽内。

⊙18

## 叁拾 下帘

30 ⊙18

纸帘用竹篾精制而成，涂上漆，大小根据纸的大小来定制。将帘放在帘床上，压好帘尺，分开捉好帘床的上下，两人同时向下插入槽笼进行操作。

## 叁拾壹 捞纸

31 ⊙19～⊙21

（1）两人将帘子下槽后，前后摆荡几次，将帘上的水倒掉。此时两人动作必须一致，掌帘的师傅就要看准纸的厚薄、拉力、均匀度来判断纸可好干、好晒。

（2）取帘上塔。纸塔就是由纸一张张叠起的，也有称之为帖子。将帘取出，右手拧梢，左手拧额。额头有帘竹，帘竹头有纱帽，将纱帽套好帘桩。右手拧梢时必须使帘和塔板成90°角，这样下放，使整个帘着板后，双手分开，轻按帘竹，使纸额部粘上塔板和塔，右手握帘竹取帘。注意上塔时不能过快，否则，有时会起泡，也就是盖有空气，或湿纸跟帘爬起等。这都要把握好技巧。

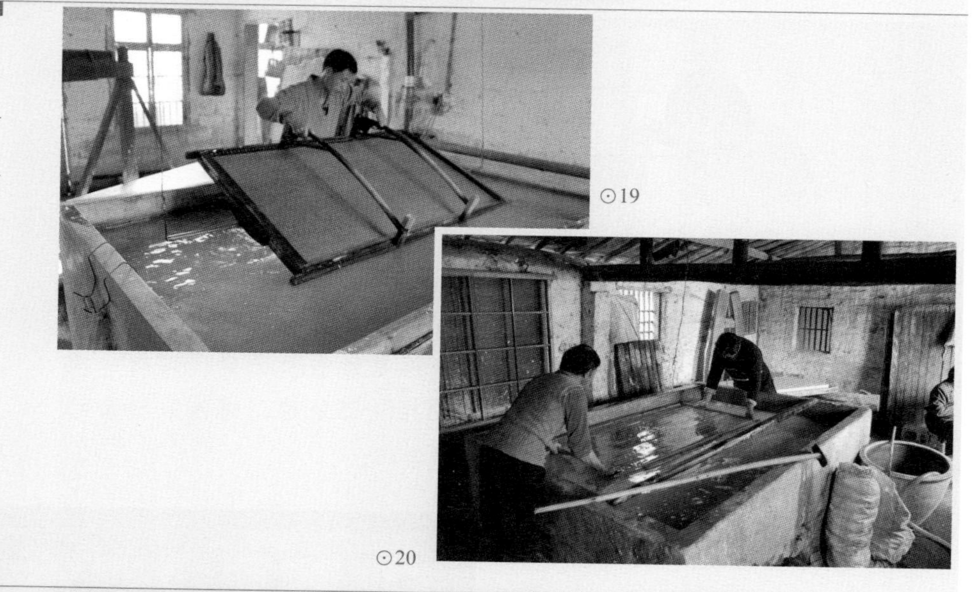

⊙19

⊙20

⊙21

⊙18 下帘 Putting the bamboo papermaking screen into water

⊙19/20 捞纸 Papermaking

⊙21 取帘上塔（樊嘉禄供图）Putting the bamboo papermaking screen on the paper pile (photo provided by Fan Jialu)

## 叁拾贰 榨帖

32 ⊙22

将捞好的帖子盖好掩塔帘，上好掩塔板，上板时必须两人抬着轻放，重放容易将塔压破（如果破了，两人一天的工夫就白费了）。这时就要看塔里面储藏的水流出的速度，速度快就说明帖子走水快，反之则慢。若快就可马上在榨驼上放上三根直驼，再在直驼上叠高横驼。再将手轻按塔的额头看是否有点变硬，如变硬，驼起榨杠，把固定的一头插进榨肩膀内。再将另一头榨绳放下，套在榨轱辘上，用榨塔棍榨。榨一会停一会，后可加快，直到塔榨干为止。

⊙22

## 叁拾叁 退塔

33 ⊙23

退塔也有技巧。在退塔之前还要将榨杠狠榨两下速退，这样退的帖子松、柔软。去掉榨杠，绕好榨绳，将驼拿下，两人从塔的额部使帖子立起抬到焙房外。

⊙23

## 叁拾肆 烧焙

34

焙是用土砖砌的。实测刘同烟家纸坊的焙长约4 m、高1.7 m，呈梯形。前面朝屋外有70 cm高的门，后面有一小门。后顶上有烟囱，中间全部是空的。每天早晨4～5点，工人将前门、后门打开，拔掉烟囱闸，把后焙前一天所烧的柴灰清理干净，再在前焙放柴点火烧焙，使土焙用手摸从前到后热度要一样匀。全部烧热后，再用4 m长的火叉将柴捆一捆送入后焙燃烧。柴烧过后，将前后焙门封死，关好烟囱闸，不要让里面的热气散出。

⊙24

## 叁拾伍 晒纸

35 ⊙24

（1）洗焙。早晨用胶桶装满清水，从前到后，从上到下，用专制的洗焙擦布擦洗干净，中午必须再用米汤擦洗一次。如果不洗，会影响纸质。

（2）整帖。抬进焙笼的帖子靠在塔板上，用刮额刀将帖子四周刮起，使周围变松，再用刮额刀从帖额下一点从右向左刮额，然后刮整个塔身。

（3）牵纸上焙。整理好塔后，将塔面上的纸一张张牵下。这时，右手拿刷夹纸一角，左手全部拿纸上焙。

（4）揭纸、刀纸。烤干的纸每张下揭时，必须先摸纸的全身是否都干了，干了才能揭下，不然容易起皱。每10张或20张折角以计算张数，到100张时将纸额两头整理齐。理好后薄纸压上刀纸棍折叠，折叠层数按纸的大小确定，厚的不用刀纸棍也可叠起。码好后用石头将额部压好。

⊙22 榨驼 Tool for squeezing the paper

⊙23 榨好的纸塔 Paper pile after squeezing

⊙24 晒纸 Drying the paper

叁拾陆

## 检　纸

36　⊙25⊙26

桑皮纸晒好后搬入检纸房。在检纸台板上打开（每次100张），每张纸的验收主要是看是否破损、匀度、杂质等，不合格的丢下台面，可重新用于制浆。

⊙25

⊙26

叁拾柒

## 切　纸

37　⊙27

（1）在切纸房内，把房间整理干净，放好切纸板，板要平整。架好磨刀石，磨刀时将刀口磨成跟剃须刀一样锋利才可。

（2）在切纸板上，将纸打开，用手把折叠的纸印擦抻平整，再按切纸的量叠放。每叠一刀必须做记号，叠好一定量后压上切纸框。压框时必须注意是额部多切还是梢部多切，还是切得一样。这时双手拿刀，脚踏纸框下切，切下的纸张要平整好看。

⊙27

切好后，跟之前一样折叠好，包上包装纸。至此，繁杂工序的纯桑皮纸制作就算完成了。

叁拾捌

## 打　包

38　⊙28

根据需要，包装箱可装5刀或10刀。放好包装箱，将桑皮纸每刀包上塑料袋再放进纸箱内，用打包机打好包装。

⊙28

⊙25/26 检纸 Checking the paper
⊙27 切纸 Cutting the paper
⊙28 包装 Packing the paper

(三)

## "紫烟"纯桑皮纸生产的工具设备

### 壹 皮甑煮锅 1

用来蒸煮桑皮料。

⊙29

### 贰 石灰浆池 2

用来沤制蒸煮好的皮料。

### 叁 洗皮篮 3

洗皮时用来盛皮的篮子。

⊙30

### 肆 皮钩 4

洗皮时用皮钩将河内的皮一把把地摆散开。

⊙31

### 伍 火叉 5

晒纸烧火时用来放入柴火。

### 陆 洗皮池 6

清洗皮料的水池，一般长3 m，宽2 m。

### 柒 土筛 7

水漂后的桑皮料用土筛挑选出不小心沾染的杂质。

⊙32

⊙29 蒸煮锅灶 Pot for steaming and boiling

⊙30 洗皮篮 Basket for cleaning the bark

⊙31 正在用皮钩摆散皮料的刘同烟 Liu Tongyan stirring the bark materials with a hook

⊙32 土筛 Sieve for filtering the bark materials

捌
## 挑皮篮
8

用来将在河水里洗好的皮料挑回纸坊。

⊙33

玖
## 皮　　碓
9

木制，用来碓打皮料。

⊙34

拾
## 拌料池
10

水泥浇筑而成，用来搅拌碓打后的皮料。

⊙35

拾壹
## 过滤池
11

水泥浇筑而成，呈长方体，用来过滤搅拌后的浆料。调查时实测刘同烟家所用过滤池尺寸为:长103 cm，宽86 cm。

⊙36

拾贰
## 料　　袋
12

清洗皮料的布袋子。

拾叁
## 袋料锤
13

捣洗皮料的木棍。

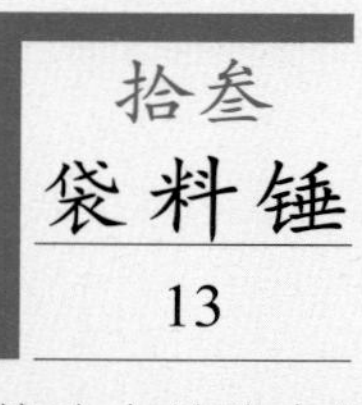

⊙37

拾肆
## 槽　　笼
14

水泥浇筑而成，捞纸用的纸槽。调查时实测刘同烟家所用槽笼尺寸为：长281 cm，宽274 cm，高85 cm。

⊙38

⊙33 挑皮篮 Baskets for carrying the bark materials

⊙34 皮碓 Pestle for beating the bark materials

⊙35 拌料池 Sink for stirring the materials

⊙36 过滤池 Sink for filtering the materials

⊙37 袋料锤 Hammer for beating and cleaning the materials

⊙38 槽笼 Papermaking trough

## 拾伍 竹帘 15

捞纸时用的纸帘。调查时实测刘同烟家所用竹帘尺寸为：长148 cm，宽77 cm。

⊙39

## 拾陆 帘床 16

捞纸时放置纸帘的帘架。调查时实测刘同烟家所用帘床尺寸为：长158 cm，宽90 cm。

⊙40

## 拾柒 植物汁缸 17

用来盛放植物纸药的缸。

## 拾捌 过滤网 18

用于过滤纸药的丝布网。

⊙41

## 拾玖 土焙笼 19

晒纸用的火墙。

⊙42

## 贰拾 晒纸刷 20

晒纸用的刷子，刷柄为木制，刷毛为松针。实测晒纸刷的尺寸为：长50 cm，宽11 cm。

⊙43

⊙39 竹帘 Bamboo papermaking screen

⊙40 帘床 Frame for supporting the papermaking screen

⊙41 植物汁缸与丝布网 Vat holding the plant papermaking mucilage and cloth net

⊙42 土焙笼 Drying wall

⊙43 晒纸刷 Brush made of pine needles

# 五 潜山县星杰桑皮纸厂的市场经营状况

## 5 Marketing Status of Xingjie Mulberry Bark Paper Factory in Qianshan County

截至调查时的2015年底，从星杰桑皮纸厂的产品售价看，四尺桑皮混料书画纸出厂价为100～120元/刀；而纯桑皮纸因品种较多，用途各异，价格差别很大，从400元/刀至5 000元/刀不等。其中，四尺拓片用纸市场价为1 800元/刀，四尺针灸艾条纸市场价为400元/刀，四尺修复用纸市场价为3 500元/刀。

⊙1

从销售渠道看，星杰桑皮纸厂进行了混合材料书画纸和纯桑皮纸的市场分割。书画纸主要在文化用品商店销售，用于古籍印刷、写字、绘画。纯桑皮纸主要是订单式生产，包括北京故宫博物院、档案馆用纸和广东针灸艾条纸的订单，较少部分用于书画家收藏。除了订单外，纸厂没有线下经销商和直营店，也没有线上网店销售，主要是因为纯桑皮纸生产是规模很小的家庭手工作坊生产方式，产量较低，难以支撑市场拓展的布局需求。

调查中了解的产销信息显示，近几年来，纸厂年产量一直在5 000刀纸左右，其中混合材料书画纸占80%左右，纯桑皮纸占20%左右，年销售额100多万元。

⊙2

⊙1 四尺针灸艾条纸 4-chi paper for acupuncture and moxibustion

⊙2 老纸坊 Old paper mill

# 六
# 潜山县星杰桑皮纸厂的品牌文化与民俗故事

# 6
# Brand Culture and Stories of Xingjie Mulberry Paper Factory in Qianshan County

## 1. 桑皮纸制作技艺的家族男性传承习俗

据刘同烟介绍，其自家纸坊生产桑皮纸到他至少已有五代的历史，至自己侄子已是第六代。按刘同烟的说法，潜山本地桑皮纸技艺传承讲究“传男不传女”，他们造纸的核心技艺六代一直传给家里的男丁，女人们只帮忙做些辅助性的工作，如今依然保持着这一传承模式。

## 2. 造纸讲究“图吉利”

据刘同烟介绍，从祖辈一直流传至今的说法是，造纸要讲究时刻，图个吉利。据说桑皮纸蒸煮时，工人需要早起，蒸煮时段内不能遇到丧事、意外伤亡等晦气事件，一旦遇到皮就会蒸不熟、蒸不好。

⊙3

⊙3 正在观察桑皮原料蒸煮火候的刘同烟
Liu Tongyan observing stove heat of the steaming and boiling step

## 3. 神丹皮的传说

坛畈村当地一直流传着这样一个传说：很早以前，当地夏季因缺乏纸药不能造纸，先辈就外出找可用的植物纸药。先辈跋山涉水接连两天都没有找到，第三天傍晚就靠在一棵大树下累得睡着了。先辈夜里做了一个梦，梦里有人告知，如果第二天早上有水滴落下，那棵树的杆子就可以做纸药。第二天早上，先辈被从树叶上滴落的水滴惊醒，于是半

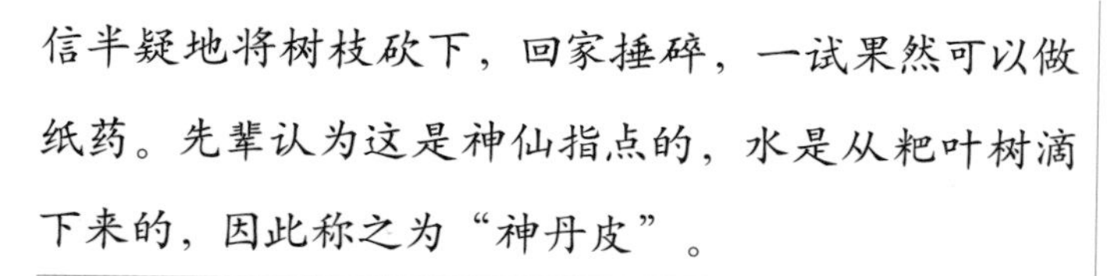

信半疑地将树枝砍下，回家捶碎，一试果然可以做纸药。先辈认为这是神仙指点的，水是从粑叶树滴下来的，因此称之为“神丹皮”。

### 4. 在继承工序的基础上改进工具

刘同烟如今拥有两项专利：一项是发明专利，以传统工艺生产桑皮纸的专用抄纸装置；二是实用新型专利，专用于桑皮纸传统生产工艺的焙笼。刘同烟在发扬传统技艺基础上注重研发和工具改进，这有利于生产效率和产品质量的提高。

⊙1



## 七 潜山县星杰桑皮纸厂的传承现状与发展思考

## 7 Current Status of Business Inheritance and Thoughts on Development of Xingjie Mulberry Bark Paper Factory in Qianshan County

⊙2

调查组通过访谈得知，星杰桑皮纸厂在发展中面临若干挑战性问题：

(1) 桑皮纸制作工序繁杂，生产周期长，耗时耗力，年轻人多不愿从事造纸行业。刘同烟已是国家级传承人，儿子宁愿做服装生意也不愿维系家族传承；侄子也是通过人到中年的老刘动员才又回到造纸作坊的，此外并无其他传人。作为国家级“非遗”保护项目的潜山桑皮纸制作技艺传承面临后继乏人的困境。

(2) 桑皮纸生产原料成本和人工成本不断上升，而产品利润空间小，造纸户造纸积极性普遍不高，像坛畈村原先有很多户造桑皮纸，调查时只剩刘同烟与侄子这一户了，经济因素对传统造纸技艺的传承和保护具有相当大的制约。



⊙1 刘同烟讲述『神丹皮』传说
Liu Tongyan telling the story of "Shendanpi"

⊙2 独自洗料的刘同烟
Liu Tongyan cleaning the materials alone

⊙3

(3) 在坛畈村造桑皮纸经常遇到因山村小河缺水而无法生产的困境，村里的石材加工产业不仅大量耗水占地，而且对山村环境破坏严重。星杰桑皮纸厂想要扩大规模和改进纸坊周边生产环境，多次寻求当地政府给予资金和政策支持都没有结果。刘同烟对此表示很无奈。

(4) 官庄镇原先有围绕桑皮纸业态的较多元的家庭手工业，如竹帘编制，2007年樊嘉禄前往官庄镇调查时专门访谈了紧邻坛畈村的纸帘制作户，当时尚处于正常生产状态，但没有几年就完全歇业了。2015年调查组访谈时，纸帘制作行当已在当地消失。

⊙4

表心感谢参加美国世界文化遗产基金会和中国北京故宫博物院合作修复紫禁城乾隆皇帝倦勤斋

In appreciation from
The World Monuments Fund and the Palace Museum
Cooperative Partnership to Restore
Emperor Qianlong's Lodge of Retirement
The Forbidden City, Beijing

⊙5

星杰桑皮纸厂生产的桑皮纸能作为故宫倦勤斋标志性修复工程用纸，说明其纸品质上佳。随着故宫修复选纸，星杰桑皮纸厂逐步形成了较高的知名度和较大的品牌影响力。面对以上问题，市场推广存在内劲严重不足的障碍，迫切需要创造以品牌影响力带动产品市场占有率的支撑环境。

目前纸坊规模小，产量有限，水源不足，加上技艺传承面临难题，潜山桑皮纸制作技艺这一国家级非物质文化遗产的传承和发展存在不小的隐患。

⊙3 污染的小溪
Polluted stream

⊙4 2007年调查时的制纸帘户（樊嘉禄供图）
Interviewing papermaking screen maker in 2007 (photo provided by Fan Jialu)

⊙5 对潜山桑皮纸参与倦勤斋修复的感谢证明
Proof of thanks to mulberry bark paper in Qianshan County participating in the restoration of Juanqinzhai

# 潜山县星杰桑皮纸厂

# 桑皮纸

Mulberry Bark Paper
of Xingjie Mulberry Bark Paper Factory
in Qianshan County

桑皮针灸艾条纸透光摄影图

A photo of mulberry bark paper for acupuncture and moxibustion seen through the light

187

# 潜山县星杰桑皮纸厂

# 纯桑皮纸 厚

Pure Mulberry Bark Paper (Thick)
of Xingjie Mulberry Bark Paper Factory
in Qianshan County

# 潜山县星杰桑皮纸厂

# 纯桑皮纸（薄）

Pure Mulberry Bark Paper (Thin) of Xingjie Mulberry Bark Paper Factory in Qianshan County

纯桑皮纸（薄）透光摄影图
A photo of pure mulberry bark paper (thin) seen through the light

# 第四节
# 岳西县金丝纸业有限公司

安徽省
Anhui Province

宣城市
Xuancheng City

岳西县
Yuexi County

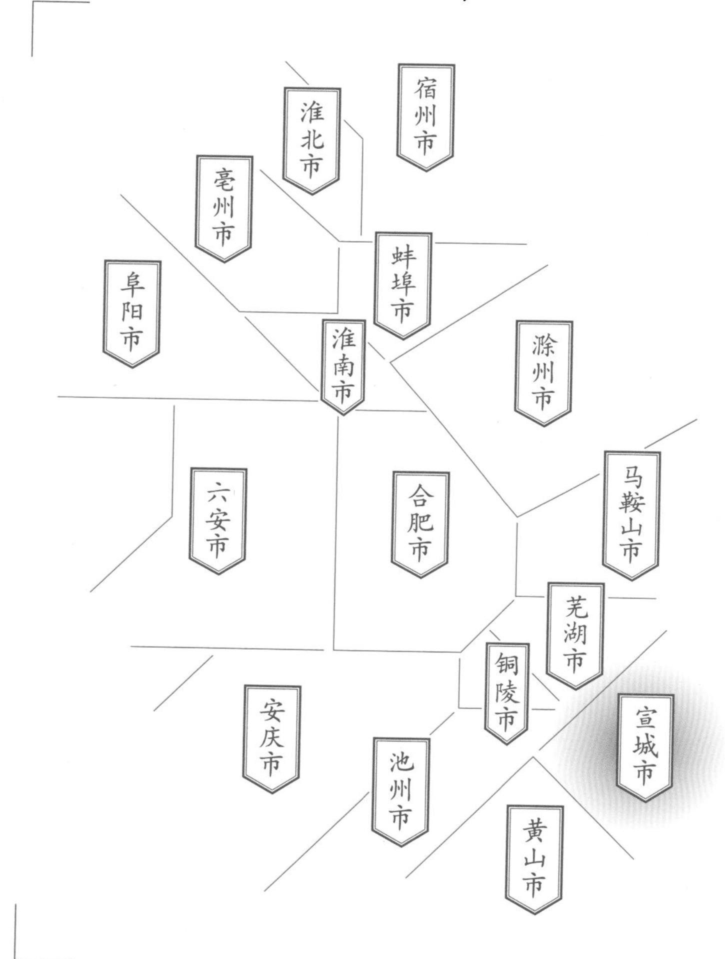


Section 4
Jinsi Paper Co., Ltd.
in Yuexi County

调查对象
毛尖山乡
岳西县金丝纸业有限公司
皮纸

Subject
Bast Paper
of Jinsi Paper Co., Ltd.
in Yuexi County
in Maojianshan Town

## 一 岳西县金丝纸业有限公司的基础信息与生产环境

## 1 Basic Information and Production Environment of Jinsi Paper Co., Ltd. in Yuexi County

岳西县金丝纸业有限公司坐落于安庆市岳西县毛尖山乡板舍行政村，地理坐标为：东经116° 26′ 43″，北纬30° 52′ 26″。桑皮纸为岳西区域内著名的传统工艺产品，据当地民间传说以及调查中的造纸人说法，它的生产始于东汉末年，因此又称为汉皮纸，并有大汉、中汉、小汉之分，其传说与相邻的潜山县桑皮纸生产区相同。如果传说有依据，那么迄今已有1 700多年的历史了，这是有标志性意义的安徽早期造纸信息，但可惜调查组目前未能发现信史及古代地方乡土文献的任何可靠记载。

金丝纸业有限公司（简称“金丝纸业”）的现任负责人和技艺传承人为王柏林。访谈中了解到的重要信息如下：王柏林祖上即以造桑皮纸谋生。2004～2005年，北京故宫博物院急需“乾隆御贡高丽纸”为倦勤斋修复通景壁画，在遍访全国各桑皮纸生产基地及朝鲜半岛的桑皮纸产区后，最终，岳西县金丝纸业和替山星杰桑皮纸厂生产的桑皮纸被指定为倦勤斋修复专用纸，一时声名大振。2008年6月，“桑皮纸制作技艺”入选第二批国家级非物质文化遗产名录。2009年6月，王柏林被评为桑皮纸制作技艺“国家级非物质文化遗产代表性项目代表性传承人”。

⊙1 坐落于山上的岳西县金丝纸业生产现场 Production scene of Jinsi Paper Co., Ltd. in Yuexi County on a mountain

2016年3月17日，调查组前往金丝纸业生产厂区进行现场考察，获知的基础生产信息如下：有一个小型厂区，厂房于2008年建立。纸厂主要生产纯桑皮纸和混合原料桑皮纸，尺寸有四尺、六尺等，根据纸品用途又可细分为大汉纸、中汉纸、小汉纸、古籍修复纸等。其中，混合原料桑皮纸是常年生产的，纯桑皮纸根据客户订单和水源状况制作。纸厂动态拥有员工8人左右。调查

路线图
岳西县城
↓
岳西县金丝纸业有限公司
Road map from Yuexi County centre to Jinsi Paper Co., Ltd. in Yuexi County

# 岳西县金丝纸业有限公司位置示意图

# Location map of Jinsi Paper Co., Ltd. in Yuexi County

考察时间
2016年3月

Investigation Date
Mar. 2016

地域名称

岳西县城
毛尖山乡
岳西县金丝纸业有限公司

A 岳西县
1 毛尖山乡
2 包家乡
3 来榜镇
4 田头乡
5 白帽镇
6 石关镇
7 头陀镇
8 店前镇

造纸点名称

岳西县金丝纸业有限公司 造纸点

位置分布

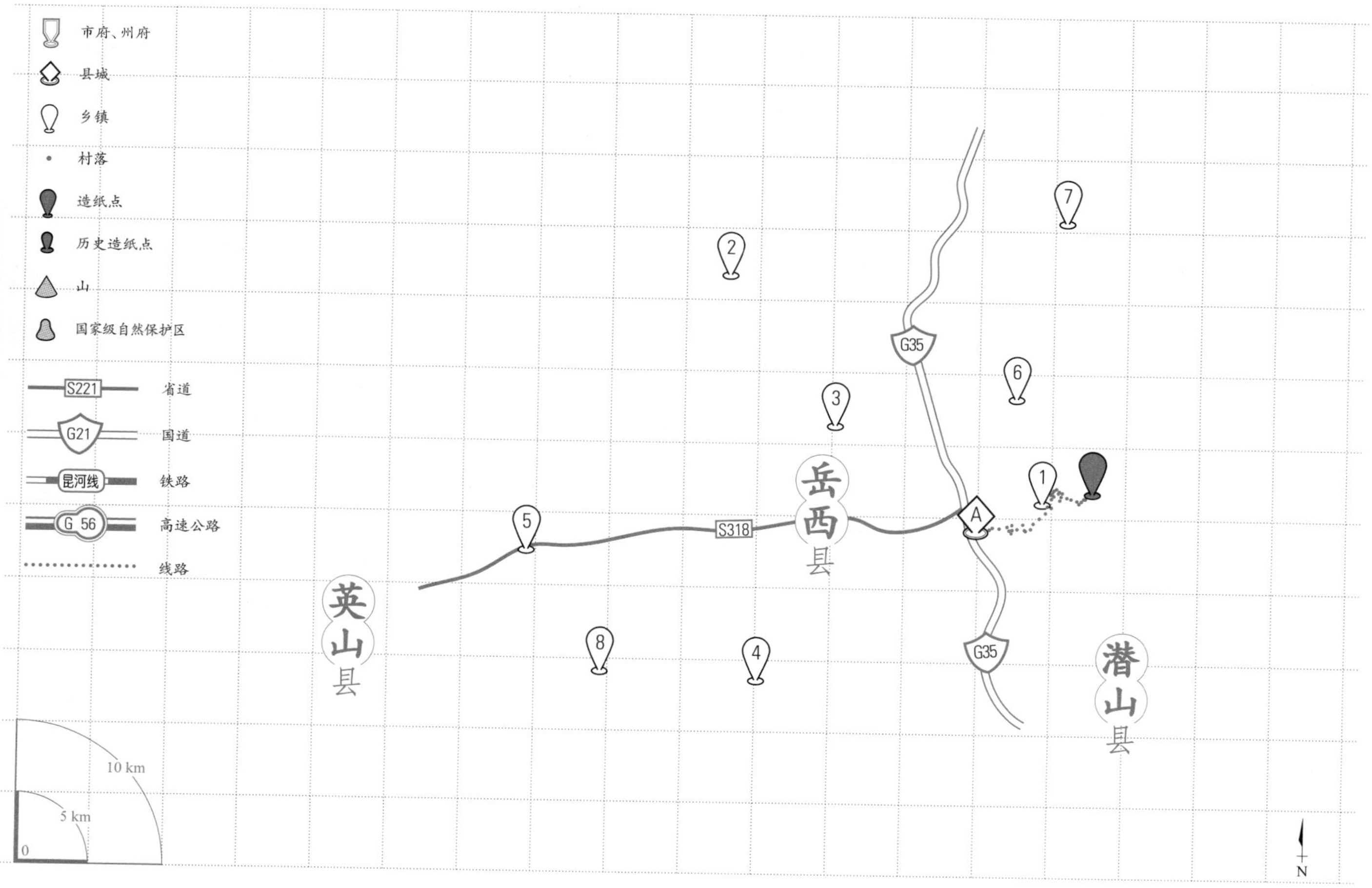

市府、州府
县城
乡镇
村落
造纸点
历史造纸点
山
国家级自然保护区
S221 省道
G21 国道
昆河线 铁路
G 56 高速公路
线路 

G35
S318
岳西县
英山县
潜山县
10 km
5 km
0
N

⊙1

时，因订单缺乏，纯桑皮纸暂时未生产，纸厂正在生产混合原料桑皮书画纸、三桠皮纸。据王柏林介绍，近五年来，金丝纸业平均年产量3 000多刀纸，年销售额约80万元，其中纯桑皮纸及桑皮混合原料纸产量占80%左右。

毛尖山乡位于岳西县东部边陲，有岳西东大门之称。内与县城、天堂镇、温泉镇、石关乡、响肠镇相衔，外接潜山县逆水乡、割肚乡。境内群山环绕，地势陡峻，山多地少，是一个集山区、库区于一体的贫困乡。2015年的检索数据显示：全乡辖12个村、133个村民组、3 390户，总面积96 $km^2$。山场总面积50 $km^2$，人均3 334 $m^2$；耕地面积3.9 $km^2$，人均280 $m^2$。





## 二
## 岳西县金丝纸业有限公司的历史与传承情况

2
History and Inheritance of Jinsi Paper Co., Ltd. in Yuexi County

古往今来，安徽地区用桑皮造纸主要集中在今潜山县、岳西县一带（今岳西县毛尖山乡及其周边区域即为核心生产区之一）。岳西地域生产桑皮纸年代悠久，史料记载统称“皮纸”，以质优量大出名。毛尖山乡区域原属潜山县后北乡，1936年1月，中华民国政府析潜山、太湖、霍山、舒城四县边陲新置岳西县，“岳西桑皮纸”名称从那时起沿用至今。

新版《安庆地区志》“造纸 印刷”条目记载：“岳西、潜山、太湖、贵池等地有生产土汉

⊙1
路口的板舍村宣传牌
Billboard of Banshe Village in an intersection

⊙1

皮纸、谱纸、白麻纸、书画纸的历史。……1949年岳西县有纸槽191座，产量为71吨。”[1]新版《岳西县志》记载：“本县造纸业历史悠久。民国二十五年（1936年）九月，全县有纸槽155座。二十八年，来榜河、和尚庄、菖蒲河等地有纸槽40座，生产汉皮、银皮等纸。……三十二年，纸槽增至166座，后受战争影响，生产萎缩。三十五年，纸槽降至150座。三十七年底，纸槽锐减至80座。建国后，造纸业手工操作与机器生产并轨发展。……1951年，有纸槽114座，从业人员6 900人，年产皮纸3万刀……农村纸槽以毛尖山乡最多，冬闲投产高峰期达300余座。”[2]

以上所引地方志史料表明：岳西县出产的桑皮纸（土汉皮纸）从民国年间就在当地具有规模化的影响力和较高的知名度。20世纪50年代，当地的皮纸生产进入繁荣态，最高峰时仅毛尖山乡一隅之地便有纸槽300余座，从业人员高达近7 000人，手工捞纸成为当地人一门重要的养家产业。但遗憾的是，到2015年前后，岳西县境内的手工造纸已萎缩到屈指可数的户数。岳西县金丝纸业有限公司正式注册成立于2001年，新厂发展至今已走过近20年的历程，而纸厂老厂房开始生产桑皮纸至今已有几代人的历史。据王柏林访谈时介绍，老厂房从祖辈流传至今，族谱有记载可查的到他这一代已有七代的传承史，从村中留存的《王氏族谱》可查证的历代传承人有：王家采（可回溯的第一代）、王知广（第二代）、王文有（第三代）、王孔怀（第四代）、王德禄（第五代）、王有贤（第六代）、王柏林（第七代），至今已历200余年。调查时，王柏林是岳西县仅有的桑皮纸制作技艺传承人，王柏林有三个女儿，都不从事造纸行业。王柏林的小舅子会造纸，时常会在厂里帮忙；另有大舅子也会造纸，调查时在北京德承贡纸坊帮工造纸。

⊙2

2009年6月，王柏林被评为“国家级非物质文化遗产代表性项目代表性传承人”，但该国家级技艺项目因客观原因，由岳西县文化馆牵头保护，因此2010年以后王柏林与保护责任单位岳西县文化馆合作较为紧密。

⊙3

⊙1
板舍村中的小河
River in Banshe Village

⊙2
王柏林
Wang Bailin

⊙3
王柏林（中）、岳西县文化馆官员（右）与调查组负责人
Wang Bailin (middle), officer from the Yuexi County Cultural Centre (right) and the head of researchers

[1] 安庆市地方志编纂委员会.安庆地区志[M].合肥：黄山书社，1995.

[2] 岳西县地方志编纂委员会.岳西县志[M].合肥：黄山书社，1996.

## 三
## 岳西县金丝纸业有限公司的代表纸品及其用途与技术分析

## 3
## Representative Paper and Its Uses and Technical Analysis of Jinsi Paper Co., Ltd. in Yuexi County

### (一)
### 代表纸品及其用途

金丝纸业生产的纸品包括纯桑皮纸、桑皮原料混合书画纸、三桠皮纸等，其中以纯桑皮纸最具代表性。纯桑皮纸根据尺寸、用途的不同可细分为好几类，如大汉纸、中汉纸、小汉纸、针灸艾条纸、文物修复纸、拓印纸。普通书画纸主要用于练习书法、绘画。

据陪同调查的岳西县文化馆汪淳介绍，民间说桑皮纸的制作源于汉末，蔡伦发明造纸术之后，人们利用野生桑皮制作出的纸叫桑皮纸，也称汉皮纸，分大汉纸、中汉纸、小汉纸，后统称桑皮纸、皮纸。大汉纸的尺寸为80 cm×60 cm，中汉纸的尺寸为70 cm×50 cm，小汉纸的尺寸为60 cm×40 cm。

⊙4

岳西纯桑皮纸品质优异，特色鲜明。纸张通常呈米白色，质地纤维细密，纹理清晰，手感棉柔，光而不滑，吸水性强，不腐不蠹，墨韵层次鲜明，纸面平整，无褶折，无洞眼和撕裂口，亦无其他附着物。纯桑皮纸主要作为书画名家用纸（因为价格很贵）、博物馆文物典籍修复纸、中国传统建筑内檐棚壁糊饰工艺内墙纸，以及匾额、对联、隔扇装裱修复及大型书画装裱修复等用纸。

### (二)
### 代表纸品的性能分析

测试小组对采样自金丝纸业的纯桑皮纸所做的性能分析，主要包括厚度、定量、紧度、抗张力、抗张强度、撕裂度、湿强度、色度、耐老化度下降、吸水性、伸缩性、纤维长度和纤维宽度

⊙4
已晒干的三桠皮原料
Dried mitsumata bark

等。按相应要求，每一指标都需重复测量若干次后求平均值，其中厚度抽取10个样本进行测试，定量抽取5个样本进行测试，抗张力（强度）抽取20个样本进行测试，撕裂度抽取10个样本进行测试，湿强度抽取20个样本进行测试，色度抽取10个样本进行测试，耐老化度下降抽取10个样本进行测试，吸水性抽取10个样本进行测试，伸缩性抽取4个样本进行测试，纤维长度测试了200根纤维，纤维宽度测试了300根纤维。对金丝纸业纯桑皮纸进行测试分析所得到的相关性能参数见表4.4。表中列出了各参数的最大值、最小值及测量若干次所得到的平均值或者计算结果。

⊙1

表4.4 金丝纸业纯桑皮纸相关性能参数
Table 4.4 Performance parameters of pure mulberry bark paper in Jinsi Paper Co., Ltd.

| 指标 | | 单位 | 最大值 | 最小值 | 平均值 | 结果 |
|---|---|---|---|---|---|---|
| 厚度 | | mm | 0.099 | 0.091 | 0.094 | 0.094 |
| 定量 | | g/m² | — | — | — | 32.9 |
| 紧度 | | g/cm³ | — | — | — | 0.350 |
| 抗张力 | 纵向 | N | 36.3 | 17.9 | 26.6 | 26.6 |
| | 横向 | N | 22.0 | 9.4 | 16.8 | 16.8 |
| 抗张强度 | | kN/m | — | — | — | 1.447 |
| 撕裂度 | 纵向 | mN | — | — | — | 32.8 |
| | 横向 | mN | — | — | — | |
| 撕裂指数 | | mN · m²/g | — | — | — | 32.8 |
| 湿强度 | 纵向 | mN | 1 100 | 800 | 950 | 950 |
| | 横向 | mN | 800 | 500 | 600 | 600 |
| 色度 | | % | 55.9 | 54.9 | 55.5 | 55.5 |
| 耐老化度下降 | | % | — | — | — | 1.3 |
| 吸水性 | | mm | — | — | — | 5 |
| 伸缩性 | 浸湿 | % | — | — | — | 0.25 |
| | 风干 | % | — | — | — | 0.73 |
| 纤维 皮 | 长度 | mm | 15.27 | 0.19 | 7.31 | 7.31 |
| | 宽度 | μm | 35.0 | 3.0 | 12.0 | 12.0 |

⊙1
纯桑皮纸（汪淳供图）
Pure mulberry bark paper (photo provided by Wang Chun)

由表4.4中的数据可知，金丝纸业纯桑皮纸最厚约是最薄的1.09倍，经计算，其相对标准偏差为0.002 4，纸张厚薄较为一致。所测金丝纸业纯桑皮纸的平均定量为32.9 g/m$^2$。通过计算可知，金丝纸业纯桑皮纸紧度为0.350 g/cm$^3$，抗张强度为1.447 kN/m，抗张强度值较大。所测金丝纸业纯桑皮纸撕裂指数为32.8 mN · m$^2$/g，撕裂度较大；湿强度纵横平均值为775 mN，湿强度较大。

所测金丝纸业纯桑皮纸平均色度为55.5%。色度最大值是最小值的1.018倍，相对标准偏差为0.305，色度差异相对较大。经过耐老化测试后，耐老化度下降1.3%。

所测金丝纸业纯桑皮纸吸水性纵横平均值为5 mm，纵横差为2.0 mm。伸缩性指标中浸湿后伸缩差为0.25%，风干后伸缩差为0.73%，说明金丝纸业纯桑皮纸伸缩性差异不大。

金丝纸业纯桑皮纸在10倍、20倍物镜下观测的纤维形态分别见图★1、图★2。所测金丝纸业纯桑皮纸皮纤维长度：最长15.27 mm，最短0.19 mm，平均长度为7.31 mm；纤维宽度：最宽35.0 μm，最窄3.0 μm，平均宽度为12.0 μm。金丝纸业纯桑皮纸润墨效果见图⊙2。

★1

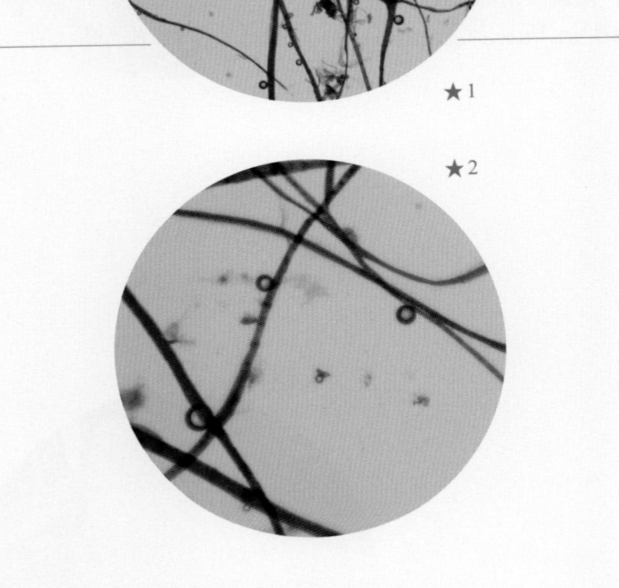

★2

⊙2

★1 金丝纸业纯桑皮纸纤维形态图（10×）
Fibers of pure mulberry bark paper in Jinsi Paper Co., Ltd. (10× objective)

★2 金丝纸业纯桑皮纸纤维形态图（20×）
Fibers of pure mulberry bark paper in Jinsi Paper Co., Ltd. (20× objective)

⊙2 金丝纸业纯桑皮纸润墨效果
Writing performance of pure mulberry bark paper in Jinsi Paper Co., Ltd.

# 四
# 岳西县金丝纸业有限公司纯桑皮纸生产的原料、工艺流程与工具设备

## 4
## Raw Materials, Papermaking Techniques and Tools of Pure Mulberry Bark Paper of Jinsi Paper Co., Ltd. in Yuexi County

据调查时王柏林介绍，金丝纸业生产的纯桑皮纸均采用手工捞制方式，原料为100%的野生桑树皮。

### (一)
### 金丝纸业纯桑皮纸生产的原料

#### 1. 主料

岳西县的野桑树与相邻的潜山县相似，通常生长在海拔500 m以上、1 000 m以下，是可再生植物，根部再生能力很强。今年将其杆枝砍去，来年又可从根部生出新苗，肥沃之地当年就可以长到2 m多高，有大拇指粗细。其叶形状与枫叶相似，野蚕食其叶（家蚕也可食用），古人称之为野桑。王柏林表示，很多人会认为造桑皮纸用的桑皮是由家桑树皮制作的，但家桑树剥下来的皮薄、纤维短、粗糙，不论是用碓打，还是机打都易成米粒状，缺乏拉力；而适龄野桑剥下的皮厚、纤维长，外表壳薄，碓打易分散，拉力很强。

#### 2. 辅料

（1）纸药。金丝纸业传统用的纸药为杨桃藤汁，但由于成本过高，后多选择化学纸药聚丙烯酰胺。化学纸药从县城当地可很方便地购买到。

（2）水源。金丝纸业生产桑皮纸使用的是板舍村外纸坊附近的山涧水。山涧水主要用来捞纸、洗料。调查组成员实测所用水pH约为6.99，非常接近中性。

⊙1 剥下来的野桑皮料 Wild mulberry bark materials

⊙2 装在袋中的纸药 Papermaking mucilage in a bag

⊙3 板舍村旁的山涧 Stream by the Banshe Village

## (二)

## 金丝纸业纯桑皮纸生产的工艺流程

王柏林介绍：制作纯手工桑皮纸工艺繁杂，工序繁多，从第一道工序到成品至少需要3个月的时间。此外，还要具备一定的有利条件，其中季节和环境是最主要的，不同的季节和当时的环境如湿度、温度的不同，都会影响桑皮纸的成品品质和出品时间。古代即流传72道工序之说。

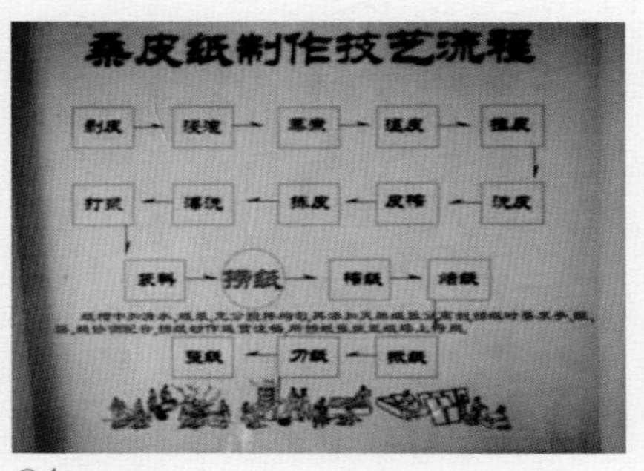


⊙4 岳西县文化馆墙上绘制的生产流程图
Production flow chart on a wall of the Yuexi County Cultural Centre

调查组于2016年3月17日对金丝纸业纯桑皮纸生产工艺进行了实地调查和访谈，综合岳西县文化馆馆长汪淳后续多次的补充介绍和提供的乡土文献参考，总结岳西金丝纸业纯桑皮纸生产的工艺流程主要包括：

| 壹 | 贰 | 叁 | 肆 | 伍 | 陆 | 柒 | 捌 | 玖 |
| --- | --- | --- | --- | --- | --- | --- | --- | --- |
| 剥皮 | 浸泡 | 蒸煮 | 沤皮 | 揉皮 | 洗皮 | 榨皮 | 拣皮 | 漂洗 |

| 拾 | 拾壹 | 拾贰 | 拾叁 | 拾肆 | 拾伍 | 拾陆 | 拾柒 |
| --- | --- | --- | --- | --- | --- | --- | --- |
| 打浆 | 袋料 | 捞纸 | 榨纸 | 揭纸 | 焙纸 | 刀纸 | 整理 |

## 壹 剥皮

### 1

⊙1

采集原材料的最佳时期为春季的晴天，手工剥取3年生桑树皮，晒干后在干燥处保存，防止霉变。农历“惊蛰”后，植物就开始行汁，这时的桑树从根部通过骨枝向上、外发汁，它的表皮和骨杆之间比较疏松，去掉细枝，修成光杆，将皮分开，脚踩杆部，手拉树皮很易剥下。要点是：小而嫩的桑树，皮质薄，没有拉力，也制不出多少纸浆；太老的桑树皮虽厚，但纤维太粗无法打碎，制出的纸质粗糙，杂质较多。杆部枝丫太多的桑树皮，每逢枝丫都有疤痕，就有一定的老茧难以除尽，后期出浆率不高。3～4龄的桑树，杆直好剥、皮质较厚，纤维适中，又无老茧，造出的纸质最好。在毛尖山乡，“清明”后桑树发出新芽，人们就停止采剥，这时的树汁散至枝叶，树皮会很难剥。

⊙1

## 贰 浸泡

### 2

⊙2

放入清洁的河水中自然漂洗数日，定时翻动桑皮，使其浸泡充分后呈现鲜皮状态为止。

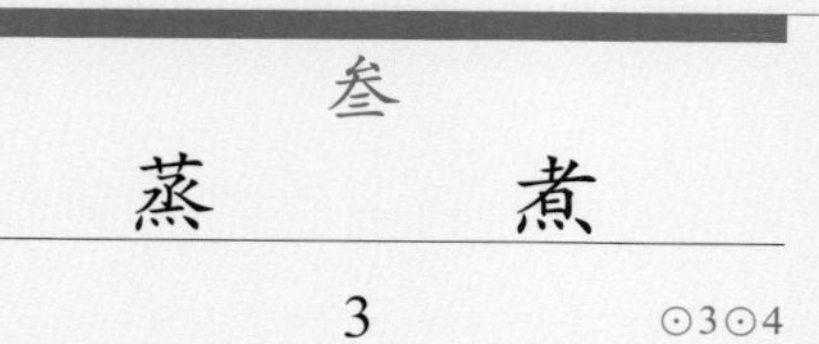

⊙2

## 叁 蒸煮

### 3

⊙3⊙4

⊙3

⊙4

桑皮1 kg左右为一扎，捆扎好后上纸甑蒸煮一日，使其变软发黑。

## 肆 沤皮

### 4

⊙5

将蒸煮好的皮料放入加石灰水的浆皮池或其他容器中浸泡沤制，使桑皮腐化，皮质纤维分离。

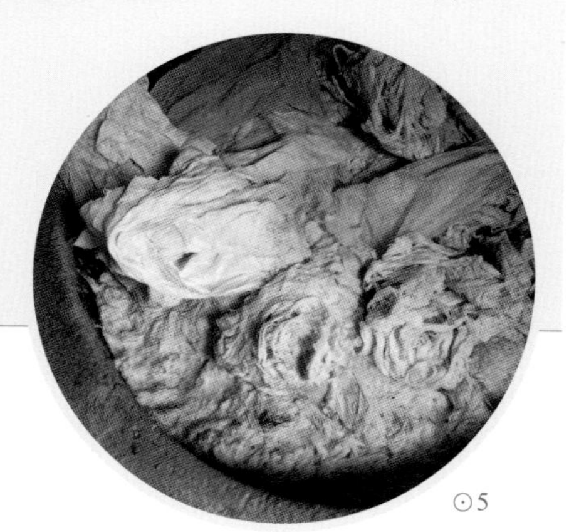

⊙5

⊙1 已剥好晒干的桑皮料 Dried mulberry bark materials

⊙2 漂洗后的皮料 Bleached mulberry bark materials

⊙3/4 蒸煮皮料的纸甑与木柴 Papermaking utensil and firewood for steaming and boiling the bark materials

⊙5 沤皮（在陶缸中沤制） Soaking the bark (in a bag pot)

## 伍

## 揉　　皮

5　⊙6

放入清洁的河水中浸泡漂洗，然后在干净的石板或木板上搓揉桑皮，使老桑皮脱落。

⊙6

⊙7

## 陆

## 洗　　皮

6　⊙7

在清洁的河水或泡料池的流水中漂洗3次，洗净石灰水等杂质。

## 柒

## 榨　　皮

7　⊙8

桑皮1kg左右为一扎，先挤去部分水分，堆放整齐后以皮榨加大挤压力度，压干水分阴干后待用。

⊙8

⊙9

## 捌

## 拣　　皮

8　⊙9

将榨干后的桑皮摊放在拣皮台上，采用人工细致地挑选，去除粗梗等杂质。

## 玖

## 漂　　洗

9　⊙10

传统方式是将挑拣好的皮料放置到清洁的河水中日夜漂洗，这期间要经过反复压榨、清洗、拣选工序，逐步遴选出洁白纯净的桑皮纤维。调查时因未赶上在山溪流水中操作的流程，王柏林在料池中做了漂洗演示。

⊙10

⊙11

## 拾

## 打　　浆

10　⊙11

分别上石碓、皮碓打浆。石碓是第一步粗加工，皮碓是第二步细加工。充分锤击桑皮，将桑皮捣成茸絮状，成为捞纸用的纸浆。

⊙6 揉皮 Rubbing the bark

⊙7 洗皮 Cleaning the bark

⊙8 榨皮 Pressing the bark

⊙9 拣皮 Picking and choosing the bark

⊙10 漂洗 Bleaching the bark

⊙11 皮碓 Pestle for beating the materials

拾壹

## 袋　　料

**11** ⊙12

将纸浆放入纯棉布做的料袋中，在干净的水中反复捣，清洗纸浆杂质，直到水清为止。

⊙12

拾贰

## 捞　　纸

**12** ⊙13⊙14

向纸槽中添加清水，放入纸浆后充分搅拌均匀，还需添加植物纸药杨桃藤汁或化学纸药聚丙烯酰胺。捞纸时要求纸帘挂浆厚薄均匀，捞好后对准纸塔起手部位，缓慢放下纸帘，轻按起手部位，让上下纸层相连后再轻掀纸帘，使纸浆脱离纸帘一次成纸。工作完毕后必须清洗纸架上残留的纸浆，纸槽的水要一周更换一次。

⊙13

⊙14

拾叁

## 榨　　纸

**13** ⊙15

在纸塔上以皮榨逐渐加压，在纸堆最上面覆盖护纸纱布，每隔几分钟逐步施压，沥干纸堆水分。

⊙15

拾肆

## 揭　　纸

**14** ⊙16

纸堆干后靠放在焙塔上，以水润湿，揉捻纸边使其松散易揭。

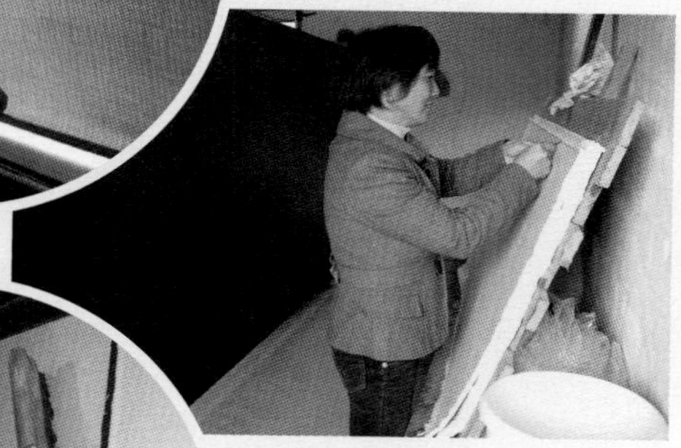

⊙16

拾伍

## 焙　　纸

**15** ⊙17

以明火加温焙墙至60 ℃左右，焙墙受热要均匀。纸堆干后靠放在焙塔上，以水润湿，揉捻纸边使其松散易揭，将揭下的纸贴上纸焙，用松刷将纸平展紧贴在焙墙上，纸面不得刷破。

⊙17

⊙18

拾陆

## 刀　　纸

**16** ⊙18

100张纸为1刀成品纸。

拾柒

## 整　　理

**17** ⊙19

将纸按三等份折叠，按客户市场要求包装待售。

⊙19

⊙12
袋料
Putting paper pulp in a cloth bag and beating it in water

⊙13/14
王柏林在捞纸（汪淳供图）
Wang Bailin making the paper (photo provided by Wang Chun)

⊙15
王柏林在榨纸（汪淳供图）
Wang Bailin pressing the paper (photo provided by Wang Chun)

⊙16
揭纸（汪淳供图）
Peeling the paper down (photo provided by Wang Chun)

⊙17
焙纸
Drying the paper

⊙18
桑皮纸成品
Mulberry bark paper products

⊙19
桑皮纸外包装样式
Package of mulberry bark paper Jinsi Paper Co., Ltd. in Yuexi County

(三)

## 金丝纸业桑皮纸生产的工具设备

### 壹 竹帘 1

用于捞纸，竹丝编织而成，表面很光滑平整，帘纹细而密集。金丝纸业所用的竹帘分大、中、小三种。实测大号尺寸为：长139 cm，宽70 cm；中号尺寸为：长117 cm，宽63.5 cm；小号尺寸为：长57 cm，宽47.5 cm。

⊙20

### 贰 帘床 2

捞纸时放置纸帘的架子，多用竹子或木头制成。

⊙21

### 叁 纸刷 3

用于晒纸时将纸刷上晒纸墙，刷柄为木制，刷毛为松针。实测金丝纸业所用纸刷的尺寸为：长48.5 cm，宽12 cm。

⊙22

### 肆 料筛 4

用于手工拣选皮料后将皮料中的杂物以及不合格的皮料筛选下去，留下能够达到要求的皮料。实测金丝纸业所用料筛的尺寸为：小号直径66 cm，大号直径104 cm。

⊙23

### 伍 拌料池 5

将制好的原料与水混合，形成合格纸浆的设施。实测金丝纸业所用拌料池的尺寸为：长264 cm，宽160 cm，深90 cm。

⊙24

⊙20 竹帘 Bamboo papermaking screen
⊙21 帘床 Frame for supporting the papermaking screen
⊙22 纸刷 Brush
⊙23 料筛 Sieve
⊙24 拌料池 Sink for stirring the materials

陆

捞纸池

6

捞纸时用来盛浆，传统为石板制成的石槽。调查时金丝纸业所用的捞纸池系水泥浇筑而成。

柒

过滤池

7

用于净化制造桑皮纸过程中产生的污水。金丝纸业使用两个过滤池，用来进行分级净化。

⊙25

⊙26

## 五 岳西县金丝纸业有限公司的市场经营状况

## 5 Market Status of Jinsi Paper Co., Ltd. in Yuexi County

金丝纸业年产量5 000多刀纸，其中桑皮纸占80%左右，桑皮纸主要根据市场需求和客户订单来生产，年销售额80多万元。

截至调查时，从售价看，皮料混合书画纸出厂价为100～120元/刀，而纯桑皮纸根据原料等级和加工方式，价格从500至5 000元/刀不等，其中按照故宫倦勤斋修复标准的四尺纯桑皮纸通常售价为4 500元/刀左右。纯桑皮纸主要是订单式生产，包括北京故宫博物院、中国国家图书馆用纸，少部分用于书画家收藏。除了订单外，纸厂有线下经销商和直营店（北京），也有线上网店销售，但因为是以家庭式很小规模的手工作坊进行生产的，产能与产量都无法使公司在渠道上铺货销售。

王柏林制造的纯桑皮纸，因为故宫博物院倦

⊙25 捞纸池 Papermaking trough

⊙26 过滤池 Filtering trough

⊙27

勤斋通景画修复的典型事件而一举成名，岳西桑皮纸已经成为当代中国桑皮纸最著名的品牌。因此，金丝纸业除了开拓出中国国家图书馆等新的标志性客户外，已陆续有韩国、日本客户前来考察，国内主要市场是深圳、北京、上海、天津等。

⊙28

⊙29

## 六 岳西县金丝纸业有限公司的品牌文化与民俗故事

## 6 Brand Culture and Stories of Jinsi Xuan Paper Co., Ltd. in Yuexi County

### (一) 被遗忘的“汉皮纸之乡”

1996年版《岳西县志》记载：“……农村纸槽以毛尖山乡最多，冬闲投产高峰期达300余座。”[1]《岳西县乡镇简志》记载：“尤以毛尖山境内板舍地区为甚，时称‘皮纸之乡’，当地民间流传方冲的柳条，板舍的纸槽……”[2]

由乡邦文献记载可见，从20世纪30年代到中华人民共和国成立初期，毛尖山乡特别是板舍村是“汉皮纸”生产高度聚集之地，从业人员和纸槽数量之多都是相当壮观的。调查中，王柏林及

⊙27 调查组成员与王柏林交流市场情况 Researchers exchanging market information with Wang Bailin

⊙28 《岳西县志》（汪淳供图） *The Annals of Yuexi County* (photo provided by Wang Chun)

⊙29 调查组成员前往捞纸槽的途中 Researchers on the way to papermaking trough

[1] 岳西县地方志编纂委员会.岳西县志[M].合肥：黄山书社，1996.

[2] 岳西县乡镇简志编纂领导小组.岳西县乡镇简志[M].合肥：黄山书社，2001.

汪淳介绍，实际上从20世纪70年代至80年代中后期，毛尖山乡及板舍村的家庭式皮纸作坊还是遍布乡间的日常业态，20世纪90年代后才开始式微的。

岳西的"皮纸之乡"虽然县志有记，但在安徽省内少有传播及口碑，他地提及更为鲜见，人们谈到安徽的手工纸自然想到泾县；谈到中国的桑皮纸，首论河北迁安桑皮"高丽纸"，其次会想到新疆维吾尔族桑皮纸，再次会想到山东、山西的桑皮纸，很少有人会提到安徽的桑皮纸。

如果没有故宫博物院倦勤斋大修在国内外海选桑皮纸及后续主流媒体的密集报道，就没有随后的快速入选第二批国家级非物质文化遗产代表性项目名录，"汉皮纸之乡"实际上已经在21世纪初只剩星星之火了，毛尖山乡及板舍村只有零零星星的造纸户在半信半疑地边造边看祖业能否维持得下去。

倦勤斋的流行故事很有探秘性，引发了官方和民间对"汉皮纸之乡"的持续关注和研究保护，岳西桑皮纸（当然也包括原为一体的潜山桑皮纸）一跃成为中国桑皮纸的掌上明珠。

## （二）倦勤斋修复专用纸的当代演绎

北京故宫博物院的倦勤斋是乾隆皇帝晚年退隐当太上皇时的专用憩息之地，殿内的大面积通景画是借鉴著名传教士画家郎世宁绘制的故宫敬胜斋内饰画稿，由郎世宁的徒弟、宫廷画家王幼学绘制完成的。乾隆皇帝去世后，倦勤斋基本处于关闭状态。

1994年，故宫专家聂崇正在宫廷部工作人员王宝光的指引下，发现了已荒废多年的倦勤斋内的巨幅通景画，通过研究，又发现，该画不仅是故宫内唯一存世的通景画，而且是全国现存最大的内饰通景画，并且全部为名家手绘，十分珍贵。2002年，故宫与世界建筑文物保护基金会（World Monuments Fund）合作对倦勤斋大修，2004年正式动工。修复组发现，殿内通景画的褙纸全部为乾隆当年御贡的高丽纸，由100%桑皮手工精制。抢修工程急需用与当年进贡桑皮褙纸品质相同或相近的纸。

从2004年开始，由原故宫科技部负责人曹静楼牵头的寻访小组跑遍了全国的桑皮纸历史产区与现存产地，还专门去了朝鲜半岛原御贡的高丽纸产地。经过国内外的大面积寻访和比较，2005年，岳西县的王柏林纸坊与相邻潜山县的刘同烟纸坊的100%纯桑皮纸脱颖而出，成为闻名全国的"倦勤斋修复专用纸"。一时新闻报道连篇累牍，从而强力带动了岳西桑皮纸业态的绝境复苏，品牌远播。

原故宫科技部主任曹静楼、世界建筑文物保护基金会叶美宁女士、颐和园纸绢类文物保护专家王敏英等都对岳西桑皮纸给予了较高评价。据王柏林介绍，2004年9月，曹静楼在岳西县毛尖山乡金丝纸坊找到手工桑皮纸制作六代单传艺人王柏林时，曾表示：王柏林生产的纯手工桑皮纸，完全可与通景画最初使用的乾隆年间的高丽纸相媲美。

调查组通过文献研究也采集到了另一种说法：据说在ZAKER新闻记载的采访中，他们先是接触到潜山县专门从事汉皮纸销售的叫余一富的农民，在余一富的帮助下，最终生产出了耐折度达到5 000次以上的纯桑皮纸，与倦勤斋通景画的桑皮裱褙纸相近。

# 七
# 岳西县金丝纸业有限公司的传承现状与发展思考

# 7
# Current Status of Business Inheritance and Thoughts on Development of Jinsi Paper Co., Ltd. in Yuexi County

依托金丝纸业有限公司活态传承生产技艺，岳西县文化馆作为责任主体负责国家级非物质文化遗产的保护，至调查时的2016年春天，岳西桑皮纸传承方面虽然保持着活态的生产，可是只有王柏林一户在造纸，调查现场的"非遗孤岛"效应非常明显。调查时比较强烈的感受是，岳西桑皮纸虽是国家级非物质文化遗产保护项目，又有故宫"倦勤斋修复专用纸"的品牌美誉度，但传承现状确实存在隐忧。

调查组通过对王柏林和汪淳等人的访谈，发现金丝纸业传承发展上面临以下突出问题：

(1) 桑皮纸制作工序繁杂、生产时间长，耗时耗力，本地乡村里的年轻人多不愿从事手工造纸行业，造纸技艺传承面临较大的困难。以王柏林纸坊为例，头顶国家级"非遗"项目和代表性传承人的荣誉，而且确实每年都会有从国家到地方一定数额的津贴，尚且没有找到年轻的技艺传承人，只有自己和小舅子两人在传承祖业，而两人的年龄都已很大；三个女儿按习俗也不会继承造纸，而且确实也没学过这门手艺，现状已是后继无人。

(2) 纯桑皮纸生产所需的野生桑树皮的原料成本和采集加工中的人工成本不断上升。虽然金丝纸业的桑皮纸名声很大，但由于用户面很窄，目前创新开拓不足，用户群规模很小，渠道也少，没有办法摊薄成本，导致桑皮纸的保本型售价相当高。产品利润空间并不大，因此造纸户造纯桑皮纸的积极性不高，反倒是造中低端的混合原料纸成为无奈的选择，这对纯桑皮纸正宗技艺的传承和保护具有一定挑战。

(3) 品牌与纸的名声很大，但不容易挣到钱，说明市场推广存在一定问题，需要拓宽桑皮纸的市场销售渠道，以优质的品牌影响力带动产品的市场占有率。目前纸坊规模很小，从业者只有一个半人（王柏林算一个，小舅子算半个），产能很有限，加上技艺传承面临难题，纸厂想要扩大规模和改进纸坊周边生产环境（主要是高品质桑皮纸漂洗袋料时需用到充足的流水）都力不从心。对此，仅靠当地政府予以一定的政策和资金扶持还不能解决这一问题。

⊙1

⊙1 文化馆『非遗』牌匾下留影的汪淳
Wang Chun by the Intangible Heritage Certificate plaque in the Cultural Centre

# 岳西县金丝纸业有限公司桑皮纸

Mulberry Bark Paper of Jinsi Paper Co., Ltd. in Yuexi County

纯桑皮纸透光摄影图
A photo of pure mulberry bark paper seen through the light

# 岳西县
# 金丝纸业有限公司
# 三桠皮纸

*Edgeworthia chrysantha* Lindl. Paper
of Jinsi Paper Co., Ltd. in Yuexi County

三桠皮纸透光摄影图
A photo of *Edgeworthia chrysantha* Lindl. pape seen through the light

# 第五节

# 歙县深渡镇棉溪村

安徽省
Anhui Province

黄山市
Huangshan City

歙县
Shexian County

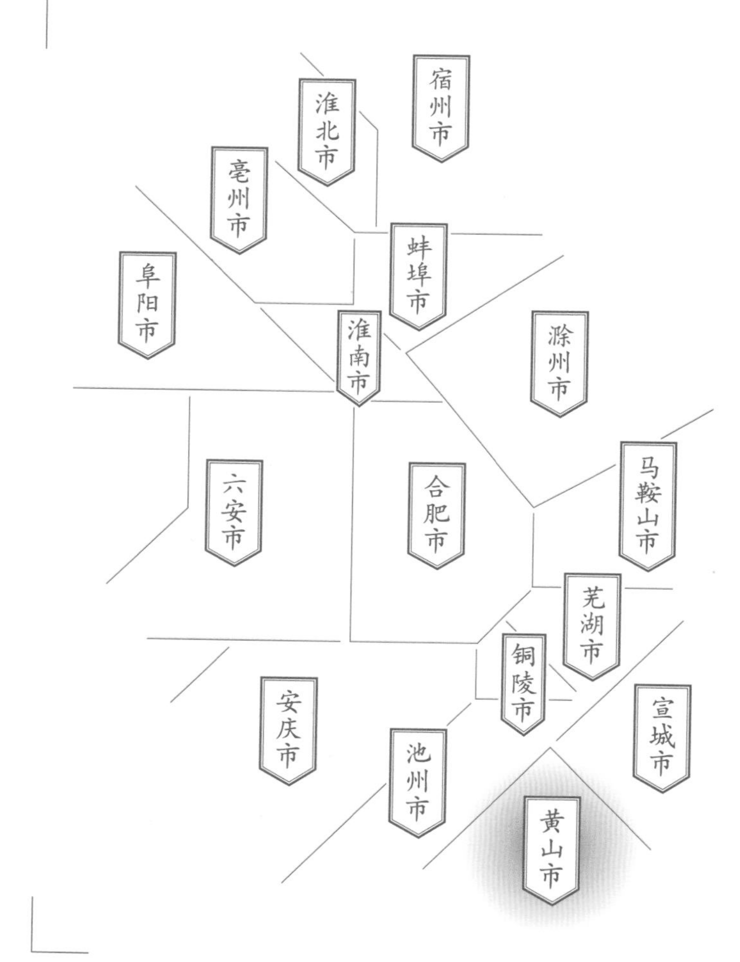




调查对象

深渡镇
棉溪村
构皮纸

## Section 5
## Mianxi Village in Shendu Town of Shexian County

Subject

Paper Mulberry Bark Paper in Mianxi Village of Shendu Town

## 一
## 歙县深渡镇棉溪村构皮纸的基础信息与生产环境

1
Basic Information and Production Environment of Mulberry Paper in Mianxi Village of Shendu Town in Shexian County

歙县深渡镇棉溪行政村坐落于安徽省黄山市歙县东部，地理坐标为：东经118° 35′ 58″，北纬29° 50′ 41″。濒临新安江支流淮源河畔，处在南深公路（歙县南源口镇—深渡镇）与棉五（棉溪村—五渡村）公路交汇点上。

2016年3月23日，调查组对歙县深渡镇棉溪村的手工造纸现存业态进行了调查，所获得的基础信息如下：棉溪村造纸原料为当地生长的构树皮，历史上的主要用途包括糊茶叶箱、包雪梨（地方名特产）、包糕点、包蚊香、做纸伞，也可在老人去世后用来垫棺材、包石灰等。调查时，村里曾经较普遍分布的造纸作坊基本停产，只有江祖术一户处于断续维持造纸状态。

⊙1

⊙1 棉溪村边的新安江 Xin'an river by the Mianxi Village

路线图
歙县县城
↓
棉溪村
Road map from Shexian County centre to Mianxi Village

# 歙县深渡镇棉溪村位置示意图

# Location map of Mianxi Village in Shendu Town of Shexian County

考察时间
2016年3月

Investigation Date
Mar. 2016

地域名称

歙县县城
深渡镇
棉溪村

A 歙县
1 三阳乡
2 金川乡
3 杞梓里镇
4 霞坑镇
5 溪头镇
6 上丰乡
7 深渡镇
8 石门乡

造纸点名称

棉溪村 造纸点

位置分布

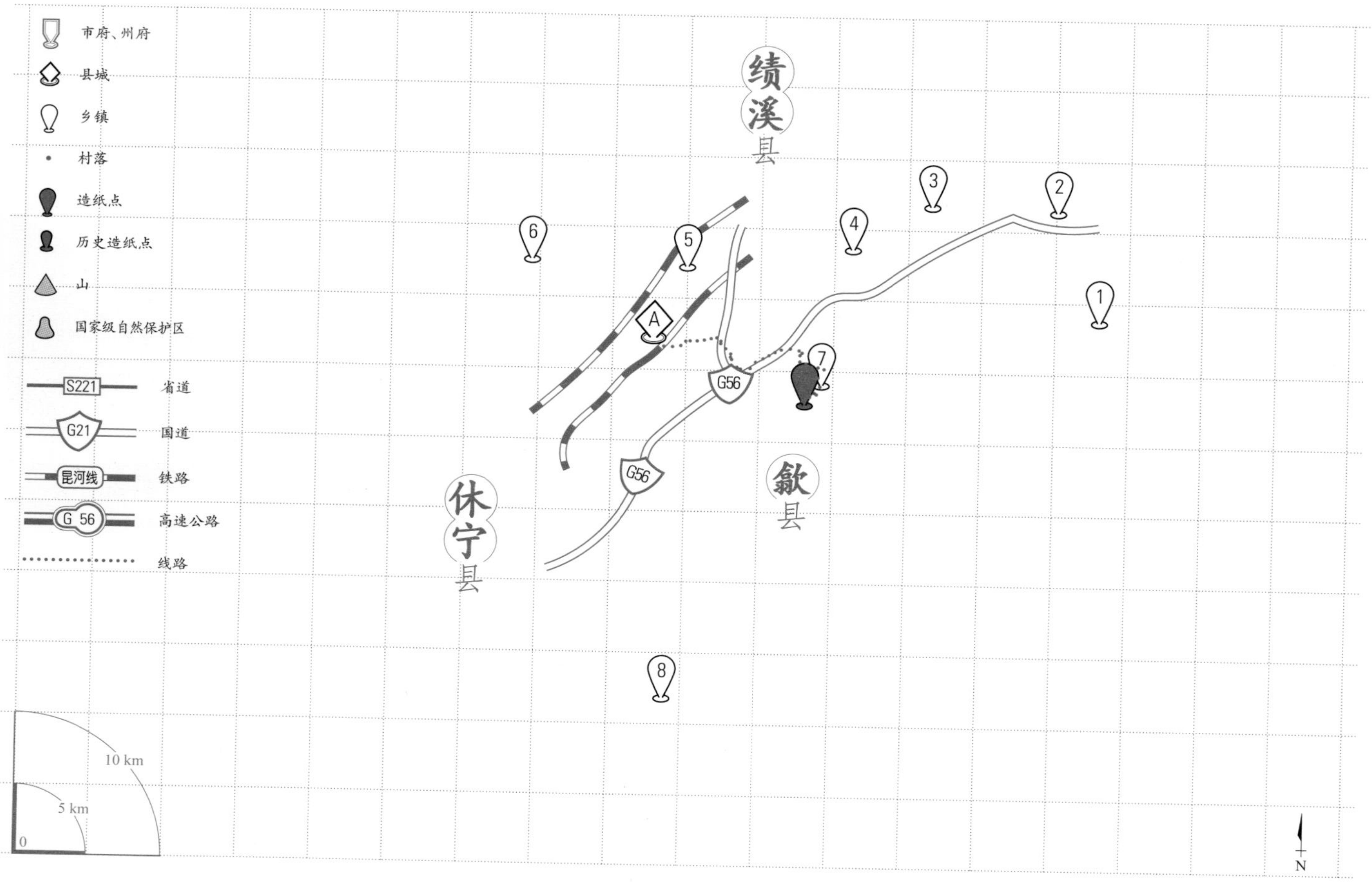

市府、州府
县城
乡镇
村落
造纸点
历史造纸点
山
国家级自然保护区
S221 省道
G21 国道
昆河线 铁路
G56 高速公路
线路

绩溪县
休宁县
歙县
G56
10 km
5 km
0
N

⊙1

⊙2

据棉溪村村主任汪成祺介绍，2015年底棉溪村共有137户、443人，以汪姓和江姓为主，两姓人数占比大致相同。

棉溪村造纸的历史到底有多长，还没有发现文献的记载，而造纸老人江祖术的说法是：听过去村里老人说江氏在始迁篁墩（现黄山市徽州区的一个镇）时就开始造纸了，那时相当于五代到北宋初期。当然，歙县的江氏从篁墩携造纸技艺迁往棉溪村是在第21代，时间大约在明清之交，算起来，大约有近400年的历史。

⊙3

## 二 歙县深渡镇棉溪村构皮纸生产的人文地理环境

## 2 Cultural and Geographical Environment of Mulberry Bark Paper in Mianxi Village in Shendu Town of Shexian County

歙县位于安徽省最南部，为古代著名的徽州府驻地，徽州一府六县的中心区域，现隶属于黄山市，下辖28个乡镇，面积2 122 km$^2$。东北与宣城市绩溪县和浙江省临安市毗邻，东南与浙江省淳安县、开化县交界，西南和西北分别与黄山市休宁县、黄山区等相连。歙县为历史文化底蕴深厚之地，1986年，被授予“国家历史文化名城”称号。在民间说法中，歙县与云南丽江、山西平遥、四川阆中古城并称为保存完好的中国四大古城。2014年，歙县古徽州文化景区入选国家5A级旅游景区。

深渡镇是著名古镇，为古代新安江由徽州前往杭州水路的重要商业中转枢纽，位于今黄山市歙县与杭州市淳安县的交界处。新安江水库和千岛湖旅游景区是两县的分界水域，也是深渡镇重

⊙1/2 冬天『猫冬』休闲的棉溪村民 Relaxing Mianxi Villagers in winter

⊙3 深渡镇区景观 View of Shendu Town

要的渔业与旅游资源。

深渡镇域的人文聚集形成于唐，聚落兴建于宋，唐至清属孝女乡，民国二十一年（1932年），首次设深渡镇建制。1952年8月，设深渡区与薛阳区，下辖深渡、大茂村、定潭、阳产、外河坑、中坑源、漳潭、绵潭、洪村、杨三、九砂、漳坑、漳岭山、漳村湾、绵潭坑等。1955年12月合并为深渡、绵潭二乡。1958年10月属北岸公社，1961年成立定潭、棉溪、漳潭三公社，隶属于深渡区。1972年3月，从定潭公社析出佛坑、深渡、凤池、大茂大队成立深渡公社。1984年7月深渡乡转为镇，1992年2月定潭乡、棉溪乡并入，2004年12月漳潭乡并入。

据2005年的数据，深渡镇辖2个居委会、29个行政村、145个自然村、187个村民小组，共9 814户25 999人，其中农业7 847户22 645人。居民绝大多数为汉族，另有苗族、布依族、傈僳族、土家族、黎族、白族共40人。镇域总面积99 km$^2$，镇区面积4.5 km$^2$，人口2.7万，辖14个村（居）委员会。

棉溪村隶属于深渡镇，占地面积2.25 km$^2$，特色果园农业是棉溪村的重要资源，村中有枇杷园和柑橘园538 693 m$^2$，林地面积786 537.3 m$^2$，耕地面积28 001 m$^2$，水浸地面积120 006 m$^2$，水域面积66 670 m$^2$。人均耕地极少，传统上一直以枇杷、茶叶和油料作物种植与销售为主要生活来源。

⊙2

⊙1

⊙3

⊙1
深渡镇政府
Shendu Town Government

⊙2/3
棉溪村外盛开的油菜花田
Rape flowers outside Mianxi Village

三

## 歙县深渡镇棉溪村构皮纸的历史与传承情况

3

## History and Inheritance of Mulberry Bark Paper in Mianxi Village of Shendu Town in Shexian County

据2016年前后的调查信息，当代歙县造纸点主要在深渡镇棉溪村。虽然该村手工造纸在20世纪80年代以前一直非常红火，80年代时棉溪村还家家户户"打皮纸"，但奇怪的是，在本地地方志和民间文献当中几乎找不到相关资料，民间流传的信息也很少寻访到。

调查中棉溪村村主任汪成棋介绍：自己家原先也从事造纸，2016年他59岁，父亲叫汪聚茂，已经整整100岁了。记忆中的老父亲从10岁开始做纸，一直做到70岁，做了60年，主要从事原料粗加工；自己学校毕业后开始造纸，一共做了19年，由于市场疲软难挣到钱而放弃了造纸。调查时，棉溪村尚在造纸的是江祖术家，原来住在村子里面，现在建了新屋，住在河对面公路旁的新居里。

通过汪成棋的热心介绍和领路，调查组成员在一幢典型的徽派民居中见到了江祖术，老人当时79岁。

江祖术介绍了家族沿袭和造纸传承：今年79岁，是江氏第33世裔。歙县江姓这一支系从歙县篁墩先迁到歙县的北岸村，次迁到本县的五渡村，再从五渡村迁到棉溪村。族谱上记载江氏21世迁到棉溪村时是兄弟三人。当地人说得很形象："三个和尚到棉溪。"

江祖术回忆道：听过去村里老人说江氏在篁墩就开始造纸了，年轻时不懂事，把自家保存的一套祖宗谱打纸浆做了手工皮纸。后来，江祖术又用自己做的手工皮纸抄了部分祖谱，但是抄的内容不全，现在很后悔。

江祖术介绍：以前棉溪村是个造纸村，整个村80%以上的农户造纸，现在只有一两户人在造。自己在1955～2008年从事造纸工作，1955年跟父亲江承恩（江承恩当年五十几岁，67岁去世，从事造纸40多年）学习造纸技术，父亲是跟爷爷江荣林学的造纸技术。算起来，江祖

⊙4

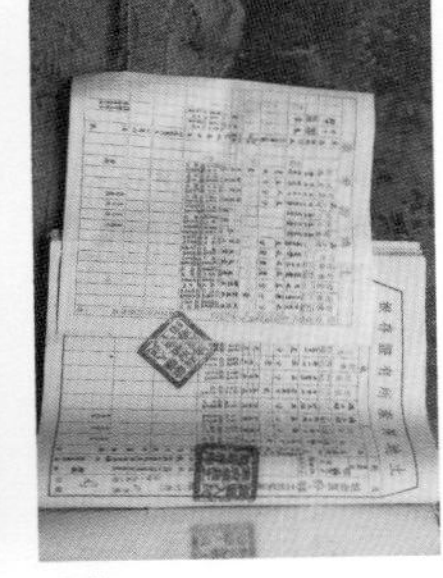
⊙5

⊙6

⊙ 4 / 5 调查组成员在棉溪村村委会查找资料
Researchers searching for information in the Mianxi Village Committee Office

⊙ 6 在江祖术新居院内访谈
Interviewing at the new house of Jiang Zushu

⊙1

⊙2

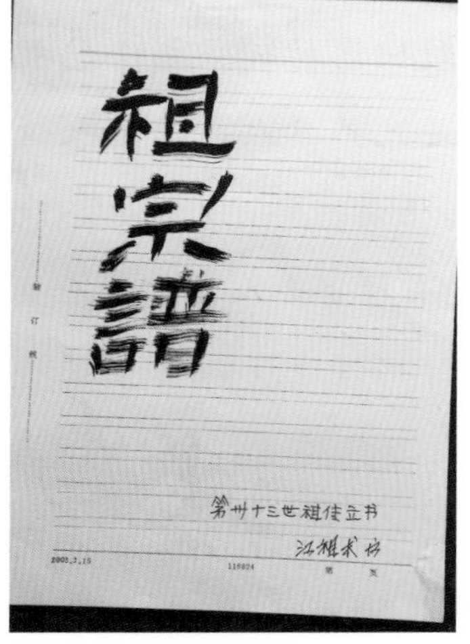


⊙3

⊙ 1 / 2
江祖术家旧宅
Old house of Jiang Zushu

⊙ 3
江祖术在机制信纸上复抄的祖谱
Jiang Zushu's family genealogy transcribed on a machine-made paper

⊙4

⊙5

⊙6

⊙7

术有明确记忆的家庭造纸历史也有百年以上了。

江祖术有兄弟三人，哥哥江祖名在浙江省金华市的百货公司做生意，弟弟江祖胜在父亲60多岁时跟父亲学习过造纸技术。江祖术的妻子汪寿花，前几年去世了，同县洪济村人，19岁嫁来从事刷纸工作。汪寿花与江祖术一共有五个孩子，其中三个儿子和两个女儿：大儿子江德辉（1958年出生），大媳妇汪月好；二儿子江德煌（1965年出生），二媳妇汪亚妹；三儿子江德成（1970年出生），三儿媳妇汪立菊；大女儿江美英（1959年出生），大女婿胡来寿；二女儿江美琴（1967年出生），二女婿汪利正。五个孩子都是十六七岁时跟江祖术学造纸的，其中汪利正娶了江祖术二女儿后也从事造纸工作。全家人人会造纸，堪称造纸之家。

据江祖术的三儿媳妇汪立菊介绍，近两年棉溪村造纸的只有他们一家，自己20岁嫁过来就做手工皮纸，一直到现在。通常是每年做茶叶季节结束，就开始打皮纸了，特别是8～10月更忙。女人的工作就是把抄好的纸往墙壁上刷，等干了再撕下来，50张1刀，一天要刷5 000～6 000张。

⊙4
江祖术与汪寿花旧日合影
An old photo of Jiang Zushu and Wang Shouhua

⊙5/6
调查组成员访谈江祖术
Researchers interviewing Jiang Zushu

⊙7
江祖术、三儿子江德成和三儿媳妇汪立菊
Jiang Zushu, his third son Jiang Decheng and daughter-in-law Wang Liju

# 四

# 歙县深渡镇棉溪村构皮纸生产的原料、工艺流程与技术分析

# 4

# Raw Materials, Papermaking Techniques and Technical Analysis of Mulberry Bark Paper in Mianxi Village of Shendu Town in Shexian County

## (一)

## 棉溪村构皮纸生产的原料

### 1. 主料

调查时汪成棋介绍，当地制作手工纸的原料为山上的“栗（歙县方言发音“li”）树”，因为不懂方言，所以调查组成员让他带着上山寻找这种“栗树”。汪成棋介绍以前村子附近的山坡上到处都有，到现场却一棵也没有发现，一直爬了一个多小时的山，跟着爬山的调查组成员浑身都是汗时，才在一处山坡上找到了这

⊙1

⊙2

⊙3

⊙4

⊙5

⊙1/5
汪成棋带队上山寻找『栗树』
Wang Chengqi leading research team to "Li" tree

种"栗树"。

棉溪村造纸户所称的"栗树"，其实是一种近地面生长的细细的藤构。与乔木型构树不同，这种藤构树皮韧性很大，纤维很长，很适合做手工皮纸。据江祖术介绍，"栗树"皮在当地又称乌鸦皮，以前50 kg皮收购价是20元，现在已经涨到100元了，现在没有人专门去做皮了，只能自己砍树自己剥树皮。

2. 辅料

（1）纸药。据江祖术介绍，当地捞纸使用的纸药为香叶，学名"山苍子"，一般采摘后浸泡2个月即可使用，目的是增加绵度，控制厚薄。20世纪80年代江祖术家曾经在歙县小洲村（音）购买，价格为50～60元/kg，村外山上也有生长的，现在自己在附近山里采摘。

（2）水源。据江祖术介绍，当地生产手工纸的水为河水。调查时，实测河水pH约为6.82，呈偏微酸性。

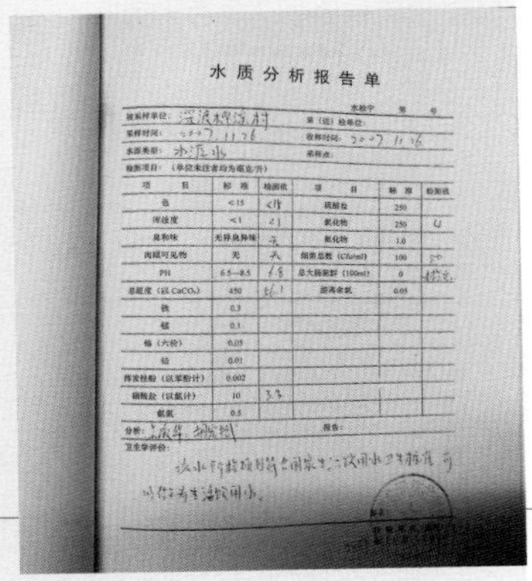

⊙6

## （二）
## 棉溪村构皮纸生产的工艺流程

由于调查时棉溪村构皮纸处于停产歇业状态，调查组成员未能完整看到工艺流程，因此据部分现场和部分工艺演示，结合江祖术、汪成棋等人的介绍，总结棉溪村构皮纸生产的工艺流程为：

| 壹 | 贰 | 叁 | 肆 | 伍 | 陆 | 柒 | 捌 |
|---|---|---|---|---|---|---|---|
| 剥皮 | 打皮 | 浸泡、烧皮 | 去壳、浸泡 | 清洗、晒干 | 踩皮 | 拣选 | 浸泡、晒干 |

| 拾伍 | 拾肆 | 拾叁 | 拾贰 | 拾壹 | 拾 | 玖 |
|---|---|---|---|---|---|---|
| 收纸、整理 | 刷纸 | 压榨 | 捞纸 | 踩皮 | 清洗 | 粉碎 |

⊙6 棉溪村2007年送检的水质报告
The water quality report of Mianxi Village in 2007

## 壹 剥皮

1 ⊙1~⊙3

⊙1

一般在农历三四月采茶季节时采剥当年的构树皮，剥好后捆成把。

## 贰 打皮

2

将采回来剥好的树皮用锤子敲打，目的是将纤维打散。

## 叁 浸泡、烧皮

3 ⊙4⊙5

将打好的树皮用清水在锅里烧煮。

⊙2 ⊙3 ⊙4 ⊙5

⊙1
构树皮
Mulberry bark

⊙2
汪成棋示范剥皮流程
Wang Chengqi showing how to strip the bark

⊙3
汪立菊示范捆皮
Wang Liju showing how to bind the bark

⊙4
煮皮的铁锅
Iron pot for boiling the bark

⊙5
蒸煮皮料的灶膛
Stove for steaming and boiling the bark materials

肆
## 去壳、浸泡
4

第二天，将烧好的皮用人工赤脚踩，将皮壳去掉；再用石灰水浸泡1天，然后放置一边堆放2天。堆放好的皮加清水烧煮1天。

伍
## 清洗、晒干
5

将烧煮好的皮用脚踩洗干净，放在河边堆放均匀、晒干。

陆
## 踩皮
6

晒干的皮进一步踩踏，将皮料中残余的皮壳脱去。

柒
## 拣选
7

通过人工拣选皮料方式，去除杂质和不合格的皮料。

捌
## 浸泡、晒干
8

将拣选好的皮料用清水浸泡2天，然后在河边晒干。

玖
## 粉碎
9 ⊙6

用粉碎机将晒好的皮粉碎（江祖术介绍：1984年后用粉碎机，1984年前用脚踏脚碓粉碎皮料，一般10 kg皮需要碓半小时）。粉碎好的浆料一般会加一些废纸边进去，也常常添加草纸、竹纸、水泥袋、毛竹浆等原料（毛竹浆是用嫩竹制作的，在嫩竹发枝时，将其砍断，按照一池15 kg左右石灰与其混合沤泡2个月后，粉碎成浆。江祖术表示，做好纸时加15 kg石灰，差纸则要加25 kg）。普通纸的成分是50%皮浆+50%废纸边，但是做纯皮纸时不加这些。如果做厚纸则要加桑皮，但是桑皮不能与新闻纸放在一起，否则捞不起来。如果做红色纸则需要在浆里加入品红（从浙江金华购买的化学染料）；如果做白色纸则需要用漂白粉进行漂白（以前用烧碱）。

⊙6

拾
## 清洗
10

将粉碎后的皮料装到皮桶里面搅拌，然后装到麻袋（后改用尼龙袋）里，在水里边挤边洗，将皮料清洗干净。

拾壹
## 踩皮
11

将清洗好的皮料从袋子里拿出，放入木桶里加入清水用脚踩踏，又称"踩纸巾"。

⊙6 石臼 Stone mortar

## 拾贰 捞纸

12 ⊙7～⊙12

将踩好的皮料放入纸槽捞纸，一般是一个人捞，一个捞纸槽一次放料一般捞200～300张浦市纸（一种棉溪纸名）。据江祖术介绍，1983年开始采用机器打浆工艺，以前用竹棍搅，然后将水放掉，再送入纸槽。一般50 kg皮可以做13 kg质量好的纸，或者15 kg普通纸。如果浆料中皮料多，只需要捞一遍水；如果皮料不太多，需要多捞几次，具体捞几遍水需要根据皮料多少决定。

江祖术表示，捞纸是个体力活，没有劲推不动，捞的纸也会厚薄不一。大集体时村里有个叫汪金付的，他年纪轻、身体好、有劲头，推过去荡过来有响声，浪法好，荡帘技术好，做的手工皮纸特别好，纸抄得均匀，那时候他做的手工皮纸是大家学习的榜样。

⊙7

⊙8

⊙9

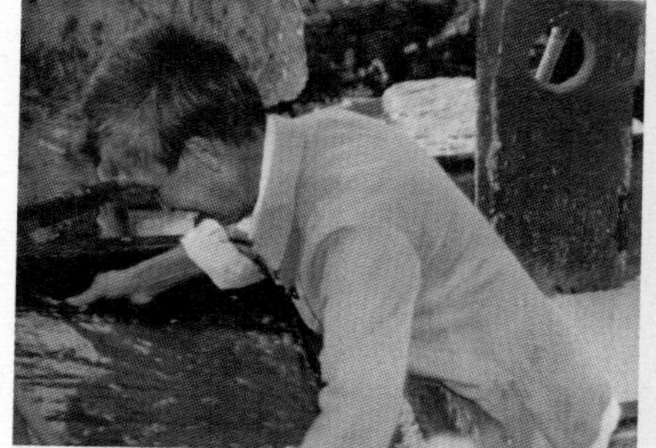

⊙10

⊙11

⊙12

⊙7 江祖术示范捞纸动作 Jiang Zushu showing the papermaking procedures

⊙8 调查组成员与汪成棋交流捞纸动作 Researchers communicating with Wang Chengqi about papermaking procedures

⊙9 汪成棋示范捞纸动作 Wang Chengqi showing the papermaking procedure

⊙10/11 江祖术捞纸（历史照片翻拍） Jiang Zushu making the paper (retook from an old photo)

⊙12 江祖术放纸（历史照片翻拍） Jiang Zushu piling up the paper (retook from an old photo)

## 拾叁 压榨

13 ⊙13⊙14

通常捞好的纸平摊在平整的石头上，用一块木板压着，一头不停地增加相当重量的石头。半小时加一块石头，开始水多时要慢一点，快干了就可以重一些。压快了纸张会变形或有裂痕，不好用，只能循序渐进。一般压榨1.5～2小时，一次压榨约1 000张纸。

⊙13

⊙14

## 拾肆 刷纸

14 ⊙15～⊙17

将压榨好的纸用松毛刷平铺地向两侧刷在墙上，一般一面墙可刷10张左右，0.5～1小时后可揭下。

⊙15

⊙16

⊙17

## 拾伍 收纸、整理

15 ⊙18

将晒好的纸揭下整理在一起，一般按50张/刀收好。

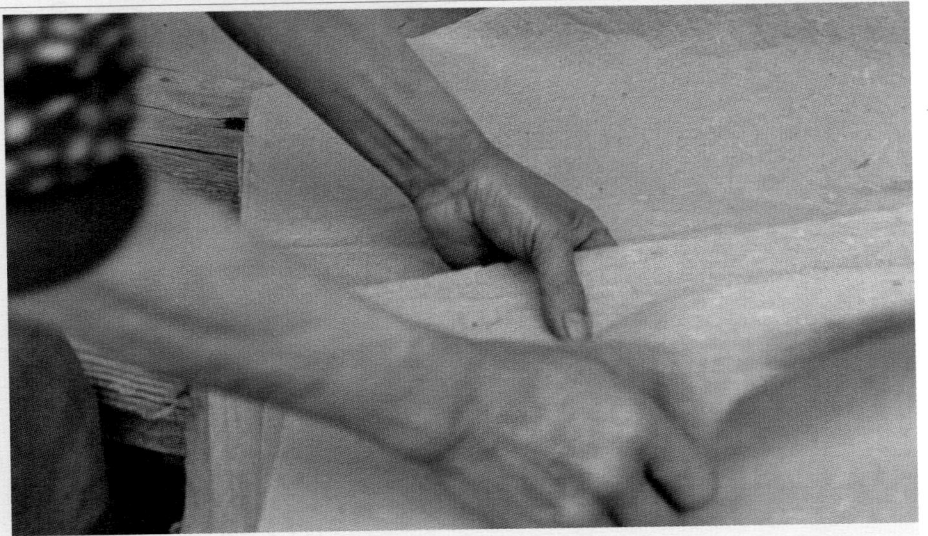

⊙18

⊙13 压榨用的旧石板 An old slate for pressing the paper

⊙14 榨床 A stone board for pressing the paper

⊙15/16 江祖术示范刷纸动作 Jiang Zushu showing how to brush the paper

⊙17 江祖术揭纸（历史照片翻拍） Jiang Zushu peeling the paper down (retook from an old photo)

⊙18 理纸示范 Showing how to sort the paper

## (三)

## 棉溪村构皮纸制作的工具设备

### 壹 纸槽 1

传统为四面石板砌成的长方体捞纸设施。

⊙1

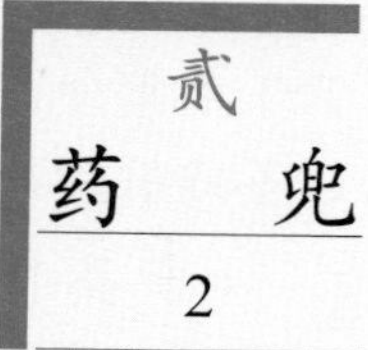

### 贰 药兜 2

洗香叶的筐子。实测江祖术家纸坊所用药兜的直径为45 cm。

⊙2

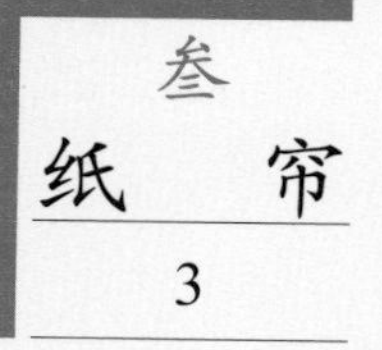

### 叁 纸帘 3

据江祖术介绍，他家的纸帘是30多年前从浙江余杭购买的，当年购买价为30元/张。实测江祖术家纸坊所用的纸帘尺寸为：长58 cm，宽50 cm，高1 cm。

⊙3

### 肆 刷子 4

晒纸的时候所用的松毛刷。据江祖术介绍，刷子是自己家做的。实测江祖术家纸坊所用的刷子尺寸为：长40 cm，宽10 cm。其中裸露在外的松毛长3 cm.

⊙4

⊙5

### 伍 压榨筐 5

压榨的时候用来放石头的筐子。实测江祖术家纸坊所用的压榨筐尺寸为：长60 cm，宽30 cm，高85 cm。

⊙6

⊙1 江祖术家的石板纸槽 Jiang Zushu's stone for papermaking trough

⊙2 江祖术家的药兜 Jiang Zushu's basket for cleaning the leaves

⊙3 江祖术家的纸帘 Jiang Zushu's papermaking screen

⊙4 江祖术家的刷子 Jiang Zushu's brush made of pine needles

⊙5 江祖术家的压榨筐 Jiang Zushu's crate for holding the stones

⊙6 压榨用的石头 Stone for pressing the paper

## (四)

## 棉溪村江祖术家构皮纸的性能分析

测试小组对棉溪村江祖术家构皮纸所做的性能分析，主要包括厚度、定量、紧度、抗张力、抗张强度、撕裂度、湿强度、色度、耐老化度下降、吸水性、伸缩性、纤维长度和纤维宽度等。按相应要求，每一指标都需重复测量若干次后求平均值，其中厚度抽取10个样本进行测试，定量抽取5个样本进行测试，抗张力（强度）抽取20个样本进行测试，撕裂度抽取10个样本进行测试，湿强度抽取10个样本进行测试，色度抽取10个样本进行测试，耐老化度下降抽取10个样本进行测试，吸水性抽取10个样本进行测试，伸缩性抽取4个样本进行测试，纤维长度测试了200根纤维，纤维宽度测试了300根纤维。对棉溪村江祖术家构皮纸进行测试分析所得到的相关性能参数见表4.5。表中列出了各参数的最大值、最小值及测量若干次所得到的平均值或者计算结果。

表4.5　棉溪村江祖术家构皮纸相关性能参数

Table 4.5　Performance parameters of mulberry bark paper in Jiang Zushu's house of Mianxi Village

| 指标 | | 单位 | 最大值 | 最小值 | 平均值 | 结果 |
|---|---|---|---|---|---|---|
| 厚度 | | mm | 0.184 | 0.092 | 0.125 | 0.125 |
| 定量 | | g/m² | — | — | — | 31.6 |
| 紧度 | | g/cm³ | — | — | — | 0.253 |
| 抗张力 | 纵向 | N | 14.4 | 10.0 | 11.2 | 11.2 |
| | 横向 | N | 7.8 | 4.4 | 5.7 | 5.7 |
| 抗张强度 | | kN/m | — | — | — | 0.564 |
| 撕裂度 | 纵向 | mN | 280 | 180 | 238 | 238 |
| | 横向 | mN | 300 | 230 | 276 | 276 |
| 撕裂指数 | | mN·m²/g | — | — | — | 7.2 |
| 湿强度 | 纵向 | mN | 200 | 180 | 190 | 190 |
| | 横向 | mN | 140 | 120 | 132 | 132 |
| 色度 | | % | 23.39 | 21.00 | 21.98 | 22.0 |
| 耐老化度下降 | | % | — | — | — | 1.2 |
| 吸水性 | | mm | — | — | — | 9 |
| 伸缩性 | 浸湿 | % | — | — | — | 0.63 |
| | 风干 | % | — | — | — | 0.68 |
| 纤维 | 长度 | mm | 8.22 | 1.07 | 3.09 | 3.09 |
| | 宽度 | μm | 35.0 | 2.0 | 13.0 | 13.0 |

由表4.5中的数据可知，棉溪村江祖术家构皮纸最厚约是最薄的2倍，经计算，其相对标准偏差为0.002 4，纸张厚薄不太一致。所测棉溪村江祖术家构皮纸的平均定量为31.6 g/m$^2$。通过计算可知，棉溪村江祖术家构皮纸紧度为0.253 g/cm$^3$，抗张强度为0.564 kN/m，抗张强度一般。所测棉溪村江祖术家构皮纸撕裂指数为7.2 mN·m$^2$/g，撕裂度较小；湿强度纵横平均值为161 mN，湿强度较小。

所测棉溪村江祖术家构皮纸的平均色度为22.0%。色度最大值是最小值的1.114倍，相对标准偏差为0.758 6，色度差异相对较大。经过耐老化测试后，耐老化度下降1.2%。

所测棉溪村江祖术家构皮纸的吸水性纵横平均值为9 mm，纵横差为8.4 mm。伸缩性指标中浸湿后伸缩差为0.63%，风干后伸缩差为0.68%，说明棉溪村江祖术家构皮纸的伸缩性差异不大。

棉溪村江祖术家构皮纸在10倍、20倍物镜下观测的纤维形态分别见图★1、图★2。所测棉溪村江祖术家生产的构皮纸的皮纤维长度：最长8.22 mm，最短1.07 mm，平均长度为3.09 mm；纤维宽度：最宽35.0 μm，最窄2.0 μm，平均宽度为13.0 μm。棉溪村江祖术家生产的构皮纸润墨效果见图⊙1。

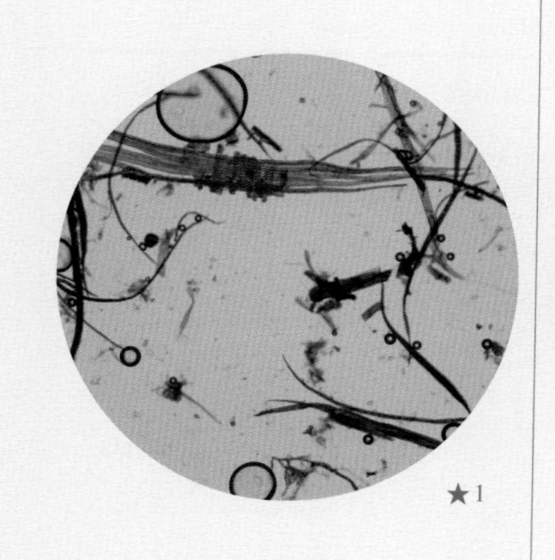

★1

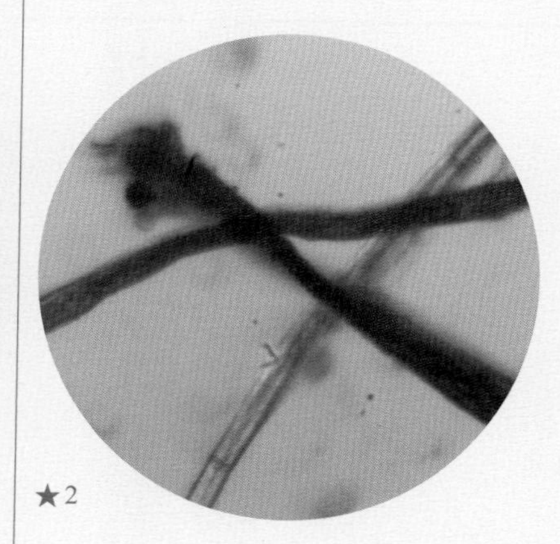

★2

⊙1

★1
江祖术家构皮纸纤维形态图（10×）
Fibers of mulberry bark paper in Jiang Zushu's house (10× objective)

★2
江祖术家构皮纸纤维形态图（20×）
Fibers of mulberry bark paper in Jiang Zushu's house (20× objective)

⊙1
江祖术家构皮纸润墨效果
Writing performance of mulberry bark paper in Jiang Zushu's house

## 五 歙县深渡镇棉溪村构皮纸的用途与市场经营状况

## 5 Marketing Status of Mulberry Bark Paper and Its Uses in Mianxi Village in Shendu Town of Shexian County

棉溪村构皮纸是用藤构制作而成的，在交流中，根据江祖术与汪成棋的介绍，调查组了解到其历史上主要用作地方的日常生产性用纸，包括糊茶叶箱、包雪梨（歙县产梨，生长期用纸严密包裹使其不见光，成梨皮色呈“雪白”颜色）、包糕点、包蚊香、做纸伞、做鞭炮引火线、包中药、包酱油缸盖、养蚕，也有用于老人去世时垫棺材、包石灰。江祖术回忆，过去还用来写阄书、契约。

棉溪村构皮纸主要有大方（规格为47 cm × 100 cm，糊茶叶箱用纸）、参皮（规格为47 cm × 50 cm）、大浦市纸（规格为60 cm × 58 cm）、浦市纸（规格为70 cm × 50 cm）、重参（厚纸，意思为好包的纸）、二伞纸（包酱油缸盖用纸）、虫桑纸（养蚕下仔用纸，较薄，规格为53 cm × 53 cm）这7种类型的纸。

⊙2

⊙3

⊙4

⊙2／4 江祖术制作的各色构皮纸 Various colors of mulberry bark paper made by Jiang Zushu

棉溪村手工皮纸制作是从什么时候开始的已经无法详细考证，汪、江两大姓氏，谁是祖传，谁是再传也难以分清，调查中看到的两姓村民也并不争论谁先造纸。

江祖术回忆：20世纪80年代棉溪村还家家户户“打皮纸”。集体造纸时（20世纪60～70年代），纸厂各种工人有120多人，其中抄纸40多人，刷纸40多人，仓库管理员2人，厂长、副厂长2人，会计2人，采购随时指派。年产量大约为20万张纸，采用发信订货形式，一天发20封信，以销定产。据江祖术介绍，20世纪60年代，集体纸厂一天造1万张纸，卖19元，当时卖1元记10个工分，造的纸和卖的钱归生产队，年底按10个工分发0.3元现金给造纸人。江祖术回忆当时小米的价格为0.16元／kg，大米为0.18元／kg，1979年后才开始不计工分。

江祖术老人至今还保管着一套有关手工皮纸的记录，有构皮纸生产销售成本、买纸客人的通信地址，以及外地来棉溪村买纸的订单，特别是订货单记录了构皮纸的去向和用途。从记录中看到棉溪村

⊙1

构皮纸销售的地区有：浙江省海宁、金华、兰溪、平湖、绍兴地区，江苏省常州地区，上海市，江西省以及安徽省的祁门、黟县、歙县本地等。

棉溪村构皮纸艺术性的书写绘画功能虽说不大，仅仅在乡村用来写契约、阄书，可在生产生活中用途十分广泛。徽州地方主要用来做茶叶出口包装箱贴纸，特别是20世纪60～80年代祁门红茶出口用量较大，调查组在江祖术家藏的账本中就发现有这样一些记录：

（1）1987年1月5日祁门县茶厂，三尺纸20 000张，660元，上交大队管理费33元，税金33元，汪德仁经手。

（2）1963年2月28日棉溪大队，我社生产蚊香需要用浦市棉纸，按月做出要货计划，4月份

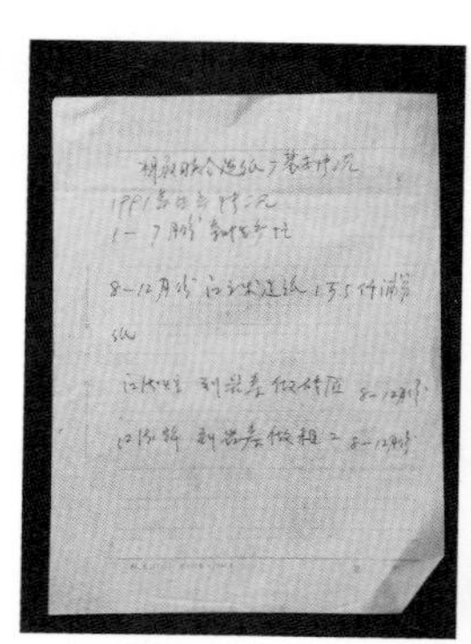

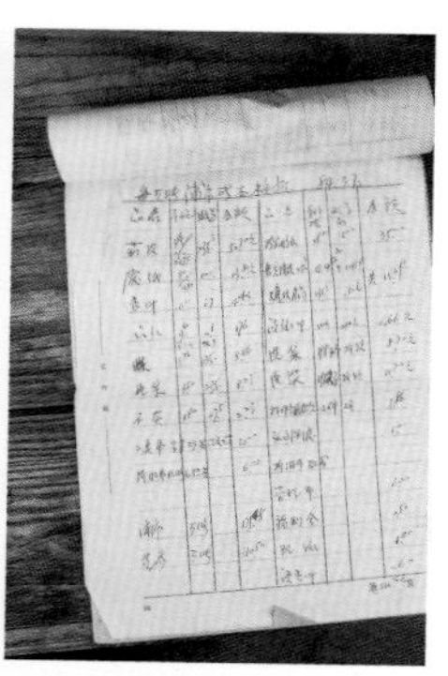

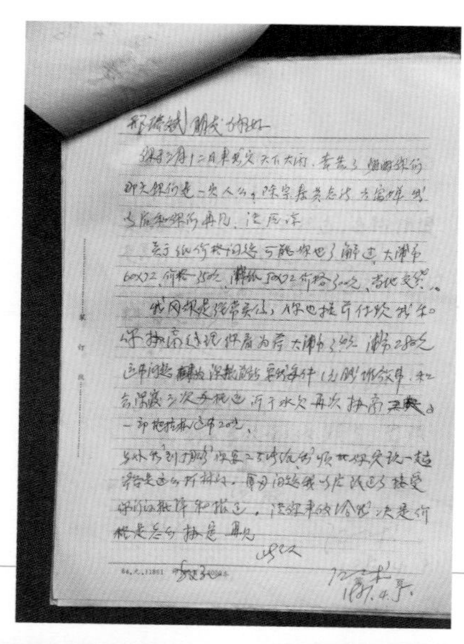

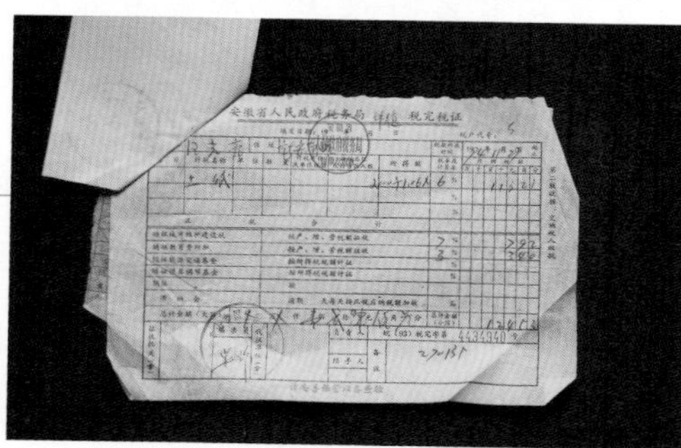

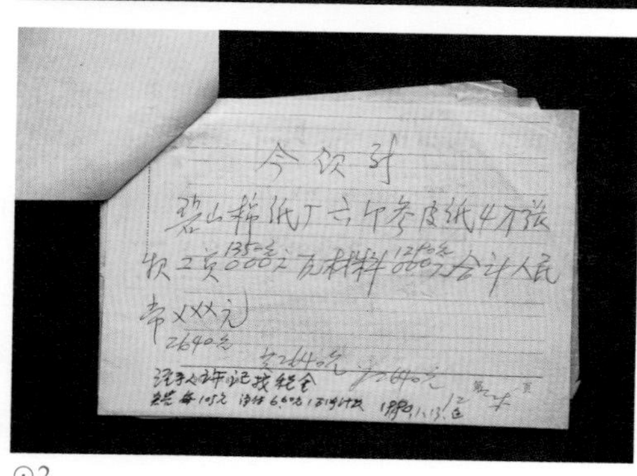

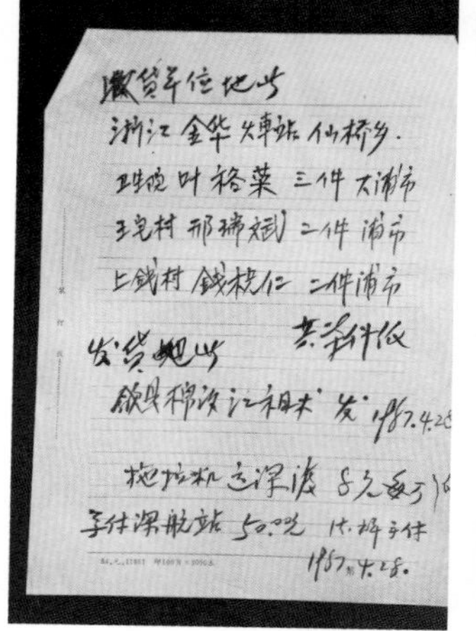

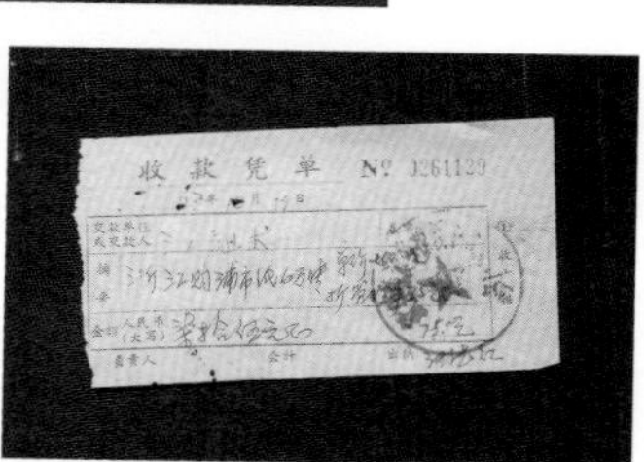

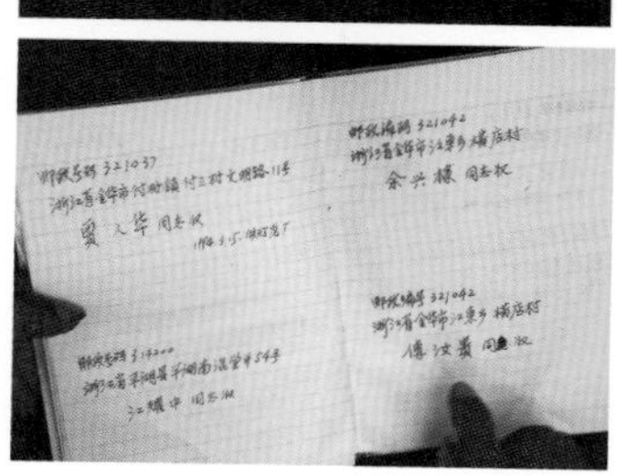

⊙2

⊙1 江祖术向调查组成员展示自己当年的记录
Jiang Zushu showing his papermaking records to a researcher

⊙2 江祖术保留的销售记录原件
Sale records kept by Jiang Zushu



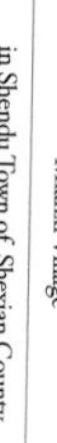

要四件，5月份要四件，6月份要四件，7月份要四件，8月份要四件，9月份要四件，共二拾（十）四件，请你大队代运，其运费由我社负担，付款办法货到汇款。浦市纸每件暂定价格160元，计4 500张。该浦市棉（纸）规格长1.5尺，阔2.15尺。歙县蚊香生产合作社。

从江祖术的记账簿中还知道：歙县老竹铺人来订纸包装老竹大方茶叶，祁门芝溪人来订纸包雪梨，金华市仙桥人来订纸做酱油封泥纸，浙江兰溪人来订纸做日用油纸伞，黟县渔亭人来订纸包食用糕点，本县更多的是养桑户来订纸给蚕宝宝产卵用，还有一些南货店铺订购代销，其用途更加广泛。

调查中江祖术的三儿子江德成补充了一些信息：今年（2016年）浙江省金华蚊香纸仍有订货单，每张4元，虽然量不大，但还是接了这单生意，准备茶季后（8～10月）生产。每年家里（江祖术随江德成小家生活）收入为卖枇杷收入10 000多元、卖茶叶收入6 000元、卖纸收入5 000～6 000元。

⊙3

⊙3 从棉溪村看新安江对岸的村落 View of the village on the other side of Xin'an River from Mianxi Village

## 六 歙县深渡镇棉溪村构皮纸的品牌文化与民俗故事

## 6 Brand Culture and Stories of Mulberry Bark Paper in Mianxi Village of Shendu Town in Shexian County

### 1. 民间传说中“澄心堂纸”得名的由来

歙县属于古徽州，历史上是著名的高端产纸区，名气最大的莫过于“澄心堂纸”。《徽州府志》记载：“黟歙间多良纸，有凝霜、澄心之号。”传说这种以楮树皮或构树皮为原料的纸原产地为歙县的搁船尖一带，搁船尖山上有天然奇观“十门九不锁”“天下第一心”，云溪穿心而过，故名澄心。南唐中主李璟和后主李煜视这种纸为珍宝，赞其为“纸中之王”，并专门建“澄心堂”藏之，取名“澄心堂纸”，供宫中长期使用。

明代诗人傅若金曾作诗称赞：“新安江水清见底，水边作纸明于水。兔白霜残晓月空，皎宫练出秋风起。”

### 2. 棉溪村造纸民谣

调查组访谈汪祖术时，汪祖术提及当地造纸的两

⊙1

⊙2

首歌谣。一首歌谣是:“打皮做纸筋,打折脚肚筋。两年吃了三年粮,三年睡了两年觉。”由于棉溪村当地造纸一般是早上4点半一直做到晚上10点,十分辛苦,因此大家口耳相传了这首歌谣来反映造纸人的辛苦。还有一首歌谣是:“棉溪杨村做纸浆,棉溪怂、挂灯笼。”意思是棉溪村做纸人淳朴老实有余,不灵活。

### 3. 棉溪村造纸工艺用方言表达的术语

据江祖术介绍,在实际生产中,有些工序和工具用当地方言,如:解开皮又称“解皮涉”,脚踩又称“礌灰皮”,拣选又称“礌皮娘”,让牛拉石碾碾原料又叫“碾纸巾”,装袋清洗又称“喷纸巾”,碓的时候添料又称“添皮”,用传统方法蒸皮又称“老法夏皮”,浸泡香叶的桶叫“药兜”,装原料的大木桶叫“湾桶”。

## 七 歙县深渡镇棉溪村构皮纸的传承现状与发展思考

## 7 Current Status of Business Inheritance and Thoughts on Development of Mulberry Bark Paper in Mianxi Village of Shendu Town in Shexian County

江祖术多年坚持不懈的记账习惯为棉溪村造纸保留了珍贵的原始交易文献,体现了古徽州民间重记述的文化传统。江祖术表示:目前找过来做纸的都是老纸客,都是冲着这手工皮纸来的,虽然村里只有1~2户造纸,但是不能不做,几十年的声誉不能丢,几百年的手艺不能忘。说着这些,面对渐行渐远的棉溪村手工皮纸,老人的神情变得暗淡。

听说江祖术老人收藏了一些准备自用的手工皮

⊙3

⊙1 调查组成员与汪成棋(前右一)合影
A group photo of researchers and Wang Chengqi (first from the right in the front row)

⊙2 向江祖术老人询问造纸术语
Inquiring Jiang Zushu about papermaking terms

⊙3 江祖术谈到传承现状
Jiang Zushu talking about current status of papermaking inheritance

⊙4

纸，我们想让他拿出来看看，几番动员之下，江祖术才小心翼翼地从屋内橱柜中搬出保存的几刀纸。面对这旧日生产的棉溪村手工皮纸，调查组成员想买回去做测试和出版的标本样纸，老人吞吞吐吐地说最多只能让出3刀。他女婿说："老爹年纪大了，这是他留着自己'老'了用的。做了一辈子纸的人，自然想与自己做的好纸为伴呢！"

棉溪村30年前还是全村造纸，但至调查时的2016年，还在造纸的农户已所剩无几，对这项快速消失的民间手工艺，村主任汪成棋表示：棉溪村现在还有一些会做手工皮纸的老人，还保存一些旧日手工皮纸的生产工具，如抄纸的竹帘、石槽，榨压皮纸的杠杆、条石，捞纸料用的竹篓等。从传承发展的措施来说，第一，我们努力继续传承发展；第二，可以结合附近的新安江山水画廊景区作为一项旅游项目来发展；第三，力争申报成为非物质文化遗产保护项目。

⊙5

⊙6

调查组认为：徽州手工造纸有着悠久的历史，唐五代时出现过具有一定历史地位的"徽纸""澄心堂纸"。而今，在整个徽州手工制纸已没有了往日的辉煌，棉溪村手工构皮纸起源时间久、与生产生活结合紧密、原材料特色鲜明，是安徽当代手工纸一个重要的研究节点。但至调查组入村时止，由于信息传播受阻等因素，棉溪村手工构皮纸尚未能列入任何层级的非物质文化遗产保护名录，需要政府相关部门立即关注，在当地造纸人的参与下，建立完整的保护和传承机制。

⊙7

⊙4
江祖术收藏自用的纸
Paper saved by Jiang Zushu for his own funeral

⊙5/6
江祖术依依不舍卖老纸
Jiang Zushu felt reluctant to sell the paper the saved for his own funeral

⊙7
调查组成员与江祖术合影
A group photo of researchers and Jiang Zushu

# 构皮纸

Paper Mulberry Bark Paper
in Mianxi Village of Shendu Town
in Shexian County

棉溪村江祖术家构皮纸透光摄影图
A photo of paper mulberry bark paper in Jiang Zushu's house of Mianxi Village seen through the light

# 染色构皮纸

Dyed Paper Mulberry Bark Paper in Mianxi Village of Shendu Town in Shexian County

棉溪村江祖术家染色构皮纸透光摄影图

A photo of dyed mulberry bark paper in Jiang Zushu's house Mianxi Village seen through the light

# 第六节

# 黄山市三昕纸业有限公司



安徽省
Anhui Province

黄山市
Huangshan City

休宁县
Xiuning County

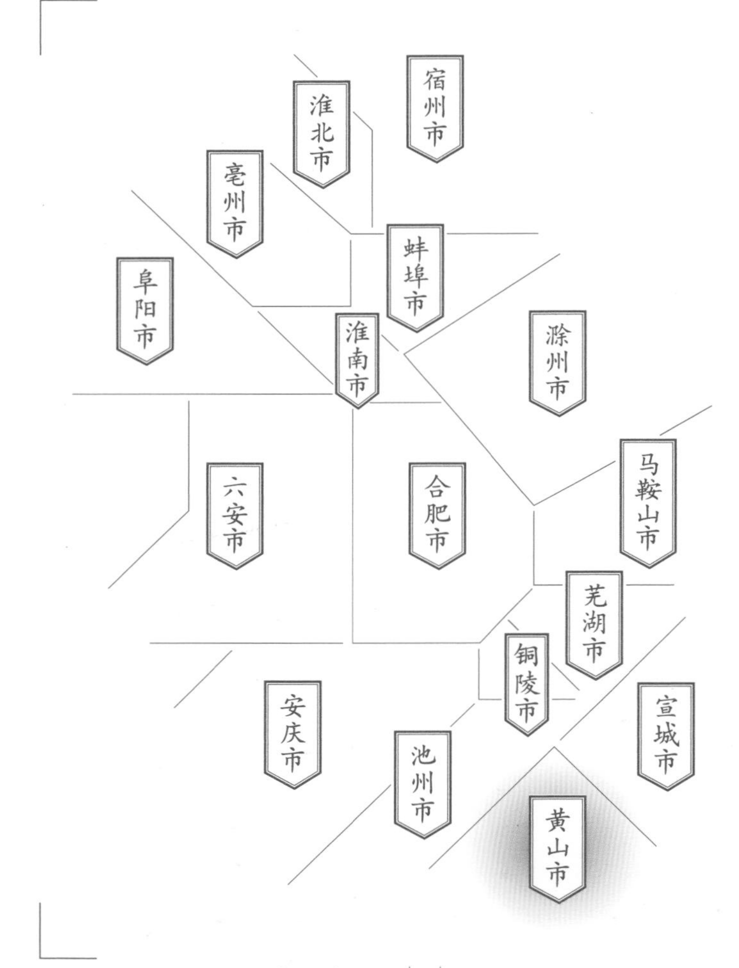


**调查对象**

海阳镇
黄山市三昕纸业有限公司
皮纸

Section 6

## Sanxin Paper Co., Ltd. in Huangshan City

**Subject**

Bast Paper
of Sanxin Paper Co., Ltd.
in Huangshan City
in Haiyang Town

## 一 黄山市三昕纸业有限公司的基础信息与生产环境

## 1 Basic Information and Production Environment of Sanxin Paper Co., Ltd. in Huangshan City

三昕纸业有限公司位于安徽省黄山市休宁县，原厂址为海阳镇玉宁街172号，地理坐标为：东经118° 10′ 44″，北纬29° 46′ 43″。公司前身为苏州市的太仓韩纸房纸业有限公司，由于大客户担忧太仓水质会影响产品品质，公司于2002年搬迁至休宁县海阳镇玉宁街。2011年，因休宁县政府整体城市规划的需要，将生产厂房迁至海阳镇晓角行政村，租用原休宁县水泥厂的旧生活区。

2016年3月24日，调查组对三昕纸业有限公司新厂区进行了第一次实地调查；2017年7月6日，又进行了第二次补充调查。截至第二次调查时，新厂房尚未开始正式生产，生产设备大部分已安装。当问到新厂恢复生产的日期比第一次调查时所定的日期延后的原因时，三昕纸业有限公司企业法人朱建新有些无奈地表示，主要是土地的购置过程中出现了波折。

2016年3月访谈朱建新时了解到的状态是：公司2002年搬至玉宁街后，曾经拥有6帘槽的手工纸生产能力，员工16人左右，生产旺盛期员工最多达到20多人。晓角村新厂房建设完成后，计划有5帘槽的手工纸产能，同时将于新厂区上线单人半自动喷浆捞纸槽，通过空气压缩机小范围使浆料流动，目前设备已试验成功，目的在于减少耗费抄纸工的体力和提高产量。据悉，新厂区还将上线机械书画纸生产设备。

三昕纸业有限公司因所有产品都出口至日本，所以产品均围绕日方客户的需求生产。至调查时，三昕纸业有限公司与阿波制纸公司建立了单一、深入、持久和稳定的合作关系，实际上相当于专门的生产单元。

⊙1

⊙2

⊙3

⊙4

⊙1/2 第一次与第二次调查时的新厂区
View of the new Sanxin Paper Factory on our first and second field investigation trips

⊙3 布满灰尘的新建平卧式焙纸台群
Newly built tables for drying the paper covered with dust

⊙4 在新厂区筹建办公室访谈朱建新
Interviewing Zhu Jianxin in the construction office of the new factory

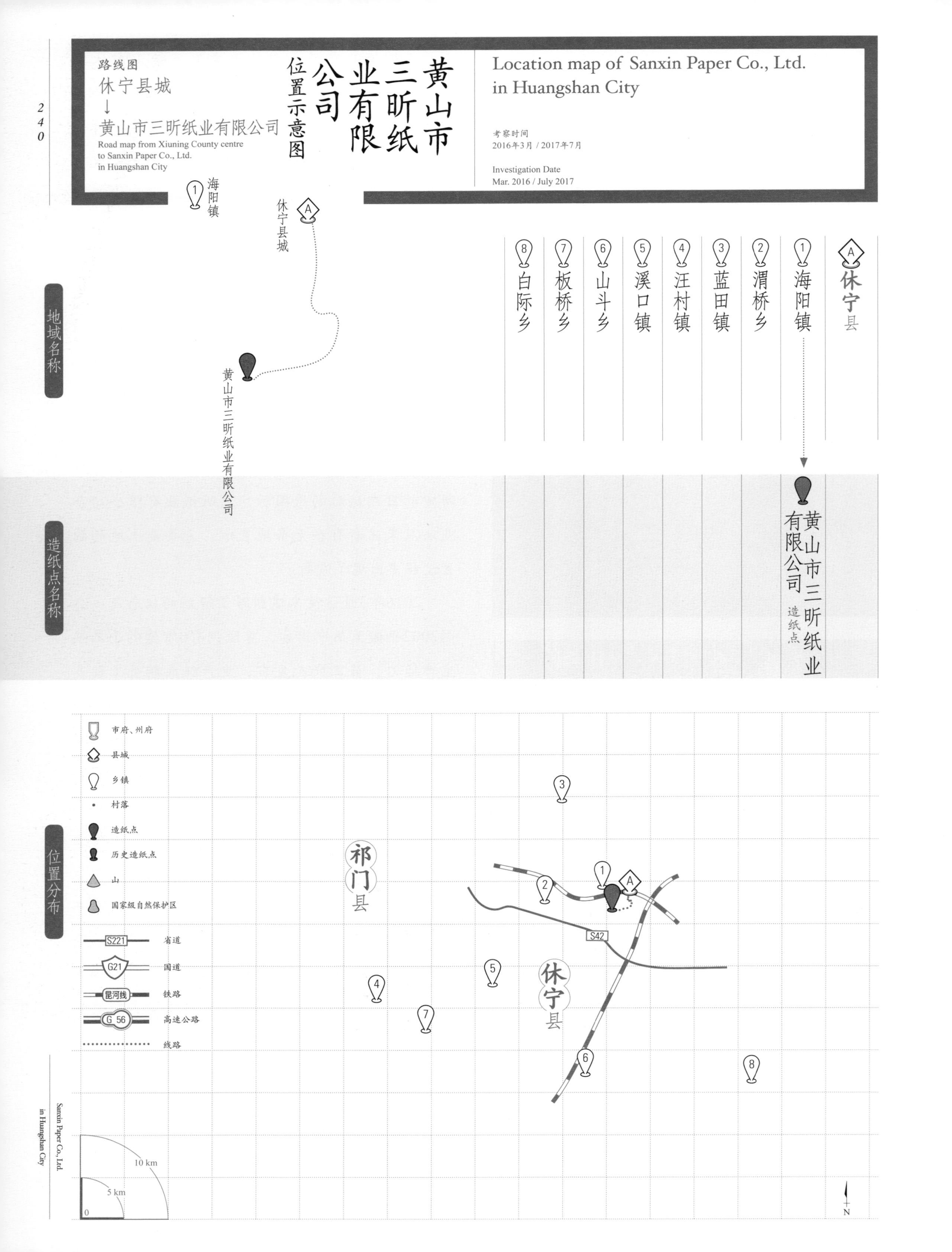

路线图
休宁县城
↓
黄山市三昕纸业有限公司
Road map from Xiuning County centre
to Sanxin Paper Co., Ltd.
in Huangshan City
位置示意图
黄山市三昕纸业有限公司
Location map of Sanxin Paper Co., Ltd.
in Huangshan City
考察时间
2016年3月 / 2017年7月
Investigation Date
Mar. 2016 / July 2017
海阳镇
休宁县城
黄山市三昕纸业有限公司
地域名称
休宁县
海阳镇
渭桥乡
蓝田镇
汪村镇
溪口镇
山斗乡
板桥乡
白际乡
造纸点名称
黄山市三昕纸业有限公司
造纸点
位置分布
市府、州府
县城
乡镇
村落
造纸点
历史造纸点
山
国家级自然保护区
S221
省道
G21
国道
昆河线
铁路
G 56
高速公路
线路
祁门县
休宁县
S42
10 km
5 km
0
N

休宁县，隶属于安徽省黄山市，位于安徽省最南端，与浙、赣两省紧邻，属古徽州“一府六县”之一。三昕纸业有限公司所在的海阳镇地处黄山脚下新安江畔，为休宁县的城关镇，镇域面积132 km²，辖4个社区17个行政村，总人口4.9万。休宁县的前身休阳县自东汉建安十三年（208年）始建，至今已有1 800多年的历史，历来以山水之美、林茶之富、商贾之多、文风之盛而蜚声海内外。

休宁县有“中国第一状元县”“中国有机茶之乡”“中国乡村旅游福地”等誉称。境内有“中国道教四大名山”之一的齐云山风景区，新安江、富春江、钱塘江三江源头的六股尖景区，中国历史名镇万安镇老街景区与古城岩景区等。

历史上，休宁县为中国名茶“屯绿”和“松萝茶”的主产地。

⊙1

⊙2

⊙1 新安江畔的万安古镇与古城岩（陈政供图）
View of Wan'an Town and ancient city by Xin'an River (photo provided by Chen Zheng)

⊙2 新厂区附近的秀阳古塔
Xiuyang Ancient Tower near the new Sanxin Paper Factory

## 二 黄山市三昕纸业有限公司的历史与传承情况

## 2 History and Inheritance of Sanxin Paper Co., Ltd. in Huangshan City

朱建新，1961年出生，江苏省太仓市人，三昕纸业有限公司创始人与法人代表，调查组调查前的10余年间，长住在黄山市休宁县。据朱建新介绍，至2016年三昕纸业有限公司已有近20年的历史。1997年朱建新进入造纸行业，因祖辈没有造纸从业传承，他是通过慢慢摸索和自我学习才将公司逐步发展壮大的。

三昕纸业有限公司正式成立于2002年，当时正值纸厂由江苏太仓搬至休宁之际，朱建新向工商部门申请注册了公司。2011年，依据休宁县城区规划，政府部门对原厂房进行了拆迁，手工纸生产也因此停了下来。因家中有老母亲要照顾，朱建新并未立即新建厂房生产。直到2015年左右，朱建新才开始选址准备继续造纸，并将纸厂由海阳镇玉宁街172号搬迁至海阳镇晓角村。第一次访谈中朱建新讲道：“纸厂预计于2016年下半年正式建成投产，2017年还将上新线生产机械书画纸。”

在2002年原造纸厂搬至休宁县之前，朱建新一直在苏州太仓从事造纸工作，三昕纸业有限公司的前身为太仓韩纸房纸业有限公司，目前太仓韩纸房纸业有限公司已不生产手工纸，只生产机

⊙1

⊙2

⊙3

⊙1 朱建新与调查组成员在建设中的抄纸车间 Zhu Jianxin and a researcher in the papermaking workshop

⊙2 新厂区已砌好的水池 Newly built water pools at the new Sanxin Paper Factory

⊙3 水质纯净的新安江上游（陈政供图） Upstream of Xin'an River (photo provided by Chen Zheng)

械书画纸。

据朱建新的回溯信息，太仓韩纸房纸业有限公司的前身则是台棉纸业有限公司，系总部在台湾南投县埔里镇的台湾绵纸厂股份公司与大陆的合资企业，1993年开始生产手工楮皮纸。台棉纸业有限公司由台湾人经营，因存在原料高价从台湾买入、产品又以低价卖出和台商两头赚差价等问题，公司很快于1995年陷入资不

抵债的境地。

朱建新原在太仓当地的农工商总公司工作，1996年，辖区的岳王镇镇政府派他去担任韩纸房纸业有限公司总经理，希望他能够改善纸厂的经营状况。由于台棉纸业有限公司当时销售渠道有限，韩纸房纸业有限公司虽拥有60帘槽的产能但无法完全发挥，一直处于亏损状态，短时间内很难拓展客户扭转经营局面。

1996年下半年，朱建新结识了大客户日本阿波制纸公司的代表，因韩纸房纸业有限公司经济效益不好，加上主持韩纸房纸业有限公司的中国台湾经理和日本人的经营理念不同，生产的偏低端产品很难满足阿波制纸公司的订单需求。基于此，1997年初，朱建新向政府提出太仓韩纸房纸业有限公司申请破产，股东清盘后，他自己购买部分设备迁出自行造纸，从那时起，开始专门生产日本阿波制纸公司所需的产品，双方建立了持久、稳定的合作伙伴关系。

⊙1

⊙2

由于在太仓造纸所用的地下水矿物质含量高，造出来的纸时间长了会因铁离子聚集在纸上而有斑点，朱建新在当地只能做带颜色的加工色纸才能获得阿波制纸公司的认可。偶然的机缘，思考着改变被动局面的朱建新在热心朋友的推荐下到休宁县进行了考察，发现休宁的优质水源和自然环境很适合造优质纸。在所取水样得到阿波制纸公司肯定后，朱建新2002年下决心将纸厂由太仓搬至休宁县海阳镇。同年，朱建新注册成立黄山市三昕纸业有限公司。

据朱建新介绍，纸厂搬迁到休宁之初，员工都是从太仓带过来的老员工，后来开始培训休宁当地员工，阿波制纸公司在原料加工、员工培训、生产管理和新产品研发上都给予了指导和直接帮助。

朱建新是家中的独子，祖辈都是太仓的农民，家族中没有造纸历史和经验传承。朱建新自述：在日本客户阿波制纸公司持续指导造纸工艺的基础上，自己靠不断探索和学习慢慢积累了造纸经验，造纸环节中所有细节问题或难题均是在多次试验的基础上才解决的。朱建新育有一儿一女，子女都未从事造纸行业。当问及传承问题时，朱建新表示：造纸传承上需要投入更多资源培养本地化的年轻人。

⊙1 新厂区正在建设中的纸槽群 Papermaking troughs under construction in the new factory

⊙2 在施工现场访谈朱建新 Interviewing Zhu Jianxin at the construction site

三

# 黄山市三昕纸业有限公司的代表纸品及其用途与技术分析

3

Representative Paper and Its Uses and Technical Analysis of Sanxin Paper Co., Ltd. in Huangshan City

## (一) 代表纸品及其用途

三昕纸业有限公司主要生产礼品包装用纸和装饰用纸，属于专向性用途很明确的类型，产品至搬迁停产时几乎全部出口日本。纸品种类主要包括楮皮纸、落水纸、强制纸、云龙纸、民芸纸（颜色纸）等，其中以楮皮纸、民芸纸和强制纸为代表纸品。同时，公司也生产少量摄影画专用纸和版画专用纸。从用途和纸品选择来看，基本上是为日本的消费需求量身打造的生产模式。

据朱建新介绍，所有纸品均采用手工捞制方式生产，其中楮皮纸、民芸纸和强制纸采用手工吊帘捞制方式，而摄影画专用纸采用手工钢丝网捞制方式。楮皮纸的主要原料为楮树皮，民芸纸和强制纸的主要原料为楮树皮和木浆。其中楮树皮从泰国进口，木浆从加拿大进口。

从规格上看，三种代表纸品的规格均为98 cm × 66 cm。从重量上看，楮皮纸可分为11 g/张、15 g / 张、19 g / 张、22 g / 张、26 g / 张、30 g / 张、38 g/张，常规品种为26 g/张；民芸纸分为19 g / 张、32 g/张；强制纸为45 g/张。

从用途上来说，楮皮纸主要用作书法用纸、门窗贴纸，也可通过加工染色后当作包装用纸。民芸纸又称“颜色纸”，是通过在浆内染色而手工捞制的各种颜色的皮纸，除了礼品包装、寺庙用纸（如烫金纸、印花纸）外，大量用于手撕画。此外，因日本十分重视小孩子练习书法和学习做纸品，民芸纸也常用于小孩手工用纸。强制纸又称“皱纹纸”，是在民芸

⊙3

⊙4

⊙3/4 纸品样册 Sample book of decorative papers

纸基础上通过人工搓揉而成的花纹纸，它的用途比较广泛，可用于相册封面、沙发垫、衣服内衬、墙纸和包装纸。

## （二）代表纸品的性能分析

### 1. 代表纸品一：楮皮纸

测试小组对采样自三昕纸业有限公司生产的楮皮纸所做的性能分析，主要包括厚度、定量、紧度、抗张力、抗张强度、撕裂度、湿强度、白度、耐老化度下降、尘埃度、吸水性、伸缩性、纤维长度和纤维宽度等。按相应要求，每一指标都需重复测量若干次后求平均值，其中厚度抽取10个样本进行测试，定量抽取5个样本进行测试，抗张力（强度）抽取20个样本进行测试，撕裂度抽取10个样本进行测试，湿强度抽取20个样本进行测试，白度抽取10个样本进行测试，耐老化下降抽取10个样本进行测试，尘埃度抽取4个样本进行测试，吸水性抽取10个样本进行测试，伸缩性抽取4个样本进行测试，纤维长度测试了200根纤维，纤维宽度测试了300根纤维。对三昕纸业有限公司楮皮纸进行测试分析所得到的相关性能参数见表4.6。表中列出了各参数的最大值、最小值及测量若干次所得到的平均值或者计算结果。

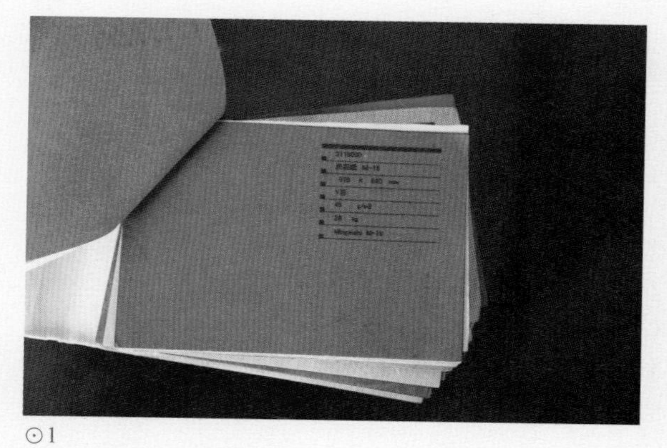

⊙1

⊙2

表4.6　三昕纸业有限公司楮皮纸相关性能参数
Table 4.6　Performance parameters of mulberry bark paper in Sanxin Paper Co., Ltd.

| 指标 | | 单位 | 最大值 | 最小值 | 平均值 | 结果 |
|---|---|---|---|---|---|---|
| 厚度 | | mm | 0.100 | 0.080 | 0.090 | 0.090 |
| 定量 | | g/m² | — | — | — | 25.8 |
| 紧度 | | g/cm³ | — | — | — | 0.287 |
| 抗张力 | 纵向 | N | 14.5 | 10.5 | 12.9 | 12.9 |
| | 横向 | N | 7.4 | 6.4 | 7.0 | 7.0 |
| 抗张强度 | | kN/m | — | — | — | 0.663 |
| 撕裂度 | 纵向 | mN | 150 | 120 | 136 | 136 |
| | 横向 | mN | 480 | 400 | 438 | 438 |
| 撕裂指数 | | mN•m²/g | — | — | — | 13.6 |
| 湿强度 | 纵向 | mN | 3 600 | 2 100 | 1 150 | 1 150 |
| | 横向 | mN | 1 800 | 1 000 | 760 | 760 |

⊙1 民芸纸样品 Minyun paper samples

⊙2 强制纸样品 Qiangzhi paper samples

续表

| 指标 | | 单位 | 最大值 | 最小值 | 平均值 | 结果 |
|---|---|---|---|---|---|---|
| 白度 | | % | 71.77 | 70.21 | 71.34 | 71.3 |
| 耐老化度下降 | | % | — | — | — | 12.6 |
| 尘埃度 | 黑点 | 个/m² | — | — | — | 24 |
| | 黄茎 | 个/m² | — | — | — | 0 |
| | 双浆团 | 个/m² | — | — | — | 0 |
| 吸水性 | | mm | — | — | — | 11 |
| 伸缩性 | 浸湿 | % | — | — | — | 0.75 |
| | 风干 | % | — | — | — | 0.68 |
| 纤维 | 长度 | mm | 6.82 | 1.61 | 3.69 | 3.69 |
| | 宽度 | μm | 19.0 | 2.0 | 10.0 | 10.0 |

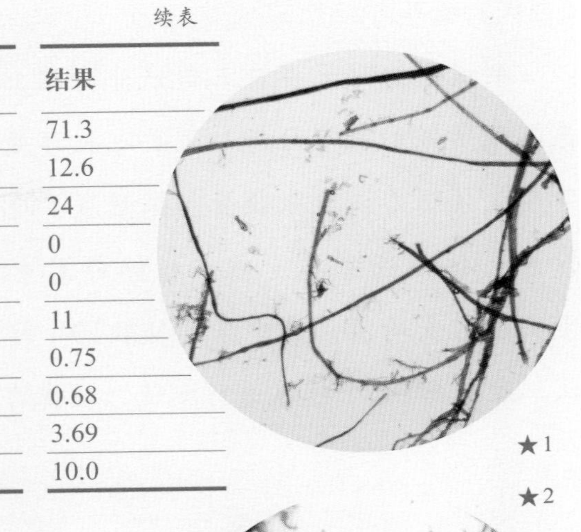

★1

★2

由表4.6中的数据可知，三昕纸业有限公司楮皮纸最厚约是最薄的1.25倍，经计算，其相对标准偏差为0.007，纸张厚薄较为一致。所测三昕纸业有限公司楮皮纸的平均定量为25.8 g/m²。通过计算可知，三昕纸业有限公司楮皮纸紧度为0.287 g/cm³，抗张强度为0.663 kN/m，抗张强度值较大。所测三昕纸业有限公司构皮纸撕裂指数为13.6 mN·m²/g，撕裂度较大；湿强度纵横平均值为955 mN，湿强度较大。

所测三昕纸业有限公司楮皮纸平均白度为71.3%，白度较高，是由于其加工过程中有漂白工序。白度最大值是最小值的1.022倍，相对标准偏差为0.473，白度差异相对较小。经过耐老化测试后，耐老化度下降12.6%。

所测三昕纸业有限公司楮皮纸尘埃度指标中黑点为24个/m²，黄茎为0个/m²，双浆团为0个/m²。吸水性纵横平均值为11 mm。伸缩性指标中浸湿后伸缩差为0.75%，风干后伸缩差为0.68%，说明三昕纸业楮皮纸伸缩性差异不大。

三昕纸业有限公司楮皮纸在10倍、20倍物镜下观测的纤维形态分别见图★1、图★2。所测三昕纸业楮皮纸皮纤维长度：最长6.82 mm，最短1.61 mm，平均长度为3.69 mm；纤维宽度：最宽19.0 μm，最窄2.0 μm，平均宽度为10.0 μm。三昕纸业有限公司楮皮纸润墨效果见图⊙3。

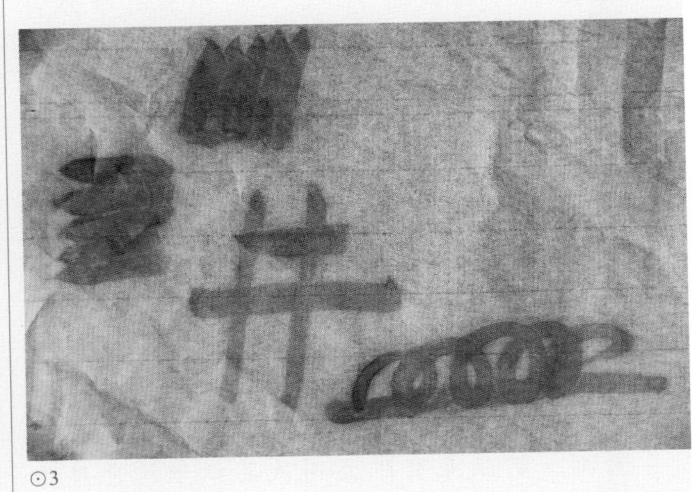

⊙3

★1 三昕纸业有限公司楮皮纸纤维形态图（10×）
Fibers of mulberry bark paper in Sanxin Paper Co., Ltd. (10× objective)

★2 三昕纸业有限公司楮皮纸纤维形态图（20×）
Fibers of mulberry bark paper in Sanxin Paper Co., Ltd. (20× objective)

⊙3 三昕纸业有限公司楮皮纸润墨效果
Writing performance of mulberry bark paper in Sanxin Paper Co., Ltd.

2．代表纸品二：民芸纸

测试小组对采样自三昕纸业有限公司生产的民芸纸所做的性能分析，主要包括厚度、定量、紧度、抗张力、抗张强度、撕裂度、色度、吸水性等。按相应要求，每一指标都需重复测量若干次后求平均值，其中厚度抽取10个样本进行测试，定量抽取5个样本进行测试，抗张力（强度）抽取20个样本进行测试，撕裂度抽取10个样本进行测试，色度抽取10个样本进行测试，吸水性抽取10个样本进行测试。对三昕纸业有限公司民芸纸进行测试分析所得到的相关性能参数见表4.7。表中列出了各参数的最大值、最小值及测量若干次所得到的平均值或者计算结果。

表4.7　三昕纸业有限公司民芸纸相关性能参数
Table 4.7　Performance parameters of Minyun paper in Sanxin Paper Co., Ltd.

| 指标 | | 单位 | 最大值 | 最小值 | 平均值 | 结果 |
|---|---|---|---|---|---|---|
| 厚度 | | mm | 0.15 | 0.14 | 0.147 | 0.147 |
| 定量 | | g/m² | — | — | — | 49.8 |
| 紧度 | | g/cm³ | — | — | — | 0.339 |
| 抗张力 | 纵向 | N | 23.6 | 16.6 | 18.08 | 18.08 |
| | 横向 | N | 11.8 | 10.2 | 11.07 | 11.07 |
| 抗张强度 | | kN/m | — | — | — | 0.972 |
| 撕裂度 | 纵向 | mN | 530 | 420 | 478 | 478 |
| | 横向 | mN | 610 | 500 | 553 | 553 |
| 撕裂指数 | | mN·m²/g | — | — | — | 10.18 |
| 色度 | | % | 8.01 | 7.40 | 7.73 | 7.73 |
| 吸水性 | 长度 | mm | — | — | — | 极难吸水 |
| | 宽度 | mm | — | — | — | 极难吸水 |

由表4.7中的数据可知，三昕纸业有限公司民芸纸最厚约是最薄的1.071倍，经计算，其相对标准偏差为0.005，纸张厚薄较为一致。所测三昕纸业有限公司民芸纸的平均定量为49.8 g/m²。通过计算可知，三昕纸业有限公司民芸纸紧度为0.339 g/cm³，抗张强度为0.972 kN/m，抗张强度值较小。所测三昕纸业有限公司民芸纸撕裂指数为10.18 mN·m²/g。

所测三昕纸业有限公司民芸纸平均色度为7.73%。色度最大值是最小值的1.082倍，相对标准偏差为0.855，色度差异相对较小，且极难吸水。

3．代表纸品三：强制纸

测试小组对采样自三昕纸业有限公司生产的强制纸所做的性能分析，主要包括厚度、定量、紧度、抗张力、抗张强度、撕裂度、湿强度、白度、耐老化度下降、尘埃度、吸水性、伸缩性等。按相应要求，每一指标都需重复测量若干次后求平均值，其中厚度抽取10个样本进行测试，定量抽取5个样本进行测试，抗张力（强度）抽取20个样本进行测试，撕裂度抽取

10个样本进行测试，湿强度抽取20个样本进行测试，白度抽取10个样本进行测试，耐老化下降抽取10个样本进行测试，尘埃度抽取4个样本进行测试，吸水性抽取10个样本进行测试，伸缩性抽取4个样本进行测试。对三昕纸业有限公司强制纸进行测试分析所得到的相关性能参数见表4.8。表中列出了各参数的最大值、最小值及测量若干次所得到的平均值或者计算结果。

表4.8　三昕纸业有限公司强制纸相关性能参数
Table 4.8　Performance parameters of Qiangzhi paper in Sanxin Paper Co., Ltd.

| 指标 | | 单位 | 最大值 | 最小值 | 平均值 | 结果 |
|---|---|---|---|---|---|---|
| 厚度 | | mm | 0.325 | 0.270 | 0.297 | 0.297 |
| 定量 | | $g/m^2$ | — | — | — | 695.3 |
| 紧度 | | $g/cm^3$ | — | — | — | 0.023 |
| 抗张力 | 纵向 | N | 24.4 | 19.8 | 22.23 | 22.23 |
| | 横向 | N | — | — | — | |
| 抗张强度 | | kN/m | — | — | — | 1.394 |
| 撕裂度 | 纵向 | mN | 820 | 620 | 714 | 714 |
| | 横向 | mN | — | — | — | |
| 撕裂指数 | | $mN\cdot m^2/g$ | — | — | — | 1.86 |
| 湿强度 | 纵向 | mN | 5 700 | 5 200 | 5 440 | 5 440 |
| | 横向 | mN | — | — | — | |
| 白度 | | % | 59.01 | 58.08 | 58.59 | 58.59 |
| 耐老化度下降 | | % | — | — | — | 0.94 |
| 尘埃度 | 黑点 | 个/$m^2$ | — | — | — | 24 |
| | 黄茎 | 个/$m^2$ | — | — | — | 0 |
| | 双浆团 | 个/$m^2$ | — | — | — | 0 |
| 吸水性 | | mm | — | — | — | 24 |
| 伸缩性 | 浸湿 | % | — | — | — | 1.63 |
| | 风干 | % | — | — | — | 0 |

由表4.8中的数据可知，三昕纸业有限公司强制纸最厚约是最薄的1.204倍，经计算，其相对标准偏差为0.018，纸张厚薄较为一致。所测三昕纸业有限公司强制纸的平均定量为695.3 $g/m^2$。通过计算可知，三昕纸业有限公司强制纸紧度为0.023 $g/cm^3$，抗张强度为1.394 kN/m，抗张强度值较大。所测三昕纸业有限公司强制纸撕裂指数为1.86 $mN\cdot m^2/g$，撕裂度较大；湿强度纵向值为5 440 mN，湿强度较大。

所测三昕纸业有限公司强制纸平均白度为58.59%，白度较高，这是由于在加工过程中有漂白工序。白度最大值是最小值的1.016倍，相对标准偏差为0.328，白度差异相对较小。经过耐老化测试后，耐老化度下降0.94%。

吸水性纵横平均值为24 mm。伸缩性指标中浸湿后伸缩差为1.63%，风干后伸缩差为0%，说明三昕纸业有限公司强制纸伸缩性差异不大。

# 四 黄山市三昕纸业有限公司代表纸品生产的原料、工艺流程与工具设备

# 4 Raw Materials, Papermaking Techniques and Tools for Making Representative Paper in Sanxin Paper Co., Ltd. of Huangshan City

## (一) 代表纸品生产的原料

### 1. 主料

(1) 楮皮。三昕纸业有限公司所用的楮皮皮料均从泰国北部的清迈进口，购买的是漂白后的皮料半成品。据朱建新介绍，泰国皮料分A、B和C三个等级。其中A级是楮皮皮料剥下后的中间部位，皮质最好，价格最高，一般为1 400～1 500元/30 kg；B级是楮皮皮料剥下后的头部部位，皮质中等，价格适中；C级是楮皮皮料根部位置，皮质最差，价格也最便宜。三昕纸业用的是B级楮皮皮料，2016年价格为600元/30 kg。与国内楮皮皮料相比，泰国楮皮皮料处理得更加干净，出浆率也比国内高。一般来说，泰国30 kg皮料能有9 kg干料，而国内只有7 kg，出浆率泰国楮皮皮料比国内高20 %左右。

(2) 木浆。三昕纸业有限公司使用的木浆原料是从加拿大进口的木浆浆板，木种为针叶林。截至2016年3月调查时，木浆浆板价格为6 000元/t。

### 2. 辅料

(1) 纸药。三昕纸业有限公司用的纸药是化学分散剂聚丙烯酰胺。调查时朱建新表示，与别的造纸厂家不同，三昕纸业有限公司采用的是日本和纸的纸药使用方式，纸槽浆内分散剂浓度较高，目的是使捞出来的纸更加均匀。一般来说，若分散剂浓度太高，则正常抄出的纸就太薄，叠放在一起不易沥水，而三昕纸业有限公司的抄纸工人能根据经验灵活捞出均匀度较佳的纸。化学分散剂均从日本进口，使用方法由日本阿波制纸公司传授。

(2) 水源。三昕纸业有限公司在老厂区时使

用的水源是自来水，调查组实测新厂区自来水pH为7.24，呈弱碱性。

⊙1

⊙2

## （二）代表纸品生产的工艺流程

### 1. 楮皮纸的生产工艺

据调查组成员对朱建新访谈的信息，总结其工艺流程为：

| 壹 | 贰 | 叁 | 肆 | 伍 | 陆 | 柒 | 捌 | 玖 |
|---|---|---|---|---|---|---|---|---|
| 洗料 | 榨干 | 拣皮 | 打浆 | 捞纸 | 压榨 | 烘纸 | 检验 | 包装 |

#### 壹 洗料

1

购买后的皮料半成品需要用清水进行清洗，洗去残留的漂白液和杂质。

#### 贰 榨干

2

清洗后的皮料需要进行压榨，该环节主要是将洗好的皮料中的水分榨干。

#### 叁 拣皮

3

榨干后的皮料需要进行人工挑拣，主要挑拣出皮料中的黄茎、杂质和未经完全漂白的皮料。

#### 肆 打浆

4

挑拣后的皮料接下来进行打浆。三昕纸业有限公司使用的是打浆机。

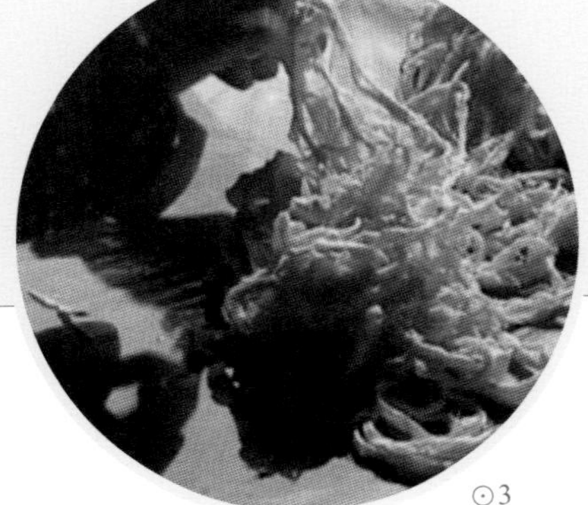
⊙3

⊙1 新厂区内的小水库 Reservoir in the new factory

⊙2 朱建新在老厂 Zhu Jianxin in the old factory

⊙3 拣皮（朱建新供图） Picking the bark (photo provided by Zhu Jianxin)

## 伍 捞纸

### 5

⊙4⊙5

三昕纸业有限公司采用单人手工吊帘方式捞制楮皮纸，吊帘捞制皮纸方式系由日本阿波制纸公司直接传授。因浆内分散剂浓度较高，浆料加入纸槽内需要打料15分钟左右，使分散剂在浆内分布均匀。据朱建新介绍，重量19 g/张及以下的楮皮纸都要求过"四道水"，即纸帘前后下水捞纸各两次。工人将纸帘放至帘床后，先自外侧向内侧将纸帘倾斜下水捞纸，然后端平纸帘使浆料均匀分布于纸帘上，再自内侧向外侧将纸帘倾斜下水捞纸，然后端平纸帘，重复此动作两次。纸张的均匀度和重量完全靠捞纸工的经验来把握。一般而言，一个捞纸工一天能捞约300张楮皮纸，捞纸工人每天工作9小时，每月工作26～27天，周日休息。

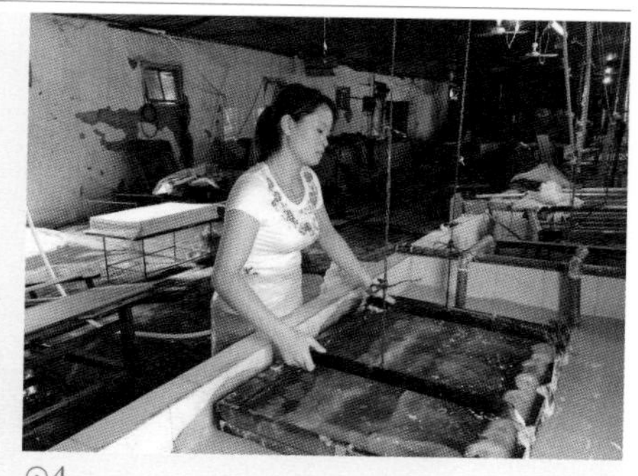

⊙4

⊙5

## 陆 压榨

### 6

与国内绝大多数厂家不同，三昕纸业有限公司压榨用的不是千斤顶，而是自动液压机，目前国内造纸厂很少在用。该压榨工艺从日本阿波制纸公司传入，自动液压机一次最多能压1 500张楮皮纸。值得注意的是，因纸浆内分散剂浓度高，纸帖渗水很慢，当天捞的纸要到第二天才能压榨，一般需要压榨1小时。

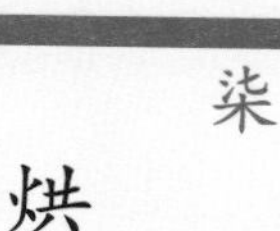

## 柒 烘纸

### 7

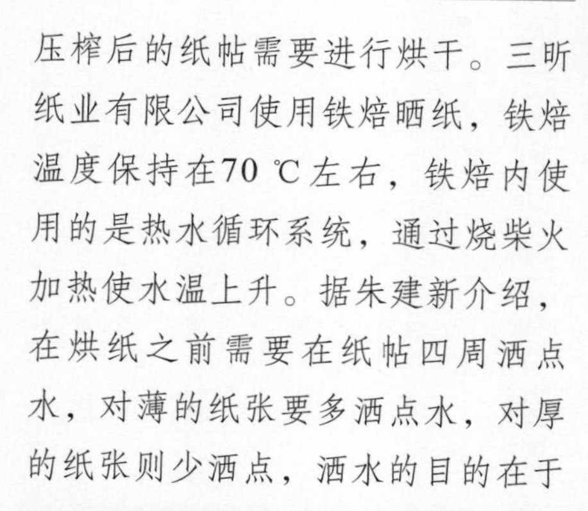

压榨后的纸帖需要进行烘干。三昕纸业有限公司使用铁焙晒纸，铁焙温度保持在70 ℃左右，铁焙内使用的是热水循环系统，通过烧柴火加热使水温上升。据朱建新介绍，在烘纸之前需要在纸帖四周洒点水，对薄的纸张要多洒点水，对厚的纸张则少洒点，洒水的目的在于便于揭纸。揭纸后，晒纸工人用松针刷将纸张平整刷上铁焙，因铁焙温度较高，刷上去的纸张几十秒后便可烘干收下。烘出的纸张一般要求四周毛边整齐，纸张无明显褶皱和缺口。

## 捌 检验

### 8

烘干后的纸张需要进行人工检验，挑选出有褶皱和缺口的纸张，这些有瑕疵的纸张一般用于手撕画。

## 玖 包装

### 9

检验合格的纸张需要进行包装。三昕纸业有限公司的出口包装箱具有伸缩性，一般装1 000～1 500张。打包好的纸箱便可通过进出口公司发至日本。

⊙4/5 捞纸（朱建新供图）
Papermaking (photos provided by Zhu Jianxin)

2. 民芸纸的生产工艺

据调查组成员对朱建新访谈的信息，民芸纸生产的原料为楮皮和木浆，纸药为化学分散剂，水源使用的是自来水，配料使用的是化学粉剂染料。染料分为不同的颜色，均是国产的，从天津购入。三昕纸业有限公司会根据日本客户给的民芸纸纸样要求，购买染料后自主进行染料调配，生产样品纸后发给阿波纸业公司的客户看，确认达到要求后再进行批量生产。

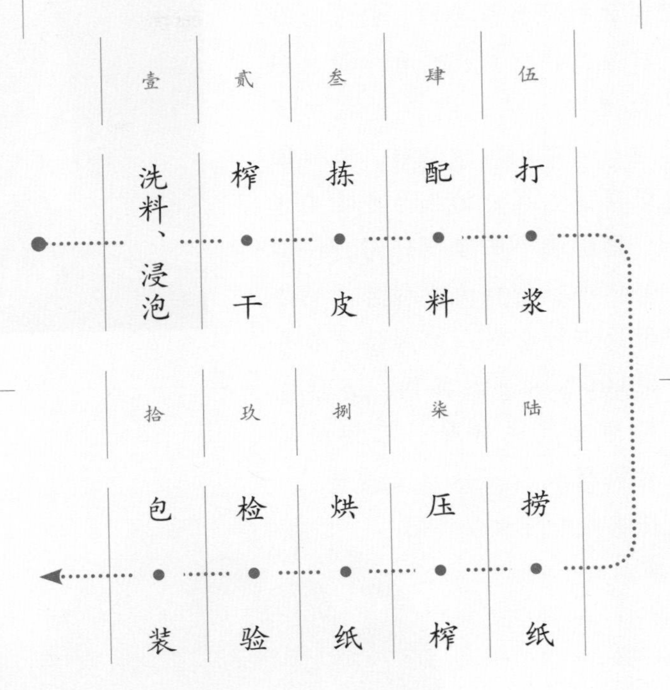


### 壹 洗料、浸泡

1 ⊙6

购买后的皮料半成品需要用清水进行清洗，洗去残留的漂白液和杂质。木浆浆板则需要浸泡，使其充分软化易于打浆。

⊙6

### 贰 榨干

2

清洗后的皮料需要进行压榨，目的是将洗好的皮料中的水分榨干。同时，浸泡后的木浆浆板也需要沥干。

### 叁 拣皮

3

榨干后的皮料需要进行人工挑拣，主要挑拣出皮料中的黄茎、杂质和未完全漂白的皮料。

### 肆 配料

4 ⊙7

民芸纸在打浆前需要按比例进行配料，楮皮皮料一般占80 %，木浆浆料占20 %。

⊙7

⊙6 洗料（朱建新供图）
Cleaning the materials (photo provided by Zhu Jianxin)

⊙7 存放待配料的染料容器（朱建新供图）
Containers for holding dye (photo provided by Zhu Jianxin)

## 伍 打浆

### 5 ⊙8

配料后的原料需要进行打浆，利用打浆机将楮皮皮料和木浆浆料进行混合打浆，打浆过程中同时加入一定比例的染料，染料颜色根据日本客户的纸品要求来选择。据访谈中朱建新介绍，民芸纸制作中加染料的流程为：先将需用的染料称好；按照估计的分量用100%的开水稀释，通常在塑料桶里完成稀释，家用的其他材料桶也行；然后直接倒入打浆机中与纸浆一同打浆。

⊙8

## 陆 捞纸

### 6 ⊙9⊙10

三昕纸业有限公司采用单人手工吊帘方式捞制民芸纸。因浆内分散剂浓度较高，浆料加入纸槽内需要打料15分钟左右，使分散剂在浆内分布均匀。据朱建新介绍，民芸纸在重量上分19 g/张、32 g/张两种，其中19 g/张的民芸纸要求过“四道水”，32 g/张的民芸纸要求过“三道水”。民芸纸吊帘捞制方法与楮皮纸相同。一般而言，一个捞纸工一天能捞约280张纸。捞纸工人月工资为3 000～4 000元，纸厂不以计件发工资，而是按时发放，要求每张捞出来的纸都达到质检要求。

⊙9 ⊙10

## 柒 压榨

### 7

与楮皮纸压榨类似，民芸纸也采用自动液压机压榨，一次最多能压1 500张纸。因纸浆内分散剂浓度高，纸帖渗水很慢，当天捞的民芸纸要到第二天才能压榨，一般需要压榨1小时。

## 捌 烘纸

### 8 ⊙11

压榨后的纸帖需要进行烘干。三昕纸业有限公司使用铁焙晒纸，铁焙温度保持在70 ℃左右，铁焙内使用的是热水循环系统，通过烧柴火加热使水温上升。

⊙11

⊙8 染料配浆（朱建新供图）
Mixing dye with pulp (photo provided by Zhu Jianxin)

⊙9/10 捞制民芸纸（朱建新供图）
Making Minyun paper (photos provided by Zhu Jianxin)

⊙11 刷纸上墙（朱建新供图）
Drying the paper on a wall (photo provided by Zhu Jianxin)

玖

## 检 验

9

烘干后的纸张需要进行人工检验，挑选出有褶皱和缺口的纸张。操作工艺与楮皮纸相同。

拾

## 包 装

10 ⊙12

先逐张检验纸的质量，检验合格的纸张才能按照标准进行包装。三昕纸业有限公司的出口包装箱具有伸缩性，一般装1 000～1 500张纸。打包好的纸箱便可通过进出口公司发至日本。

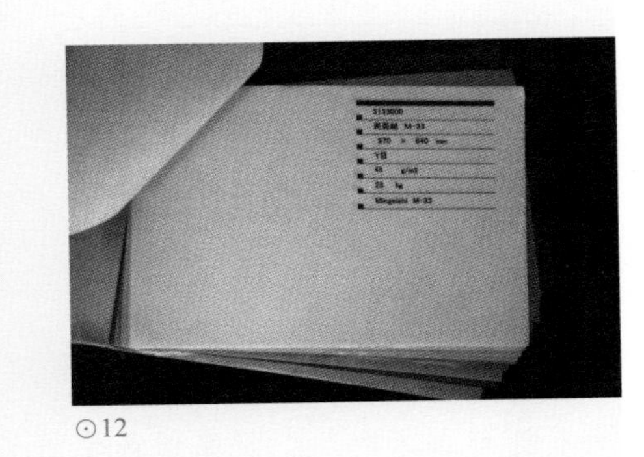

⊙12 民芸纸样品 Minyun paper samples

### 3. 强制纸的生产工艺

强制纸的原料也为楮皮和木浆，纸药为化学分散剂，用得比较多的是聚丙烯酰胺，有时也会用聚氧乙烯醚，一般从上海购买。水源是自来水，配料使用的是化学粉剂染料。强制纸的捞制技艺类似于民芸纸，并在楮皮纸与民芸纸的基础上增加"揉纸"工序。

强制纸生产的工艺流程如下：

| 壹 | 贰 | 叁 | 肆 | 伍 | 陆 |
| --- | --- | --- | --- | --- | --- |
| 洗料、浸泡 | 榨干 | 拣皮 | 配料 | 打浆 | 捞纸 |

| 拾壹 | 拾 | 玖 | 捌 | 柒 |
| --- | --- | --- | --- | --- |
| 包装 | 检验 | 揉纸 | 烘纸 | 压榨 |

壹～伍

## 洗料、浸泡至打浆

1～5

工艺与民芸纸相同。

陆

## 捞纸

6

三昕纸业有限公司采用单人手工吊帘方式捞制强制纸。因浆内分散剂浓度较高，浆料加入纸槽内需要打料15分钟左右，使分散剂在浆内分布均匀。据朱建新介绍，强制纸重量一般为42 g/张，要求过“三道水”。强制纸吊帘捞制方法与楮皮纸类似。一般而言，一个捞纸工一天能捞280张强制纸。

柒～捌

## 压榨至烘纸

7～8 ⊙13

工艺与民芸纸相同。

⊙13

拾

## 检验

10

烘干后的纸张需要进行人工检验，挑选出有褶皱和缺口的纸张。

玖

## 揉纸

9 ⊙14⊙15

烘干后的纸张需要进行人工揉纸，每张纸要揉出菱形的花纹状。

⊙14

⊙15

拾壹

## 包装

11 ⊙16

检验合格的纸张需要进行包装。三昕纸业有限公司的出口包装箱具有伸缩性，一般装1 000～1 500张纸，没有固定的数量包装标准。打包好的纸箱通过进出口公司发至日本。

⊙16

⊙13 烘好的纸张 Dried paper

⊙14/15 揉纸 Rubbing the paper

⊙16 成品纸 Paper products

（三）

三昕纸业有限公司手工纸生产的工具设备



壹
打浆机
1

用来打散、打融楮皮浆料、木浆浆料及混合浆料的设备，为伏特式构造。

⊙18

⊙17

贰
纸　槽
2

盛浆工具，旧厂由水泥浇筑而成，新厂多数已改由钢板制成。三昕纸业有限公司手工吊帘捞出的纸张规格均为98 cm × 66 cm。

⊙19

叁
纸　帘
3

用于捞纸，由竹丝编织而成，表面很光滑平整，帘纹细而密集。访谈中得知，三昕纸业有限公司使用的纸帘系从福建购买。

肆
液压机
4

用来压榨纸帖，设置一定数值后可自动压制纸帖。三昕纸业有限公司采用液压机压榨，在国内手工造纸行业中具有一定特色，此工艺是由日本客户传授而来的。

⊙17/18 打浆机 Beating machine

⊙19 金属制纸槽 Metal papermaking trough

伍
### 刷子
5

用于晒纸时将纸刷上晒纸墙，刷柄为木制，刷毛为松针。三昕纸业有限公司使用的刷子是从泾县购买的，长约26 cm。

陆
### 焙笼
6

用来晒纸，由两块长方形钢板焊接而成，中间用水蒸气加热，双面墙，可以两边晒纸。

⊙20

柒
### 压纸石
7

在晒纸工序中，当晒纸工将晒好的纸张整齐堆放在纸板上时，就用压纸石来固定纸张。

## 五 黄山市三昕纸业有限公司的市场经营状况

## 5 Marketing Status of Sanxin Paper Co., Ltd. in Huangshan City

从销售渠道看，三昕纸业有限公司生产的所有产品几乎都销往日本，大客户为日本阿波制纸公司。三昕纸业有限公司根据日本客户的订单和需求来生产各种纸品，搬迁停产前的状态实际上是阿波制纸公司在中国的定点生产商家。据朱建新的描述，三种主要产品的用途如下：一般来说，白色的楮皮纸日本客户用来装饰门窗、写书法和七色加工；民芸纸用于礼品包装、手撕画，阿波制纸公司还会进行纸品再加工，通过烫金、印花等工序生产寺庙与神社用纸；强制纸多用于

⊙20 焙笼 Drying wall

日本流行的装饰和礼品包装，日本公司通常会要求加湿强剂，这样可以做衣服内衬、沙发垫（日本公司会用柿漆再涂两遍，然后用强制纸包上填充物做垫）、墙纸等。

⊙21

⊙22

⊙23

⊙24

⊙25

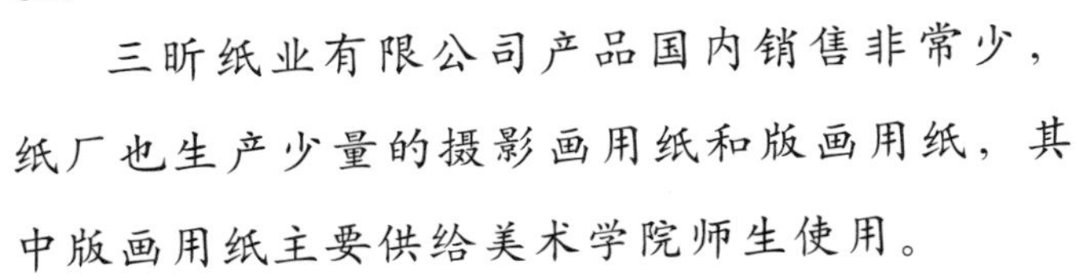

三昕纸业有限公司产品国内销售非常少，纸厂也生产少量的摄影画用纸和版画用纸，其中版画用纸主要供给美术学院师生使用。

从销售价格看，截至调查时的2016年3月，三昕纸业有限公司各种纸的市场价如表4.9所示。据朱建新介绍，纸厂年产量约40万张纸，年销售额200万元左右。因纸厂与日本客户建立了多年的合作关系，每年的产品订单很稳定，所以纸厂没有产品库存，这种以销量定产量的生产模式既节约了成本，又保障了稳定的市场需求。

⊙21 三昕纸业有限公司强制纸（白色）
Qiangzhi paper in Sanxin Paper Co., Ltd. (white)

⊙22 三昕纸业有限公司强制纸（蓝色）
Qiangzhi paper in Sanxin Paper Co., Ltd. (blue)

⊙23 三昕纸业有限公司民芸纸（黑色）
Minyun paper in Sanxin Paper Co., Ltd. (black)

⊙24 三昕纸业有限公司民芸纸（红色）
Minyun paper in Sanxin Paper Co., Ltd. (red)

⊙25 三昕纸业有限公司楮皮纸
Mulberry bark paper in Sanxin Paper Co., Ltd.

表4.9　三昕纸业有限公司纸品市场价格
Table 4.9　Market price list of paper products in Sanxin Paper Co., Ltd.

| 纸品名称 | 种类 | 市场价格 |
|---|---|---|
| 楮皮纸 | 11 g楮皮纸 | 5.8元/张 |
| | 26 g楮皮纸 | 7元/张 |
| | 30 g楮皮纸 | 7.5元/张 |
| 民芸纸 | 常规民芸纸 | 4.2～4.5元/张 |
| | 民芸纸（红色） | 4.3元/张 |
| 强制纸 | 强制纸（白色） | 8元/张 |
| | 强制纸（蓝色） | 3元/张 |

⊙1

⊙2





## 六 黄山市三昕纸业有限公司的品牌文化与民俗故事

## 6 Brand Culture and Stories of Sanxin Paper Co., Ltd. in Huangshan City

### 1. “三昕”厂名的由来和寓意

关于公司从太仓迁到休宁后为什么取名“三昕”，朱建新在访谈中讲述：“三昕”与朱建新和妻子张珊的名字有关，其中“昕”音同“新”，“三”音近“珊”。取名“三昕”第一层寓意是夫妻能够同心同德，通过共同努力将新纸厂发展壮大；第二层寓意在于，纸厂迁往新地方是因为要获得好水源，寄望于新地方建新厂，用新水源造出客户认可的好纸来。

⊙1 摄影画用纸样品 Photographic paper sample

⊙2 新厂区的临时纸库与宿舍 Temporary paper warehouse and dormitory

⊙3

## 2. 有品质才有品牌生命的文化

访谈中朱建新反复表示：日本造纸企业对产品品质的要求对他的影响很大，特别是仅仅因为造纸的水源质量达不到标准，阿波制纸公司就强烈建议另择优质水源地重新建厂。而且在他义无反顾地将厂迁往皖南山区的休宁县后，阿波制纸公司坚持全部采购三昕纸业有限公司合格的纸品，成为三昕纸业有限公司长期稳定的单一客户，为三昕纸业有限公司注重原料加工的精细和追求优质的产品质量提供了经营保障。

朱建新又举例：以楮皮纸来说，用国内厂家生产的楮皮纸写毛笔字普遍会出现白点，纸张质感效果受到干扰，这主要是因为楮皮皮料中的树汁没有完全处理干净。三昕纸业有限公司按照阿波制纸公司力求完美的标准，舍得投入资源，通过持续多次的试验，终于成功解决了楮皮树汁如何清理干净的问题，造出来的楮皮纸没有白点，书法效果较佳，在行业内具有较好的口碑。

⊙3 新安江水 Xin'an River

# 七
# 黄山市三昕纸业有限公司的传承现状与发展思考

# 7
# Current Status of Business Inheritance and Thoughts on Development of Sanxin Paper Co., Ltd. in Huangshan City

⊙1 与朱建新在小水库石坝上聊环保
Talking about environmental protection with Zhu Jianxin on a dam

## 1. 三昕纸业有限公司面临的发展问题

从调查中获得的信息来看，三昕纸业有限公司生产与销售的平衡关系维系较好，这当然与阿波制纸公司作为单一稳定大客户的产供销模式有关。与此同时，三昕纸业有限公司在发展过程中也遇到若干瓶颈问题，其中较大的问题有：

（1）环保问题。由于休宁县地处新安江源头，既是杭州重要的饮用水源——千岛湖的上游，又是国家级皖南国际旅游区的腹地，因而环保要求十分严格，造纸的污水净化处理设备是必需的投入，但这一系统成本高昂，所以三昕纸业有限公司现有生产规模和发展能力均受到很大的环保要求的压力。访谈中朱建新表示，按照当地环保标准生产当然是必须执行的，但在现阶段特别希望当地政府能够给三昕纸业有限公司这样的外来文化型小企业予以一定的扶持。

（2）工厂搬迁问题。纸厂自2002年从太仓搬迁到休宁海阳镇的老城区玉宁街后，经过7～8年的努力才实现了工厂正常生产。2011年，因县城整体连片规划需要，老厂区拆迁，中间停产数年，到2015年才在原休宁县水泥厂废弃生活区租了新的厂房，并迅速投入建设。但由于环保、转租权益等问题，新厂的建设受到一定的干扰，迫切需要获得促进性的支持。

## 2. 三昕纸业有限公司传承发展的思路

（1）三昕纸业有限公司生产的纸品几乎都销往日本，产品按照日本单一大客户需求来定制生产，最终消费在日本市场，产品的渠道建设具有鲜明的特色性。这种产销紧密配合的稳定供给定位具有渠道与市场的从容性，一定程度上避免了大众化产品带来的扭曲化竞争，三昕纸业有限公司如何达成差异化发展和个性化定制的思路值得借鉴。

（2）朱建新始终坚守精益求精的“工匠精神”，同时也按照阿波制纸公司高标准、高品质

⊙2

的造纸理念，要求工人们用心做好每张纸，确保每张纸做出来都是有用的。例如，纸厂捞纸工采用按时间的方式发放工资而非按件计算，不过分追求产品数量而是优先保障产品质量，这种薪酬制度导向值得中国手工造纸厂坊思考。

(3) 考虑到国内机械纸市场需求量大，而书画纸生产厂家普遍往低价竞争方向博弈，导致品质出色的机械书画纸比较缺乏。访谈中朱建新表示：如果新工厂建设顺利，拟于2017年上半年开始生产优质机械书画纸。

⊙3

⊙2
海阳镇老街(陈政供图)
Old street of Haiyang Town (photo provided by Chen Zheng)

⊙3
一丝不苟的『三昕』造纸人
Conscientious papermakers of Sanxin Paper Co., Ltd.

# 楮皮纸

## 黄山市三昕纸业有限公司

Paper Mulberry Bark Paper
of Sanxin Paper Co., Ltd.
in Huangshan City

手揉纸透光摄影图
A photo of wrinkled paper seen through the light

# 黄山市三昕纸业有限公司

# 民芸纸

Minyun Paper of Sanxin Paper Co., Ltd. in Huangshan City

民芸纸透光摄影图

A photo of Minyun paper seen through the light

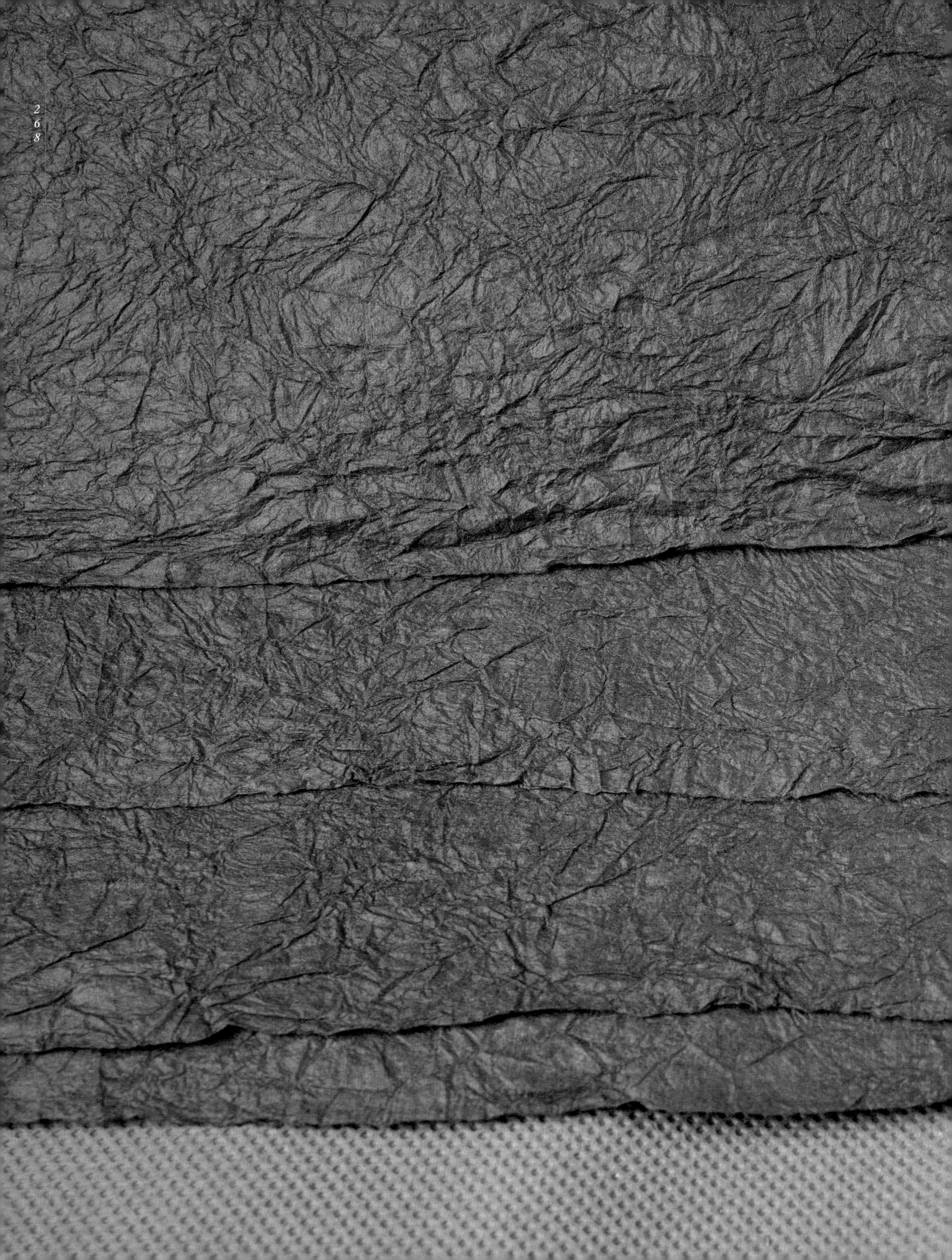

# 黄山市三昕纸业有限公司

# 楮皮纸

Paper Mulberry Bark Paper
of Sanxin Paper Co., Ltd.
in Huangshan City

染色手揉纸透光摄影图

A photo of dyed wrinkled paper seen through the light

# 第七节

# 歙县六合村

安徽省
Anhui Province

黄山市
Huangshan City

歙县
Shexian County

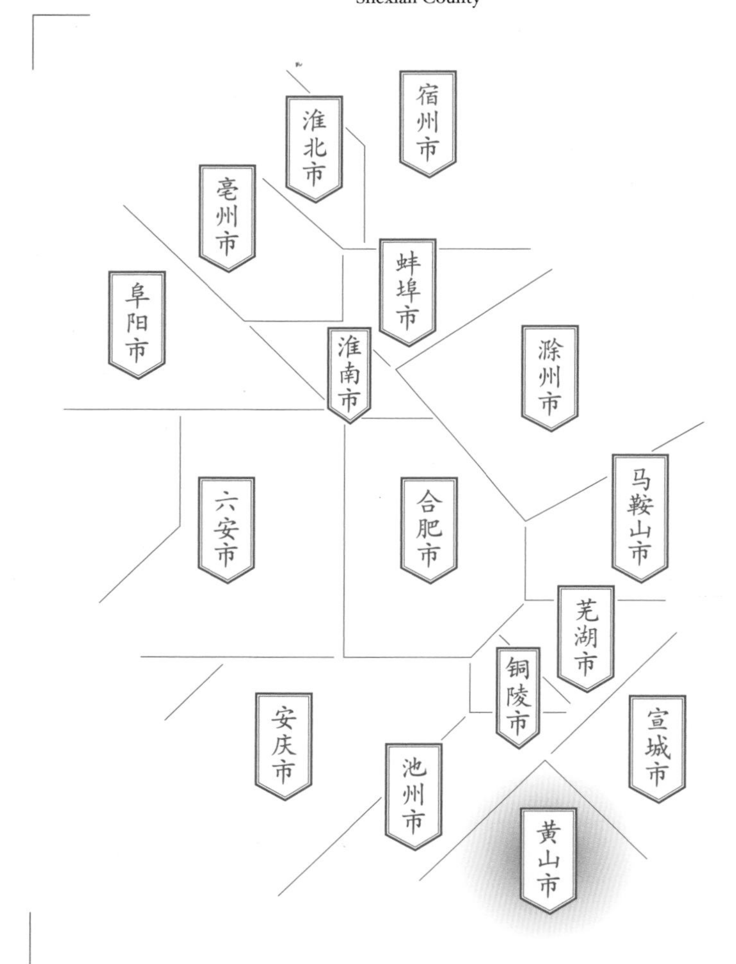


调查对象

杞梓里镇
六合村
构皮纸

Section 7

## Liuhe Village in Shexian County

Subject

Paper Mulberry Bark Paper in Liuhe Village of Qizili Town

## 一 歙县六合村构皮纸的基础信息与生产环境

## 1 Basic Information and Production Environment of Mulberry Bark Paper in Liuhe Village of Shexian County

六合村构皮纸的生产地位于黄山市歙县杞梓里镇六合村的庄坑村民组，地理坐标为：东经118° 43′ 46″，北纬29° 56′ 5″。六合村处于群山深处，从歙县县城徽城镇出发，经南源口村转向三阳镇方向，在杞梓里镇的苏村右拐进山，沿着昌源河曲曲折折的坡山而转，在唐里村再向左转方到达六合村庄坑组。2017年3月25日，调查组在黄山市非物质文化遗产管理负责人陈琪及县、镇、村三级"非遗文化人"的接力带领下，找到了庄坑，并对郑火土家造纸坊进行了重点访谈。

六合村位于群山环抱之中，黄子坑河流经村前，村外有竹林、茶园与梯田，村内种有多处小块连片的桑树。据调查中多位村民回忆的信息，六合村世代都有造纸的传统，至今有400多年的历史了。其业态为村内广泛分布的家庭手工造纸作坊，主要生产的品类是构树皮纸，当地称之为"棉纸"。不过到21世纪初，业态已经大大萎缩，村中最近一次造纸的信息是郑火土在2014～2015年制得二尺七寸的构树皮纸5 000张左右。

六合村由庄坑组等6个自然小村组合起来，故有此名。据六合村村主任介绍，六合村主要为郑姓氏族，在20世纪80年代有92户人家，最近几十年从附近山上陆续下来一些山民，2017年有百余户。六合村主要种植水稻、茶树、油菜等，也有部分农户养蚕、养猪，更多人选择外出打工赚钱。整个村里现有30余人掌握造纸技艺，最年轻的一位46岁，一般是50～60岁，主要为男性，但近5年内从事过皮纸生产的仅有郑火土一户。

⊙1

⊙2

⊙1 六合村外的山间公路 Mountain road alongside Liuhe Village

⊙2 六合村内的民居 Local residences in Liuhe Village

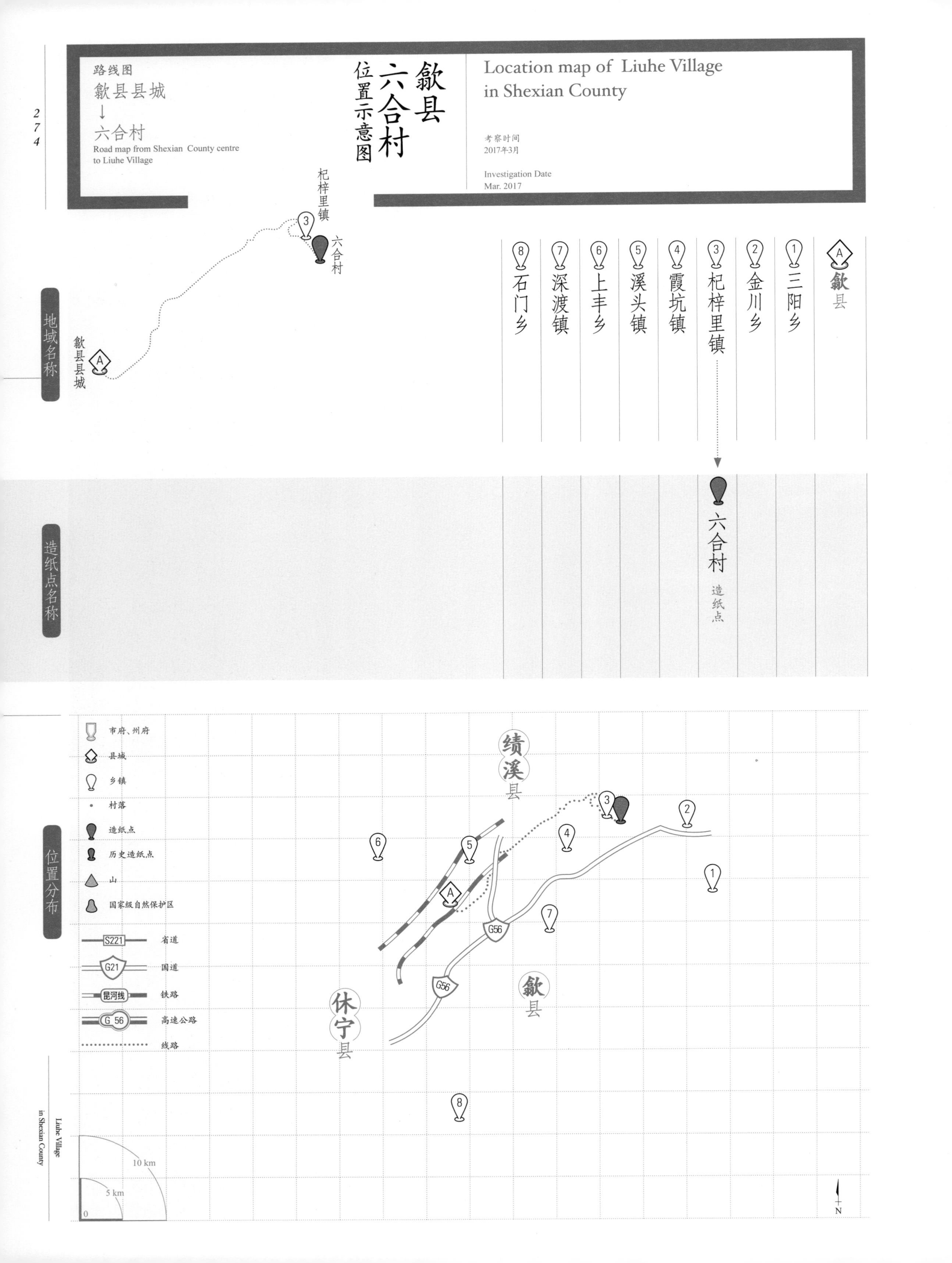

路线图
歙县县城
↓
六合村
Road map from Shexian County centre to Liuhe Village
歙县
六合村
位置示意图
Location map of Liuhe Village in Shexian County
考察时间
2017年3月
Investigation Date
Mar. 2017
杞梓里镇
六合村
歙县县城
地域名称
A 歙县
1 三阳乡
2 金川乡
3 杞梓里镇
4 霞坑镇
5 溪头镇
6 上丰乡
7 深渡镇
8 石门乡
造纸点名称
六合村
造纸点
位置分布
市府、州府
县城
乡镇
村落
造纸点
历史造纸点
山
国家级自然保护区
S221 省道
G21 国道
昆河线 铁路
G 56 高速公路
线路
绩溪县
歙县
休宁县
G56
10 km
5 km
0
N

## 二
## 歙县六合村造纸的历史与传承情况

## 2
## History and Inheritance of Papermaking in Liuhe Village of Shexian County

调查中，多位当地村民共同的说法是：六合村的造纸传统有400多年。2017年已86岁的造纸师傅郑义生老人能够清晰地回忆出，从他爷爷到现在他这三代人，都长期造过纸，更早的家族造纸历史因年代久远而失传。调查时郑金旺保存了一本《郑氏祭祀簿》。据郑金旺的记忆，祭祀簿是他爷爷按着老簿誊抄过来的，到了郑金旺已是歙县的第19世了，这样看来400多年的历史应该是有乡土记忆依据的。

根据郑火土、郑义生等人的记忆，六合村全村至少有七八十户村民都造过纸，但并非每户都独立完成造纸的所有工序，而是按村里的习俗以“换工”的形式进行工艺合作。当地人将造纸的工艺流程分为“文场”和“武场”，技术含量较高，如捞纸、晒纸，或相对比较轻松、由妇女进行的工作，如验收等，称为文场，而技术含量较低且比较辛苦的，如砍树、剥皮、煮皮、洗皮、沾石灰、踩料等，称为武场。抄纸是造纸所有工序中难度最大的一项，有人家觉得自己抄纸技术不够好，就会帮助别家进行剥皮、拣料等工作，而以换工的方式从别家请抄纸技术好的人来帮自己家抄纸。

20世纪60年代，集体经济模式的生产队造纸时期，全村按流程如打皮、踩皮、拣皮等分工，统一规划、统一安排、统一生产、统一收购，按照工作的难易和辛苦程度记不同工分，年终结算。郑金旺出生于1965年，他先做过竹编工作，后改行造纸，调查时则以茶叶的种植和加工为主业。他带着调查组成员察看了村中仍残存的旧日集中煮皮设施，那是在70年代末集体生产队建造的，当时一次可以煮500多千克皮料。据现场观察，烧火的灶台仍在，当年所用的铁锅已经完全烂掉了。

今將逐年三月十五日上北鄉
上利寺祖祠內拜掃海公墳
標掛再能公墳標掛十公墳
標掛路程開明于後
徽州府府城出北門過萬年
橋到富堨二十里沿大街出

⊙1

⊙1
《郑氏祭祀簿》（陈琪供图）
*Sacrificial Ceremony Book of the Zhengs*
(photo provided by Chen Qi)

⊙1

⊙2

⊙3

⊙4

据郑金旺回忆，当时全村在一个集中的房子里进行造纸原料的制备，该房子现已拆除，其中洗皮不是直接在河里，而是挖渠引水，以电带动大的圆形铁丝网（类似水车状），可以转动，由此进行洗皮。当时全村采取自愿原则共集合了18个纸槽，全部放置在郑家祠堂内进行集中抄纸，这些纸槽现在大多已损坏，石槽也被村民当作日用石板流失了，只有当年祠堂前那硕大的古柏树还顽强地活着，见证了庄里村造纸业的繁荣与兴衰。

⊙5

⊙6

六合村老一辈造纸工匠中的佼佼者有郑庚丁、郑仁义、郑仁山、郑仁善等人，其中抄纸

⊙1 20世纪70年代后期建造的集中煮皮设施
Facilities for boiling the bark built in the late 1970s

⊙2 放置铁锅的灶台
Kitchen range for holding the iron pot

⊙3 烧火的石头灶
Stone stove for firing

⊙4 村口老祠堂旧址前的古柏
Old cypress in front of the former ancestral hall

⊙5 郑义生（右）和郑火土（左）在郑火土家门口合影
A group photo of Zheng Yisheng (right) and Zheng Huotu (left) sitting in Zheng Huotu's doorway

⊙6 老屋墙上郑火土的岳父岳母旧照
An old photo of Zheng Huotu's father-in-law and mother-in-law on the wall

⊙7

技术最好的是郑庚丁。当访谈中问到技术好的标志是什么时，郑金旺等人表示：打的纸浆均匀，捞纸轻松，纸张厚薄平实，一般纸工达不到。村中目前年龄最大的造纸工匠是郑义生，他13岁随父学习捞纸，其家庭造纸传承谱系依次为：祖父郑灶城—父亲郑仁镇（45岁去世）—郑义生。

村里老一辈造纸工匠因年事已高不再有能力造纸，近年来还继续造纸的主要工匠为郑火土。郑火土曾用名郑惠恩，因有算命的告知命格太硬，命中缺火与土，故改为此名。

郑火土的造纸技艺传承自其岳父。一般来说，六合村传统的风俗是造纸技艺传男不传女，即使有女婿上门帮忙造纸，在关键工艺即下纸药的时候，也会有人陪着女婿喝茶，不给看。但是因为郑火土的妻弟觉得造纸太辛苦，不愿意接续这门手艺，于是老人才把造纸技艺传给了女婿郑火土。一般来说，师傅教徒弟的时候会非常严格，如果徒弟做错会被打板子。但郑火土的岳父对自己的女婿很有耐心，而郑火土也非常用心地学习，他还遵从当地“一个师傅带不出好徒弟”的风俗，到处走亲访友去学习，曾师从郑仁义、郑仁善。

郑火土的妻子郑来花也曾帮忙造纸，调查时已在外地打工。郑火土的女儿已出嫁，按照习俗没有习得造纸技艺，儿子会一些造纸的技艺，但目前没有从事和造纸相关的工作，而是在歙县开了家小饭店。

郑为民是目前村中在世的抄纸技术最好的师傅，和郑火土同岁，师从郑庚丁，也教出过四个徒弟，不过都外出打工去了，并没有留在村中继续造纸。

⊙8

⊙7 郑火土抄纸工作坊 Zheng Huotu's papermaking mill

⊙8 郑为民在自家门口展示用过的纸帘 Zheng Weimin showing papermaking screen in front of his house

# 三 歙县六合村的代表纸品及其用途与技术分析

# 3 Representative Paper and Its Uses and Technical Analysis of Liuhe Village in Shexian County

## （一）代表纸品及其用途

由于村中绝大多数造纸户均停止生产，因此选择仍处于活态生产的郑火土家为调查分析的目标对象。郑火土所造手工纸主要为构树皮纸，规格按当地的叫法有七六（长一尺六，宽一尺六）、普四（长二尺四，宽一尺四）、七三（长二尺六，宽一尺二）、七二（长二尺四，宽一尺二）和规格定制的茶箱纸。郑火土表示：现在做得最多的是七六，也有普四，七三和七二在20世纪80年代初期分产到户之前也做过，因为后来销路不好，已很久没做了。

⊙1

⊙2

⊙3

六合村构皮纸的用途主要有：

（1）书写。如写家谱、契约等。

（2）包装。如包茶叶、菊花、中药，做水泥袋的里层用纸等。

⊙1/2
郑火土展示历年所造手工纸纸样
Zheng Huotu showing handmade paper samples he made over the years

⊙3
郑火土家手工皮纸存放的地方
The container for holding handmade bast paper in Zheng Huotu's house

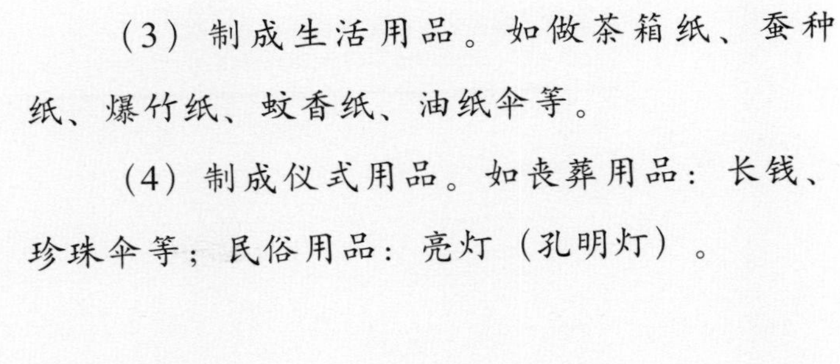

⊙4

（3）制成生活用品。如做茶箱纸、蚕种纸、爆竹纸、蚊香纸、油纸伞等。

（4）制成仪式用品。如丧葬用品：长钱、珍珠伞等；民俗用品：亮灯（孔明灯）。

## （二）郑火土家构皮纸的性能分析

经过调查访谈郑火土家造纸的实际情况，以及测试分析采样的郑火土家构皮纸，得知纸样主要由构树皮加上由杨桃藤制成的少许纸药制作而成，属于本色构皮纸。

郑火土家构皮纸的性能分析主要包括厚度、定量、紧度、抗张力、抗张强度、色度、纤维长度和纤维宽度等。按相应要求，每一指标都需重复测量若干次求平均值，通常情况下需要测量10次。调查组对采集的郑火土家构皮纸进行测试分析得到的相关性能参数见表4.10。表中列出了各参数的最大值、最小值及测量10次后所得到的平均值。

⊙5

⊙6

⊙7

⊙4 六合村现存旧日的书写用纸簿 Old writing book of Liuhe Village

⊙5 郑火土用自制手工纸做的长钱产品 Handmade joss paper products for sacrificial ceremony made by Zheng Huotu

⊙6/7 郑火土家手工构皮纸采样 Sampling handmade mulberry bark paper made by Zheng Huotu

表4.10 郑火土家构皮纸相关性能参数
Table 4.10 Performance parameters of mulberry bark paper in Zheng Huotu's house

| 指标 | | 单位 | 最大值 | 最小值 | 平均值 | 结果 |
|---|---|---|---|---|---|---|
| 厚度 | | mm | 0.138 | 0.037 | 0.086 | 0.086 |
| 定量 | | g/m² | — | — | — | 18.9 |
| 紧度 | | g/cm³ | — | — | — | 0.220 |
| 抗张力 | 纵向 | N | 10.3 | 4.8 | 7.4 | 7.4 |
| | 横向 | N | 9.3 | 4.2 | 7.2 | 7.2 |
| 抗张强度 | | kN/m | — | — | — | 0.487 |
| 色度 | | % | 39.2 | 38.1 | 38.7 | 38.7 |
| 纤维 | 长度 | mm | 7.36 | 0.52 | 2.99 | 2.99 |
| | 宽度 | μm | 73.2 | 7.9 | 18.8 | 18.8 |

由表4.10中的数据可知，所测郑火土家构皮纸最厚约是最薄的3.73倍，经计算，其相对标准偏差为0.029，差别较小。所测郑火土家构皮纸的平均定量为18.9 g/m²。通过10次数据测量，测得郑火土家构皮纸紧度为0.220 g/cm³。郑火土家构皮纸抗张强度为0.487 kN/m。

所测郑火土家构皮纸平均色度为38.7%。色度最大值是最小值的1.03倍，相对标准偏差为0.375，差别较小。

所测郑火土家构皮纸在10倍、20倍物镜下观测的纤维形态分别见图★1、图★2。所测郑火土家构皮纸纤维长度：最长7.36 mm，最短0.52 mm，平均长度为2.99 mm；纤维宽度：最宽73.2 μm，最窄7.9 μm，平均宽度为18.8 μm。

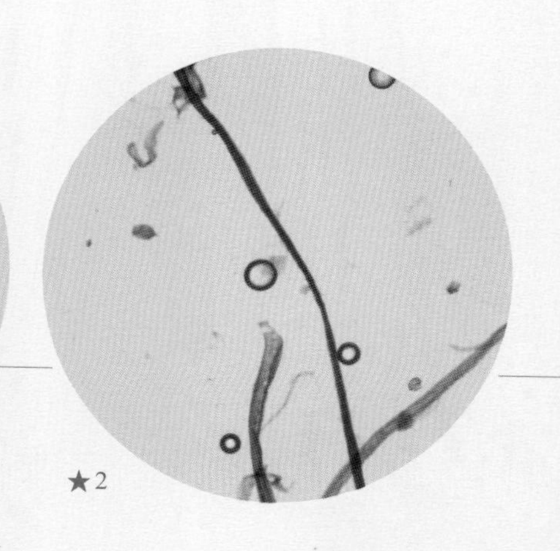

★1
郑火土家构皮纸纤维形态图（10×）
Fibers of mulberry bark paper in Zheng Huotu's house (10× objective)

★2
郑火土家构皮纸纤维形态图（20×）
Fibers of mulberry bark paper in Zheng Huotu's house (20× objective)

# 四 歙县六合村构皮纸生产的原料、工艺流程与工具设备

# 4 Raw Materials, Papermaking Techniques and Tools of Mulberry Bark Paper in Liuhe Village of Shexian County

## （一）六合村构皮纸生产的原料

### 1. 主料

郑火土家造纸所用主要原料为构树皮，当地人称为栎（音）树皮，因其果实成熟后呈红色，香甜如草莓，常被乌鸦等鸟类采食，当地人又叫老鸦皮。郑火土介绍说，歙县南乡一带常常说收老鸦皮，或把构皮纸叫老鸦皮绵纸。

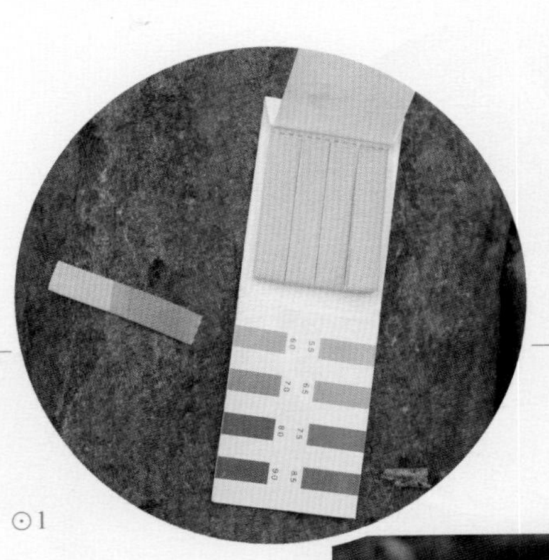

⊙1

⊙2

按照郑火土家等造纸人的说法，六合村及附近乡镇构树很多，村民会把构树枝砍下来剥皮晒干，郑火土等人会前去收购，通常一个村每年可以收购到5 000多千克干皮。其价格信息如下：20世纪60年代约8分/斤；生产队年代（70年代）约3毛（角）2分至3毛5分/斤；1982年开始分产到户后，村民造纸积极性提升，抢购造纸原料，构树皮的价格升至8毛到1块（元）/斤。而今郑

⊙1 pH实测比照 pH level test

⊙2 构树枝 Mulberry tree branch

⊙1

⊙2

⊙5

⊙3

⊙4



火土家只进行一年5 000张左右的小规模造纸，构树原料基本上都是自己采集，不再花钱收购。构皮的内料肉较厚，“取真料折头”30%～35%，即50 kg构树皮，能出造纸原料15～17.5 kg。

歙县及周边邻县自古是养蚕植桑基地，桑树皮原料充足。据郑火土回忆，在20世纪80年代，郑火土家也曾尝试以桑树皮为原料做过一年桑皮纸，到邻县绩溪、宁国、宣城等地的合作社收来桑树皮，回来自己再加工成造纸原料，但造出的桑皮纸的质量一般。桑皮纸用途与构皮纸一样，但出料率低，取真料的15%～20%，即50 kg桑树皮只能出造纸原料7.5～10 kg，不划算。尝试一年之后，郑火土家就不再做桑皮纸了。

据陪同调查的村干部和郑火土的介绍，六合村中也有其他村民家造纸时会将机制纸打碎添加作为辅料，一般会添加10%左右，不超过15%。



2. 辅料

（1）纸药。郑火土家造纸所用纸药，春冬两季以杨桃藤（猕猴桃藤）扎捆泡水为主；夏秋两季以芦皮（当地方言发音）汁水为主，芦皮即青桐（中国梧桐）树的叶、枝、杆，剥皮后泡水用。但天太热的时候，芦皮汁水很容易失效，这时便会用山苍子叶泡水1个月左右出汁作为纸药使用，或者与前两种纸药混合用。

（2）水源。郑火土家造纸水源为村口的黄子坑河，实测其pH约为6，呈偏酸性。

⊙1 未加工处理的干构树枝条 Unprocessed dry mulberry tree branches

⊙2/3 处理成造纸原料的构树皮 Processed Mulberry bark for papermaking

⊙4 六合村里种的桑树 Mulberry tree in Liuhe Village

⊙5 黄子坑河 Huangzikeng River

⊙6

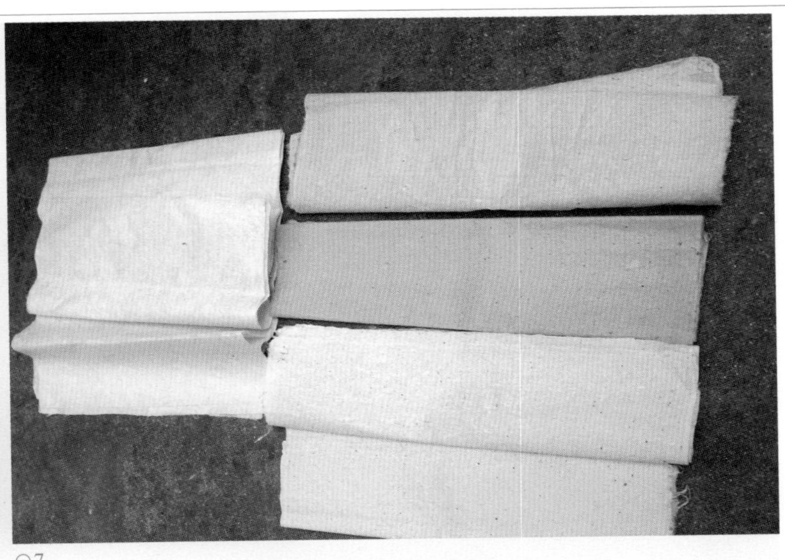

⊙7

（二）

## 六合村构皮纸生产的工艺流程

由于调查时正处于郑火土家造纸的歇业阶段，因此部分工艺工序根据郑火土等造纸人的叙述提炼而成。郑火土家造纸的工艺流程为：

| 壹 | 贰 | 叁 | 肆 | 伍 | 陆 | 柒 | 捌 | 玖 | 拾 |
|---|---|---|---|---|---|---|---|---|---|
| 浸料 | 煮料 | 踩料 | 粘皮 | 二次煮料 | 洗皮 | 煮料 | 揉皮 | 拣皮 | 打皮 |

| 拾壹 | 拾贰 | 拾叁 | 拾肆 | 拾伍 | 拾陆 | 拾柒 | 拾捌 | 拾玖 |
|---|---|---|---|---|---|---|---|---|
| 泡皮 | 袋皮 | 划槽 | 打药 | 捞纸 | 榨纸 | 晒纸 | 揭纸 | 检纸 |

⊙6/7
郑为民展示自家所造加有机制纸辅料的纸张
Zheng Weimin showing his paper made with machine-made paper materials

## 壹 浸料

1

将原料树皮在沤池里浸泡24小时。

## 贰 煮料

2

下锅煮24小时直至熟透。

## 叁 踩料

3

放在浅水潭里用牛圈踏，起到脱壳作用，然后继续人工用脚踩皮。

## 肆 粘皮

4

用石灰粘皮，堆放1周左右。

## 伍 二次煮料

5

石灰处理之后的构树皮再入锅煮24小时。

## 陆 洗皮

6

石灰皮煮好后，泡在水中，把皮料中的杂质洗净，晾干，进行做料。

## 柒 煮料

7

皮料晾干后，将豆萁灰用炭烘三天三夜后加入锅中，或石碱按比例放锅内再次煮透，以达到漂白去污的目的。煮过后再晾干。

## 捌 揉皮

8

对晾干的原料人工用脚揉皮。

## 玖 拣皮

9

拣枝，拣皮钉，拣皮壳。

⊙1

## 拾 打皮

10

用水浸泡1～2天，捞起皮料，把皮料捣打成泥膏状，使皮料中的纤维能够交织成具有一定强度的纸页。打皮是人工造纸操作中最繁重的一道工序，最早用人工脚碓，后来用像水车一样的水碓，现在已用电动打浆机替代。

## 拾壹 泡皮

11

将打好的皮料在水中浸泡。

## 拾贰 袋皮

12

将打好的皮料装进袋子，在河里清洗，用手揉搓或脚踩，除去皮料中各种杂质污质。

## 拾叁 划槽

13

将皮料放入纸槽内，用槽棍搅动形成均匀悬浮状。

## 拾肆 打药

14

加入纸药。

⊙1 电动打浆机
Electric beating machine

## 拾伍 捞纸

15 ⊙2～⊙4

即抄纸，在当地称为“淘浪”（音）。郑火土介绍，他的捞纸动作只有前后晃动，没有左右晃动。若药水（纸浆与纸药混合液）做得特别好，则下去约3 cm就可以，好捞、好晒，纸也特别均匀。一般则要深入下去10～16 cm。通常一槽可以捞100～120张六尺纸，需制作好的精料1.5～2 kg、纸药1.5～2.5 kg。

⊙2

## 拾陆 榨纸

16 ⊙5

捞一天纸，压榨一次，当天捞的纸当天压榨，每次约12小时，第二天晒纸。压榨技法要点：盖上木板后，先慢后快，前10分钟左右加1块不太大的压纸石，到3～5小时后，水榨得差不多了，就可以放上大石头了。

⊙5

## 拾柒 晒纸

17

不用火墙，而选择天气较好的时候，在房屋外墙上自然晒干。

## 拾捌 揭纸

18

将晒干的纸从墙上揭下来。由于六合村传统民居砌房屋墙体的青砖具有吸水性，一般20分钟就干了，现在的水泥砖墙则需要更长时间。如果下雨就要赶紧揭下来。

## 拾玖 检纸

19

逐张检验纸张的光滑度和完整度。

⊙3

⊙4

郑火土表示：自己年轻的时候，从早上5点捞纸到下午5点，如果捞不完，要挂马灯，没有灯就用山上采来的松明子，最多一天可以捞1 100张，现在年纪大了，每天只捞500～600张。

⊙2/4 郑火土演示并讲解捞纸动作
Zheng Huotu showing and explaining papermaking procedures

⊙5 榨纸床
Bed for pressing the paper

## （三）六合村构皮纸生产的工具设备

### 壹 打浆机 1

1982年，在生产队解散实施分产到户以后，郑火土去别人家收购了两个已经损坏不用的打浆机，又去买了一些零件，自己拼合成一个打浆机，至今仍可使用。除了打纸浆，也打过饲料。

⊙1

### 贰 脚碓 2

在调查中发现以前所用的脚碓石材很好，纹路仍在。因其功能已被打浆机替代，现已被砌入墙中作为建房材料。

⊙2

⊙3

### 叁 舂臼 3

图⊙4中舂臼一是放在图⊙5宅院中的，家中女人晚上在家里就可以舂料，为第二天做准备。舂臼一尺寸为：外径45 cm，内径33 cm，高58 cm。舂手头部是生铁，底下是木头。

图⊙6中舂臼二则在宅院外抄纸的工坊（现已闲置废弃）旁，方便临时加工处理。

⊙4

⊙5

⊙6

⊙7

⊙1 郑火土自己拼合出的『家造』打浆机
Zheng Huotu's home-made beating machine

⊙2/3 砌入墙中的脚碓石
Pestle stone in a wall

⊙4 舂臼一
Mortar (1)

⊙5 存放舂臼一的宅院
A courtyard for holding beating mortar (1)

⊙6 舂臼二
Mortar (2)

⊙7 存放舂臼二的造纸坊（已闲置废弃）
Abandoned papermaking mill for storing mortar (2)

## 肆 纸槽 4

实测郑火土家纸槽的材质为水泥，规格为：外径长138 cm，内径长124 cm；外径宽103.5 cm，内径宽88.5 cm；内径高46.5 cm，外径高54.5 cm。

⊙8

六合村现存最古老的纸槽距今已有120多年的历史，是村中郑圣禄老人（郑火土的叔叔）的母亲方嫚娣的陪嫁，采用的是远近闻名的王家店石料，号称千年不碎不裂，现已闲置，但仍能够使用。实测其规格为：外径长142 cm，内径长129 cm；外径宽105.5 cm，内径宽93 cm；内径高43 cm，外径高51.5 cm。

⊙9

⊙10

## 伍 纸帘 5

郑火土家现在所用的纸帘是20多年前从浙江余杭定制的，以家蚕丝线与竹篾制成。郑火土回忆：旧日没有电话的时候，都是写信告知对方不同规格，需要七六、普四、茶箱纸帘各多少张等，当时（1995年左右）七六的纸帘价格为130元/张，普四的为160元/张，现在价格均上涨至千余元。

⊙11

⊙12

一张纸帘可捞3万～4万张纸，其中帘额要自己装，因为每个人操作的技法不同，帘额也稍有区别，如果不合适，则搞不干净，以后晒纸很困难。郑火土装帘额一般会两边紧一点，中间松一点。实测纸帘规格为：长64 cm，宽55 cm；帘额规格为：长66 cm，宽2.6 cm。

⊙13

⊙8 郑火土家现用纸槽 Zheng Huotu's papermaking trough currently in use

⊙9/10 六合村现存最古老的纸槽 The oldest papermaking trough in Liuhe Village

⊙11 郑火土家所用纸帘 Zheng Huotu's papermaking screen

⊙12 郑火土手捏处为帘额 Zheng Huotu holding the top of papermaking screen

⊙13 六合村其他村民家的纸帘 Papermaking screens from other villagers' houses in Liuhe Village

## 陆 帘夹 6

实测郑火土家所用的帘夹规格为：长54 cm，宽7 cm。共2个，背面写有“郑惠恩办于一九八三年十月”。

⊙14

⊙15

## 柒 帘床 7

郑火土家所用的帘床规格为：长64.5 cm，宽56 cm。

⊙16

⊙17

## 捌 压纸台 8

郑火土一般在压纸台底面上放三层已破旧不能用的纸帘。

⊙18

⊙19

⊙20

## 玖 纸刷 9

材料为棕毛。晒纸时，用纸刷将纸刷在晒纸墙上。

⊙21

## 拾 纸凳 10

晒纸时，存放待晒纸的支架。

⊙22

⊙14 帘夹正面 The front view of the papermaking screen holder

⊙15 帘夹反面 The reversed view of the papermaking screen holder

⊙16 帘床 Frame for supporting the papermaking screen

⊙17 帘床、纸帘、帘夹放置图 Papermaking screen and its supporting frame and holder

⊙18 压纸木板 Board for flattening the paper

⊙19 压纸石 Stone for pressing the paper

⊙20 闲置的木榨构件 Unused wooden parts

⊙21 纸刷 Brush

⊙22 纸凳 Chair for holding the paper

## 五
## 歙县六合村构皮纸纸坊的市场经营状况

## 5
## Marketing Status of Mulberry Bark Papermaking Mill in Liuhe Village of Shexian County

在集体经营年代（20世纪60～80年代），由于造纸是技术活，比一般农活的工分高，所以村民参与造纸的干劲十足，而且造纸属于副业，可以吃商品粮，可以去县里看病。如果家里小孩多，饭不够吃，也可以先跟生产队借一些钱，第二年多干活还上。

据村里老人回忆，生产队时期一般来说六合村每年可造纸约120万张，由县里的土产公司订立合同负责收购。当年汽车已经可以通到苏村，六合村村民将制作好的纸挑到苏村，5 000张纸打一件，上下垫板用铁丝捆住，一挑共1万张纸，一般选择天晴的时候运输。力气小的村民，挑七六规格的纸，1万张约40 kg；力气大的村民，挑普四规格的纸，1万张62.5～65 kg；而茶箱纸则更重。

⊙23

⊙24

⊙25

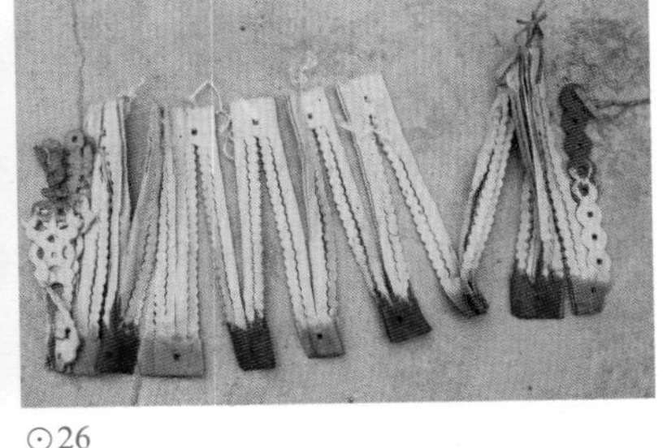
⊙26

村民并不特别清楚自己做的纸会被卖到哪里，土产公司对此保密。村民听说自己做的纸远的卖到山东、福建，近的卖到宁国、泾县和旧徽

⊙23 郑火土演示长钱的做法 Zheng Huotu showing how to make joss paper

⊙24 做纸扎的工具 Tools for making joss paper

⊙25 郑火土已经做好的纸扎 Completed joss paper made by Zheng Huotu

⊙26 郑火土用自制手工纸做的长钱 Joss paper made by Zheng Huotu

州的"一府六县"，如渔梁茶厂以前每年用纸40万张。六合村的纸一般用来包茶叶、菊花、中药，也做水泥袋的里层、茶叶箱、蚕种纸、爆竹纸等。

1976年左右，全村造纸数量达到巅峰，达到230万张。后来造纸转变为个体经营，也红火过一阵。1983年，蚕种场要做1万个蚕种，就需要1万张纸。90年代，随着机制纸的出现，以低价格冲击了手工纸的市场，手工纸的销路越来越差，村民们也不愿意继续造纸，而选择外出打工。

调查时，村里几乎只有郑火土一户近年来还坚持手工造纸，一年生产约5 000张构皮纸，最近一次造纸是2014～2015年。郑火土讲述道：近年来曾卖给过开蚊香场的曹惠生做蚊香包装纸，也曾有做油纸伞的艺人跟他买过几百张。2015年，泾川有人来买了360张纸，据说是用来放亮灯（孔明灯）。郑火土表示：现在坚持造纸不完全是为了赚钱，也为了满足当地传统习俗上对手工纸的少量需求。

郑火土曾经花了2 000多元去学了做丧葬仪式用的纸扎手艺，现在所做手工纸也多用来做丧葬仪式用品，如长钱、珍珠伞等。郑火土还到杭州参加过纸扎比赛，所做的观音衣（阴历三月十九日拜观音用），本来660元，后来涨到1 200元，曾卖出去18副。

制作一杆长钱需要3～5天，2杆算一副，一般是死者的亲属购买。3.6万个长钱耗60张纸，卖出价格为96元，大小为九尺八寸；6.6万个长钱耗80张纸，卖出价格为130元，大小为一丈八寸，耗时3天；9.6万个长钱耗88张纸，卖出价格为167元，大小为一丈二尺八，耗时3天。

⊙1

⊙2

长钱需染色，染色要放酒，白酒度数越高越好，起到固色作用。制作一把珍珠伞需要130张纸，因为成本原因，不全用自制纸，也用买的机制纸，短的珍珠伞成本约为30元，售卖价格为60元。用自制手工纸可以做得更长，但价格也更贵，销售不太好。

⊙4

⊙3

⊙1／2
长钱打开后的形态
Unfolded joss paper

⊙3
郑火土用机制纸做的珍珠伞
Pearl umbrella made of machine-made paper by Zheng Huotu

⊙4
珍珠伞打开之后的形态
Unfolded pearl umbrella

# 六
# 歙县六合村构皮纸的相关民俗及文化事项

## 6
## Folk Custom and Culture of Mulberry Bark Paper in Liuhe Village of Shexian County

### 1. 放亮灯（孔明灯）

郑火土回忆村里老辈人的说法：以前，六合村当地民俗有用所造纸制成亮灯，在正月元宵时节去高山上放出图吉利的说法。山里人放亮灯有个讲究，亮灯要扎得大小适中，灯要放得出去，飞得越高越远，福气越好越吉利。如果亮灯没有飞走，近地就落了下来是不吉利的，因此，必须重新稍扎，重新放飞，直到飞得看不见才好。因此，放亮灯应该在地势开阔的高山处，而六合村的手工棉纸轻，飞得高，受到人们的好评，歙县大部分的乡镇到这里来买。现在因为怕造成山林火灾，不允许再放。在问起以前不怕火灾的原因时，郑火土解释说，以前没有煤、沼气，也没有电，山上的小丛树木、杂草都砍掉当柴烧了，很难有山火，现在杂草太多确实易造成山火。

⊙5

⊙6

⊙5/6 村中的古巷 Old alley in the Liuhe Village

### 2. “佛”字纸符

行走在六合村中的古巷，调查组成员发现两旁的民居墙上，凡是有门有窗的地方都会贴上一张巴掌大小的土纸，纸上写有“佛”字。村民说“老”了人出丧时，灵魂会到处游荡，为避邪他们在有洞孔的地方贴上写有“佛”的纸符，可以避免邪魅侵入，同时也是为了亡魂早日找到归处。

3. 做纸扎的行规与禁忌

郑火土谨遵行规，做纸扎只赚取基本的原料和很少的手工费，绝无暴利，认为这是积阴德的事情，而且一般到了腊月三十就不做纸扎了，过年期间图个吉利，正月也要到十八以后才再开始做。一般地方做冥钱上面的孔大多为单数，而郑火土这里则用双数。

4. 收纸趣闻

郑火土回忆：听上一辈造纸老人说，生产队的时候，有时候县里土产公司库存纸很多，而生产队还要再送几十万张做好的纸，负责收纸的老宋便不太乐意，会表示拒收，这时队里就会给老宋送些村里的火腿等土特产，请他继续帮忙收纸。

5. 纸绳“拔河”

为了展示纸的结实程度，郑火土邀请调查组成员汤书昆教授和他现场使用由纸做成的绳子进行“拔河”，两人都花了很大气力，才将纸绳扯断。

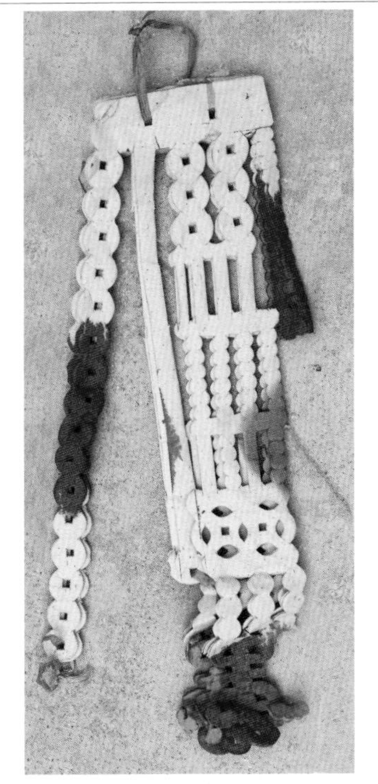

⊙1

⊙2

⊙1 双数孔纸钱 Joss paper with even numbers of hole

⊙2 纸绳『拔河』 A tug of war using paper rope

# 七
# 歙县六合村构皮纸的发展问题与未来思考

## 7
## Problems of Development and Thoughts on Future of Mulberry Bark Paper in Liuhe Village of Shexian County

根据调查和访谈的田野信息，目前六合村手工皮纸因销路较窄，收益不高，难以成为村民谋生的主业，村民造纸的积极性很低，数十年前多达80余户的造纸繁荣局面早已消失，只有郑火土一户仍在坚持造纸。而且郑火土多次表示：现在坚持造纸不完全是为了赚钱，也是为了满足当地传统习俗上对手工纸的少量需求，反正目前自己在村里也有空闲的时间、技艺和精力。

按照郑火土的说法，手工造纸已不仅仅是技艺的传承、谋生的手段，也是对六合村传统手工文化的一种守望，更是对一种渐行渐远的农耕文化的“乡愁”。

六合村的造纸工匠、技艺、工具仍基本完好，可以通过提高和改进制作技艺生产出更高档次的书写纸拓展市场。此外，距离六合村不远的坡山村是有名的摄影基地，每年前来拍摄自然景色的游人络绎不绝，或许也可以将六合村的手工造纸工艺整合成为一种体验式旅游资源，打造成为一体化区域经济。

⊙4 ⊙3

⊙3 郑火土 Zheng Huotu

⊙4 六合村附近的油菜花与山水空间 Landscape of yellow rape flower near the Liuhe Village

# 歙县六合村构皮纸

Paper Mulberry Bark Paper in Liuhe Village of Shexian County

郑火土造构皮纸透光摄影图

A photo of paper mulberry bark paper made by Zheng Huotu seen through the light

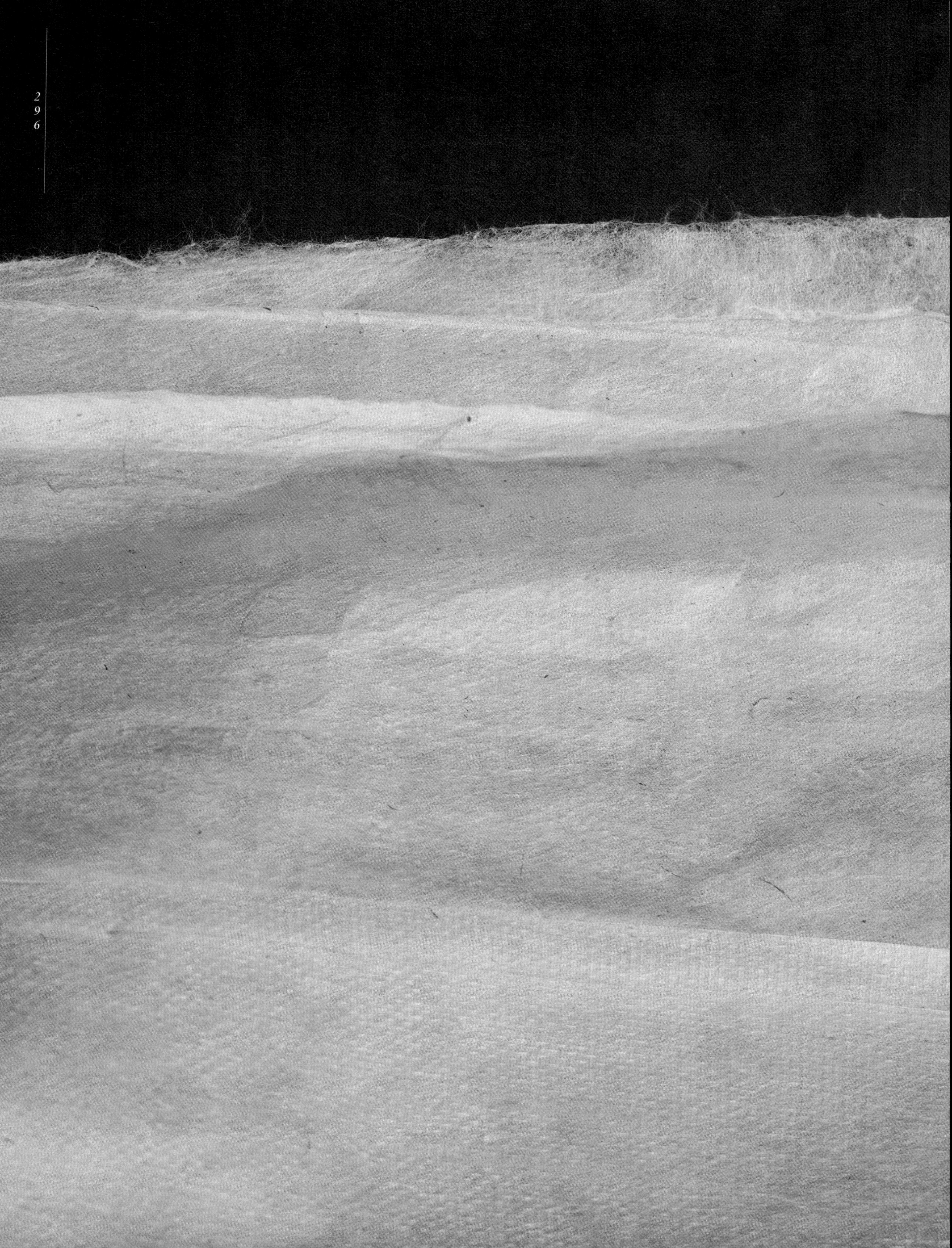

# 六合村构皮纸

Paper Mulberry Bark Paper in Liuhe Village of Shexian County

郑火土造染色构皮纸透光摄影图

A photo of dyed paper mulberry bark paper made by Zheng Huotu seen through the light

# 第五章 竹纸

# Chapter V Bamboo Paper

# 第一节

# 歙县青峰村

安徽省
Anhui Province

黄山市
Huangshan City

歙县
Shexian County

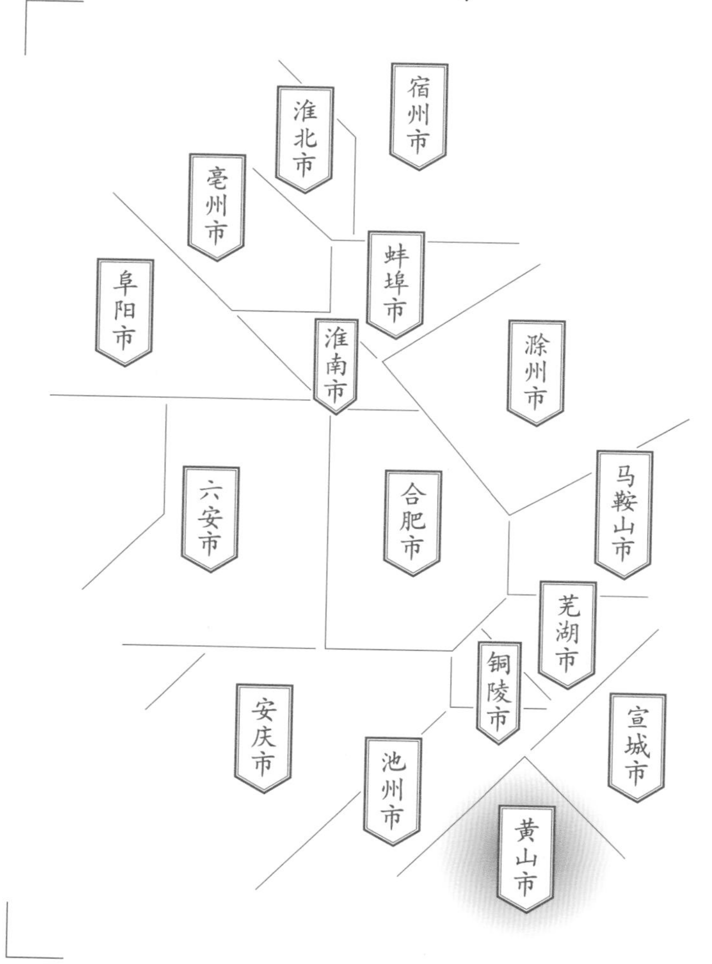


Section 1

## Qingfeng Village in Shexian County

**调查对象**

石门乡
青峰村
竹纸

Subject

Bamboo Paper in Qingfeng Village of Shimen Town

# 一
# 歙县青峰村竹纸的基础信息与生产环境

1
Basic Information and Production Environment of Bamboo Paper in Qingfeng Village of Shexian County

2017年1月22日，调查组前往黄山市歙县调查歙县当地竹纸。此次调查的造纸户名为程忠余，目的地为安徽省黄山市歙县竹岭行政村的青峰自然村，地理坐标为东经118°23′42″，北纬30°04′29″。据程忠余描述的基础信息，歙县青峰村近10年来已经不再生产竹纸，村内只有几位老人曾经从事过竹纸生产。

⊙1

⊙2

⊙1 村里的主干道路 Main road of the village

⊙2 通往程忠余家的土路 Dirt road to Cheng Zhongyu's house

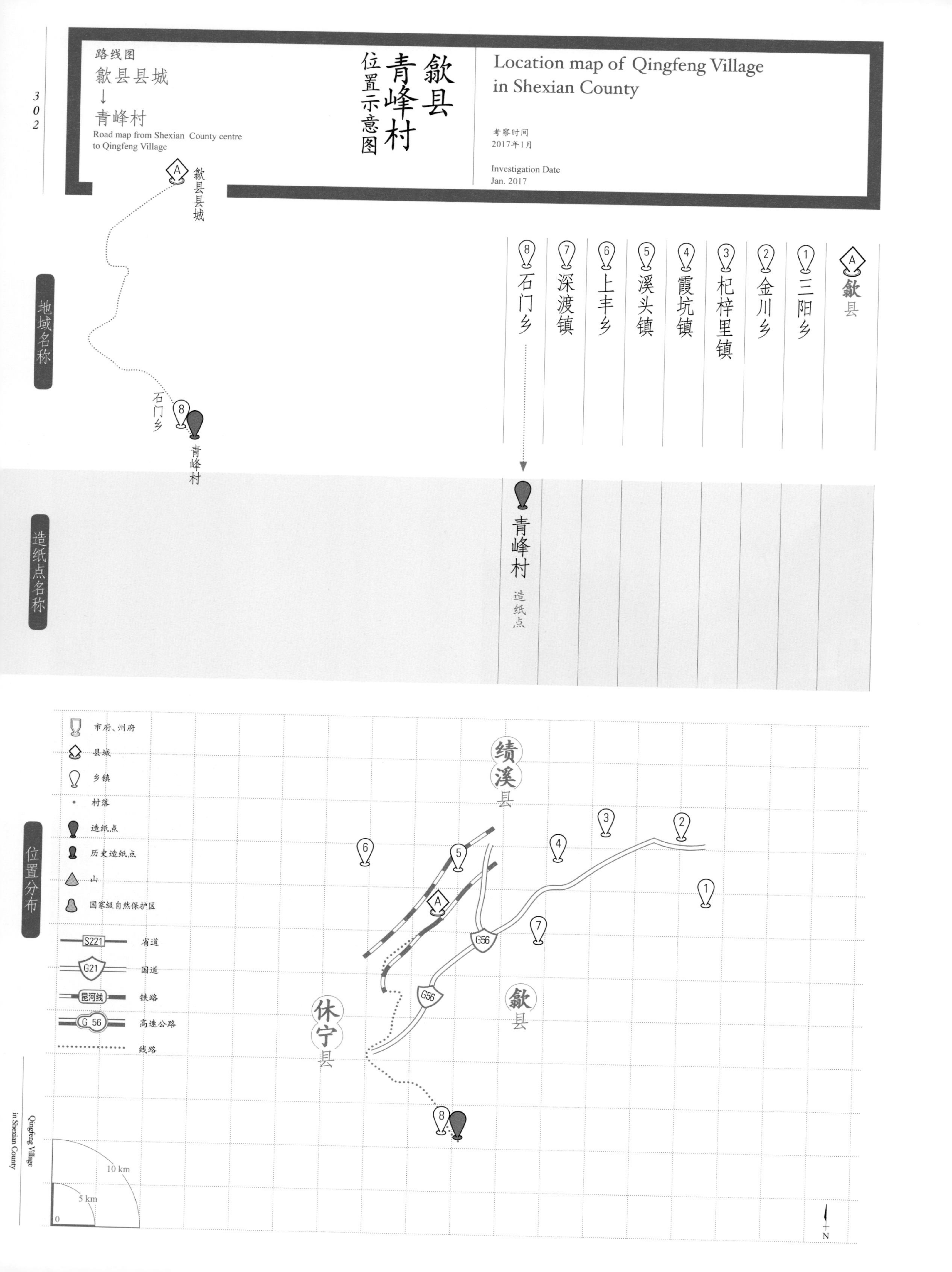

路线图
歙县县城
↓
青峰村
Road map from Shexian County centre to Qingfeng Village
歙县
青峰村
位置示意图
Location map of Qingfeng Village in Shexian County
考察时间
2017年1月
Investigation Date
Jan. 2017
歙县县城
石门乡
青峰村
地域名称
A 歙县
1 三阳乡
2 金川乡
3 杞梓里镇
4 霞坑镇
5 溪头镇
6 上丰乡
7 深渡镇
8 石门乡
造纸点名称
青峰村
造纸点
位置分布
市府、州府
县城
乡镇
村落
造纸点
历史造纸点
山
国家级自然保护区
S221 省道
G21 国道
昆河线 铁路
G 56 高速公路
线路
绩溪县
休宁县
歙县
G56
10 km
5 km
0
N
Qingfeng Village in Shexian County

## 二 歙县青峰村竹纸的历史与传承情况

## 2 History and Inheritance of Bamboo Paper in Qingfeng Village of Shexian County

青峰村以程姓为主体，属历史上著名的徽州篁墩程氏（程朱理学的二程家族）后裔，调查中所见村内现存程氏族谱载：“始祖有大经济之才，而性爱幽雅，厌城之喧嚣，喜山林之静寂，见此地为骡驼形，有骡驼献宝之吉祥风水，便由忠溪迁居此地。”青峰村原名宋家坑，明代以前就有宋姓人民居住，程氏迁入后逐渐强盛，宋氏渐渐没落与消失，但交替过程发生在什么时间和有什么原因，访谈中程姓老人们也语焉不详。

⊙1

据程忠余口述的信息，整个青峰村4个村民组一共有200多户，共有900多人，从事过造纸的不超过20人。程忠余2017年初70岁，所居住和造纸的小村当地称为外山。

⊙1 当地称为外山的造纸村 A papermaking village called Waishan by the locals

⊙1

程忠余口述的该村造纸起源与传承信息如下：这个村的造纸在1949年前就开始有了，是一位从温州逃避抓壮丁而逃难过来的造纸师傅将技术带过来的。造纸师傅名叫朱同和，儿子叫朱联志，前几年朱联志还在家中纸坊造纸，现在已经不造纸了。当地的造竹纸技术都是由朱同和这一脉系传授的，是单一来源输入的技艺发育形态。村里鼎盛时期4个生产组共有造纸槽8个，程忠余所在的二组生产力最高，有4个槽，现在还剩1个槽处在半停产状态，由程忠余的堂弟在使用。

程忠余的造纸技术是直接向朱同和学的，属于嫡传徒弟。调查中程忠余对自己的技术很自豪，他向调查组成员们讲了一个故事：老师傅朱同和年纪大了不造纸之后，程忠余主动向师傅请求，将师傅的纸槽接管过来进行造纸，第一天进行生产，就捞了700张纸，后面几天开始每天增加100张的产量，最后就和师傅产量齐平，每天在1 000张左右。

程忠余家共兄弟三人，老大陈忠余和老三从事造纸行业，程忠余的妻子汪玉琴只会晒纸工艺，老二未学过造纸。

⊙2

⊙3

⊙1
废弃的石头老纸槽
Abandoned stone papermaking trough

⊙2
程忠余回忆造纸往事
Cheng Zhongyu recalling the past stories of papermaking

⊙3
程忠余向调查组展示前些年造的纸
Cheng Zhongyu showing the paper made in previous years to the research team

## 三
## 歙县青峰村的代表纸品及其用途与技术分析

## 3
## Representative Paper Its Uses and Technical Analysis of Qingfeng Village in Shexian County

### （一）
### 代表纸品及其用途

据程忠余的说法，当地竹纸主要有刷红纸、表芯纸、五色纸三种。

刷红纸是指染成红色的竹纸，都是自己生产纸张自己染色。刷红纸主要用来包装食品，特别是当地称为“红纸包”的麻酥糖。由于纸张加工后主要用于包装食品，所以会加入明矾以起到防潮的作用。

⊙4

⊙5

⊙6

表芯纸又称黄表纸，因颜色多为偏黄的竹纸本色，故名，多用于祭祀祖先、供奉神佛，以及用作卫生用纸，青峰村一带的另一传统用法是常用于卷旱烟。

据程忠余所述，五色纸有红、蓝、黄、绿、紫五种颜色，主要用于写标语和迷信用纸，迷信用纸的用途有祭祀时的燃烧祭奠、给过世的人做寿衣等。

### （二）
### 代表纸品的性能分析

歙县青峰村竹纸的性能分析主要包括厚度、定量、紧度、抗张力、抗张强度、色度、纤维长度和纤维宽度等。按相应要求，每一指标都需重复测量若干次求平均值，通常情况下需要测量10次。调查组对采集的歙县青峰村竹

⊙4 刷红纸 Shuahong paper

⊙5 表芯纸原纸 Raw paper of Biaoxin paper

⊙6 写着乡民贺礼金额的五色纸 Five-color paper recording the gift money amount of the villagers

纸进行测试分析，得到其相关性能参数，见表5.1。表中列出了各参数的最大值、最小值及测量10次后所得到的平均值。

表5.1　歙县青峰村竹纸相关性能参数
Table 5.1　Performance parameters of bamboo paper in Qingfeng Village of Shexian County

| 指标 | | 单位 | 最大值 | 最小值 | 平均值 |
|---|---|---|---|---|---|
| 厚度 | | mm | 0.255 | 0.086 | 0.132 |
| 定量 | | $g/m^2$ | — | — | — |
| 紧度 | | $g/cm^3$ | — | — | — |
| 抗张力 | 纵向 | N | 20.5 | 12.4 | 18.8 |
| | 横向 | N | 9.0 | 5.5 | 7.5 |
| 抗张强度 | | kN/m | — | — | — |
| 色度 | | % | 22.6 | 22.4 | 24.7 |
| 纤维 | 长度 | mm | 3.63 | 0.12 | 0.80 |
| | 宽度 | μm | 47.0 | 3.0 | 12.2 |

由表5.1中的数据可知，所测竹纸最厚约是最薄的2.97倍，经计算，其相对标准偏差为0.047。所测歙县青峰村竹纸的平均定量为48.3 g/m²。通过10次数据测量，测得歙县青峰村竹纸紧度为0.366 g/cm³，纸抗张强度为0.877 kN/m。

所测歙县青峰村竹纸平均色度为24.7%。色度最大值是最小值的1.01倍，相对标准偏差为0.098，差别较小。

所测歙县青峰村竹纸在10倍、20倍物镜下观测的纤维形态分别见图★1、图★2。所测歙县青峰村竹纸纤维长度：最长3.63 mm，最短0.12 mm，平均长度为0.80 mm；纤维宽度：最宽47.0 μm，最窄3.0 μm，平均宽度为12.2 μm。

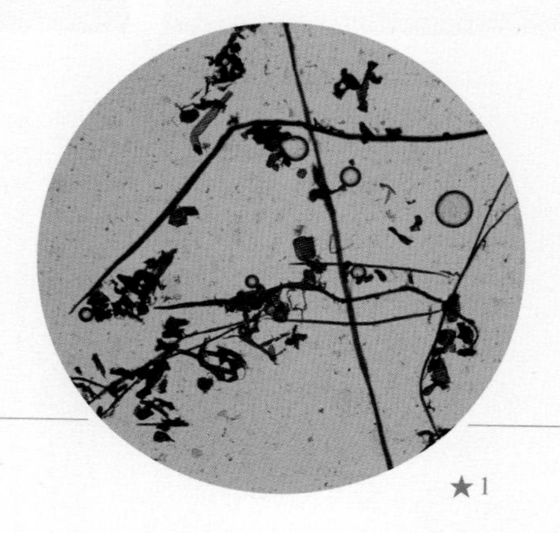

★1

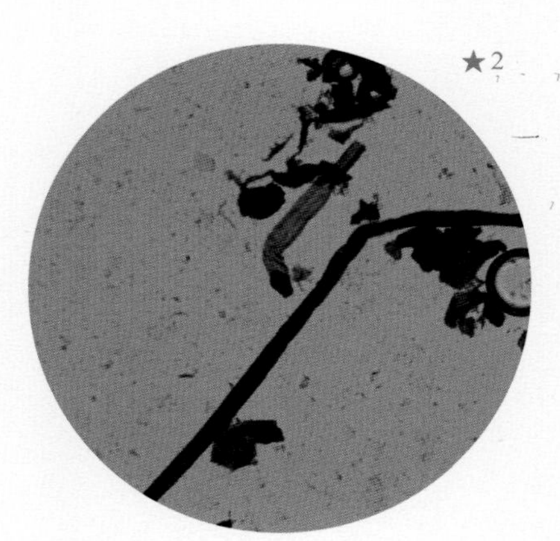

★2

★1 歙县青峰村竹纸纤维形态图（10×）
Fibers of bamboo paper in Qingfeng Village of Shexian County (10× objective)

★2 歙县青峰村竹纸纤维形态图（20×）
Fibers of bamboo paper in Qingfeng Village of Shexian County (20× objective)

# 四
# 歙县青峰村竹纸生产的原料、工艺流程与工具设备

4
Raw Material, Papermaking Techniques and Tools of Bamboo Paper in Qingfeng Village of Shexian County

## （一）
## 青峰村竹纸生产的原料

### 1. 主料

青峰村竹纸的主要原料是当地产的毛竹。据程忠余介绍，毛竹都是自己上山砍伐的当年生的嫩毛竹。

### 2. 辅料

（1）明矾、松香。只有在制作用于“红纸包”的红纸时，由于防潮的需要，在纸浆制作时才会加入明矾和松香。据程忠余介绍，两种原料均在歙县当地直接购买。

（2）纸药。青峰村造竹纸使用的纸药是当地的山苍子树的叶子，从山上采集下来后，在大锅里蒸煮出黏汁液，加入捞纸槽中与竹浆料混溶在一起，捞纸时起到使浆料在纸槽中悬浮的作用，揭纸时起到便于湿纸垛分张的作用。

（3）染料。制作刷红纸和五色纸时用于将原色纸染色。青峰村使用的染料都是从当地歙县城里的五金公司购买的，而歙县五金公司又是从黄山市府驻地屯溪区批发来的染色颜料。据程忠余向调查组成员介绍，最近几年，红色染料的价格最便宜，为90元/kg，绿色染料的价格最贵，为140元/kg，均为化工染料。

（4）水源。在竹纸的制作过程中，选用的都是当地的山溪水。

⊙1

⊙2

⊙3

⊙ 1
青峰村边的毛竹
Mao bamboo forest next to Qingfeng Village

⊙ 2
青峰村里造纸用的山溪水
Mountain stream used for papermaking in Qingfeng Village

⊙ 3
程忠余家染纸用的化工染料
Cheng Zhongyu's chemical dyes for dyeing paper

# （二）

## 青峰村竹纸生产的工艺流程

根据程忠余老人的工艺描述，综合调查组成员在纸坊实地观察到的情况，可归纳青峰村竹纸生产的工艺流程为：

| 壹 | 贰 | 叁 | 肆 | 伍 | 陆 | 柒 | 捌 |
|---|---|---|---|---|---|---|---|
| 砍竹、杀青 | 腌制、洗料 | 制浆 | 捞纸 | 压榨 | 晒纸 | 检验 | 染色 |

### 壹 砍竹、杀青

1 ⊙1

嫩毛竹的砍伐时间不能晚于毛竹发芽六七节之后，不然毛竹就会长得太老，纤维变得粗硬，不适合生产竹纸。砍伐后用刮皮刀刮去嫩毛竹的青皮和粗纤维，砍成每节1 m左右的竹段，再用木刀从中间劈成四片，捆扎起来。一捆竹料一般重25～30 kg。

⊙1

### 贰 腌制、洗料

2 ⊙2

将捆好的竹料放入塘中，加入石灰浸泡。一般一塘竹料共有3.5万～4万kg，石灰与竹料的配比在1∶10左右。

腌制约40天后，把石灰水放掉，在塘中用清水将竹料洗干净，直到竹料上没有石灰残留，一般需要用清水洗3～4次才能清洗干净。

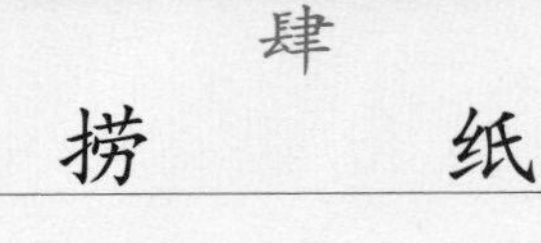

⊙2

### 叁 制浆

3 ⊙3

将获得的竹料按生产需要运到纸槽中，与山溪水以及纸药混合。纸药加多少全靠个人经验，加得多捞纸容易，但是揭纸时纸容易断裂。刷红纸纸浆制作时还要加入兑水的明矾和用开水化开的松香，一槽纸浆要加入约1 kg明矾和1～1.5 kg化开的松香。

⊙3

### 肆 捞纸

4 ⊙4⊙5

使用纸帘和帘床捞纸时，捞纸工需要用手腕晃动纸帘，将纸槽内的浆液分布到纸帘上，再将帘床左右晃动，推出多余的浆液，使纸帘上的浆液均匀分布，最后仅留一层薄薄的纸浆在纸帘上。湿纸形成后，捞纸工取下纸帘，将有纸浆的一面朝下，从离其最近的一边开始将纸帘

⊙1 程忠余的儿子演示用『木刀』将竹料劈开
Cheng Zhongyu's son showing how to split the bamboo by a wooden knife

⊙2 砍伐下来的竹料
Lopped bamboo materials

⊙3 旧日煮山苍子叶的老铁锅
Iron pot used for boiling *Litsea cubeba's* leaves

⊙4

⊙5

缓缓放置于榨纸架的榨床上，待到整个纸帘完全与纸架贴合，从离其最近的一边迅速将纸帘抬起，继续捞纸。如此反复操作，纸架上会形成越来越高的湿纸块。

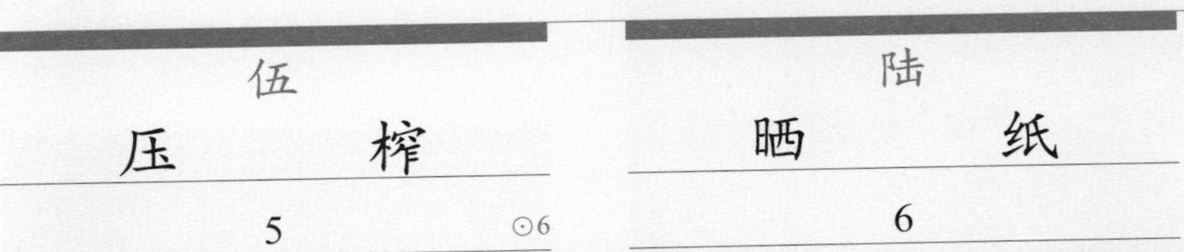

## 伍 压榨 5

⊙6

按青峰村的操作习惯，对当天捞出来的纸，到傍晚就使用压板、木榨和麻绳将其缓缓榨干，放置一个夜晚，第二天上午开始晒纸。

⊙6

## 陆 晒纸 6

将前一天晚上压榨过的纸用手逐张撕起，贴在晒纸的焙墙上，并用松毛刷在纸上迅速刷四五下，使湿纸与焙壁完全贴合。由于晒纸墙温度较高，1～2分钟后刷在墙上的湿纸就可被烘干，此时就可以一张张地将纸从焙墙上揭下了。若纸块太干，就沿着纸块的上边缘位置喷一点清水，稍微湿一点后容易撕开。晒纸过程中将破碎的纸放在一边，用于回炉打浆。表芯纸由于使用者的要求不高，不需要晒纸，只要将其放在阳光下晒干即可。

## 柒 检验 7

⊙7

把晒干的纸整理好，按100张/刀的规格数好，摆放整齐以用于染色或者直接销售。

⊙7

## 捌 染色 8

⊙8⊙9

用购买的染料将成纸染成各种颜色。值得注意的是包装麻酥糖的染色纸颜色是一面红一面白，染色时只染作为外层纸的一面。

⊙8

⊙9

⊙4/5 在半停产状态纸坊里捞纸与扣纸演示
Demonstration of papermaking procedures in the papermaking mill nearly stopped production

⊙6 半停产纸坊里的木榨
Wooden presser in the papermaking mill nearly stopped production

⊙7 数纸
Counting the paper

⊙8/9 乡民用染色纸做成的贺礼簿
Villager's Gift Brochure made of dyed paper

## (三) 青峰村竹纸生产的工具设备

### 壹 纸槽 1

盛放纸浆的工具，方形，捞纸工站在侧边工作。歙县当地的纸槽一般用松木做成，因为松木油脂大，防水性好。实测程忠余家的纸槽尺寸为：长200 cm，宽122 cm，深73.5 cm。

⊙1

### 贰 纸帘 2

捞纸工具，用于形成湿纸膜和过滤多余的水分。由细竹丝编织而成，表面刷有黑色土漆，光滑平整。实测程忠余家的旧纸帘尺寸为：长83 cm，宽57 cm。

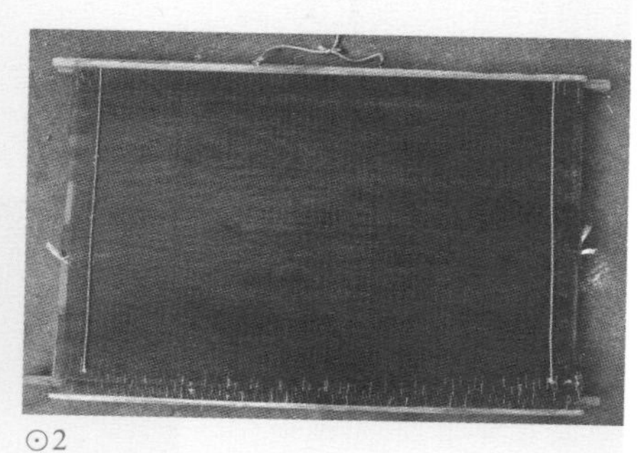

⊙2

### 叁 帘床 3

又称帘架，捞纸时放在纸帘下，起到承托纸帘和便于纸工操作的作用。

⊙3

### 肆 木刀 4

用于将砍下的竹节劈开。

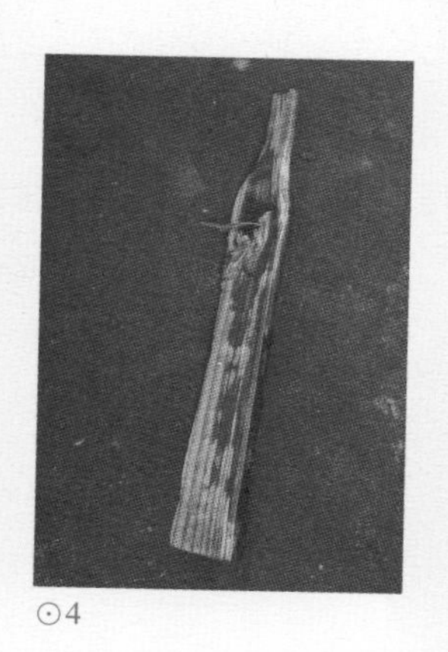

⊙4

### 伍 混料耙 5

用于搅拌纸槽中的浆料。

⊙5

⊙1 老纸坊中的纸槽 Papermaking trough in an old papermaking mill

⊙2 程忠余家的旧纸帘 Old papermaking screen in Cheng Zhongyu's house

⊙3 程忠余家的帘床 Frame for supporting the papermaking screen in Cheng Zhongyu's house

⊙4 程忠余制作的木刀 Wooden knife made by Cheng Zhongyu

⊙5 混料耙 Rake for mixing the materials

陆

## 松毛刷

6

用于晒纸时将纸刷上晒纸墙。刷柄为木制，刷毛为松毛。

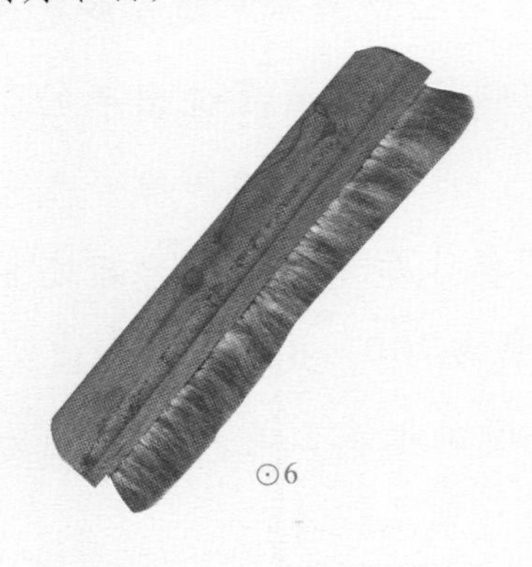

⊙6

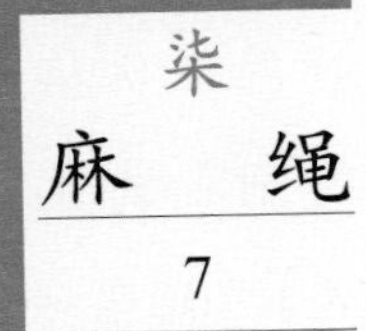

柒

## 麻　绳

7

用于压榨。

⊙7

捌

## 盖　板

8

压榨时压在纸块上的木质工具。

⊙8

玖

## 篮　子

9

生产过程中出现的残破的纸放在篮子中，用于回炉打浆。

⊙9

拾

## 捞筋叉

10

用来捞出打料制浆时未消解的竹筋块等。

⊙10

⊙6 松毛刷 Brush made of pine needles

⊙7 压榨用麻绳 Hemp rope for pressing the paper

⊙8 压榨用盖板 Wooden board for pressing the paper

⊙9 装残破纸的篮子 Basket for carrying the broken paper

⊙10 捞筋叉 Fork for picking up the residues

# 五 歙县青峰村竹纸的造纸文化与造纸故事

# 5 Papermaking Culture and Stories of Bamboo Paper in Qingfeng Village of Shexian County

⊙1

## 1. 生产队造纸的往事

调查中，程忠余回忆了数十年前生产队集体造纸时的往事：一般好的纸制作出来1刀重3.3～3.5斤，大米1斤1角2分钱的时候，1刀质量好的纸价格是2元5分钱，那时是生产队集体所有制，为生产队工作，一个槽一天最多可以生产1 700张纸。用计工分来赚钱，每生产1 000张纸，集体会多加2元的工资。

访谈中程忠余颇显自豪地表示：现在所居住的房子就是从造纸的收益中积累建造出来的。程忠余夫妻两人配合，产量最多的年份一年一个槽可以做1 300多刀纸，后来分产到户后还生产了不到5年，年产量为700～800刀，但是与生产队时期比利润要增加很多，毛利一年约为1 000～2 000元。

## 2. 抽烟点火的竹纸媒

在访谈中，因为和调查组成员一起抽烟，程忠余兴致勃勃地表示：青峰村竹纸在旧日是非常好的点土烟叶烟的纸媒材料，并现场演示制作了数根纸媒。

⊙2

⊙3

⊙4

⊙1 程忠余招待调查组成员特色粽子餐 Cheng Zhongyu entertaining the researchers with special Zongzi meal

⊙2/4 程忠余老人做点烟的纸媒与燃纸媒 Cheng Zhongyu using paper stick to light a cigarette

# 六
# 歙县青峰村竹纸的保护现状与发展思考

# 6
# Current Status of Protection and Thoughts on Development of Bamboo Paper in Qingfeng Village of Shexian County

由于青峰村竹纸多属于乡土生活民俗用纸，对品质要求不高，因此受到机制纸冲击的影响很大，几乎造成了完全的替代。加之当地若干旧日民俗用纸需求的萎缩，在20世纪90年代，青峰村竹纸的销售价格不仅已无法随人工、材料等成本上升而上浮，而且客户和市场也越来越少。

需求下降、利润太低，加上作业辛苦、环境简陋，按照程忠余的说法，村里没有年轻人愿意再从事造纸工作，而老一辈的纸工年纪太大，没有能力维系生产，只能停产。因而到2017年1月调查组入村访谈时，青峰村的竹纸生产业态已经完全中断。虽然造纸工具、设施、老一辈技术工人都在，但传承的基础和市场出路已基本被破坏。

⊙5

⊙6

⊙5 冬天村里烤火炉晒太阳的老人们 Old villagers basking in the sun by the stove

⊙6 废弃的泡料池 Abandoned pool for soaking the materials

# 歙县青峰村竹纸

Bamboo Paper in Qingfeng Village of Shexian County

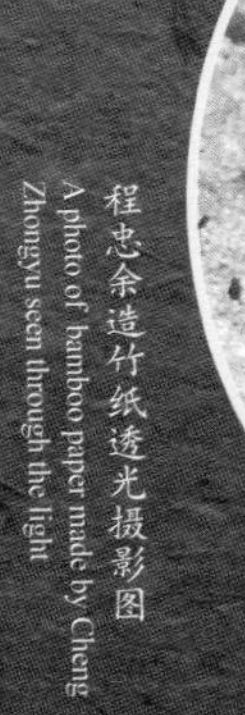

程忠余造竹纸透光摄影图

A photo of bamboo paper made by Cheng Zhongyu seen through the light

# 第二节

# 泾县孤峰村

安徽省
Anhui Province

宣城市
Xuancheng City

泾县
Jingxian County

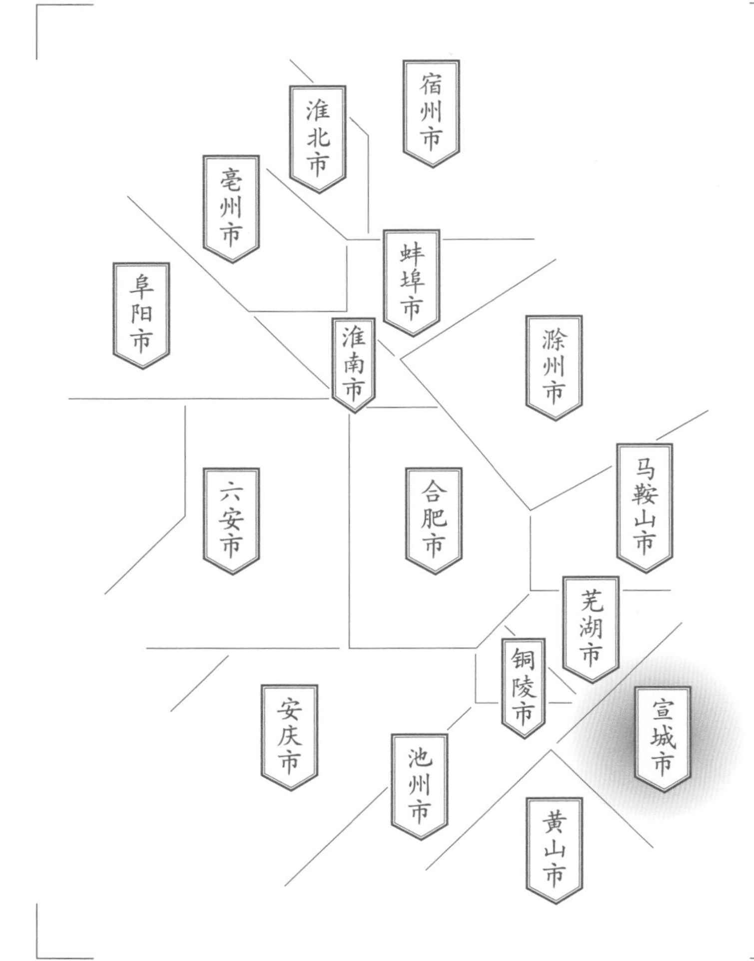




调查对象
昌桥乡
孤峰村
竹纸

## Section 2
## Gufeng Village in Jingxian County

Subject
Bamboo Paper in Gufeng Village of Changqiao Town

## 一 泾县竹纸的基础信息与分布

## 1 Basic Information and Distribution of Bamboo Paper in Jingxian County

泾县地处东经117°57′~118°41′、北纬30°21′~30°50′范围。境内山多地少，雨量充沛，光照资源丰富，竹林茂密，属于江南以茂林修竹山水景观见长的区域。泾县历史上不仅有宣纸在国内外享有盛名，而且曾经也是竹纸业态分布很广的县。

2014年3月19日，调查组在查阅相关泾县竹纸产区产地的若干文献后，来到昌桥乡孤峰行政村华侨自然村的盘坑、清水塘等竹纸产地实地考察，了解到当地手工竹纸至少已停产15年，但有不少造竹纸的遗迹尚存。同年4月，调查组又到泾川镇古坝行政村的程村组考察，在程村程家冲，发现还有腌沤竹料的麻塘旧址。根据泾县蔡村、古坝、孤峰等地的老制纸人回忆，调查组记录了泾县竹纸的部分工艺信息。

## 二 泾县竹纸制作的人文地理环境

## 2 The Cultural and Geographic Environment of Bamboo Paper in Jingxian County

考《尔雅》对字源的诠释，水直波为径，又通作“泾”。《泾县志》载：泾水湍急，流皆直下，颇无回旋，水之名泾及县，因水得名。汉武帝元封二年（公元前109年），泾县正式设置。《汉书•地理志》载：汉置泾，属丹阳郡。

泾县总面积2 059 km$^2$，自古盛产竹木，林业面积1 410 km$^2$，占土地总面积的68.5%，森林覆盖率达64.2%，绿化程度达95%以上。其中竹林面积173.16 km$^2$，毛竹蓄积量约5 000万根，年毛竹砍伐量为650万根，活木砍伐量为11.6万m$^3$。泾县的蔡村镇曾有“华夏毛竹第一镇”的乡土名号，也曾被安徽省林业厅授予“安徽竹乡”称号。

泾县竹类品种繁多，有毛竹、元竹、水竹、紫竹、红壳竹、雷竹等十几个品种，其中毛竹产量最高，元竹次之。泾县所产毛竹竹质优良、径

路线图
泾县县城
↓
孤峰村
Road map from Jingxian County centre to Gufeng Village

# 泾县孤峰村位置示意图

# Location map of Gufeng Village in Jingxian County

考察时间
2014年3月/2016年5月

Investigation Date
Mar. 2014 / May. 2016

地域名称

A 泾县
1 丁家桥镇
2 云岭镇
3 泾川镇
4 昌桥乡
5 琴溪镇
6 黄村镇

造纸点名称

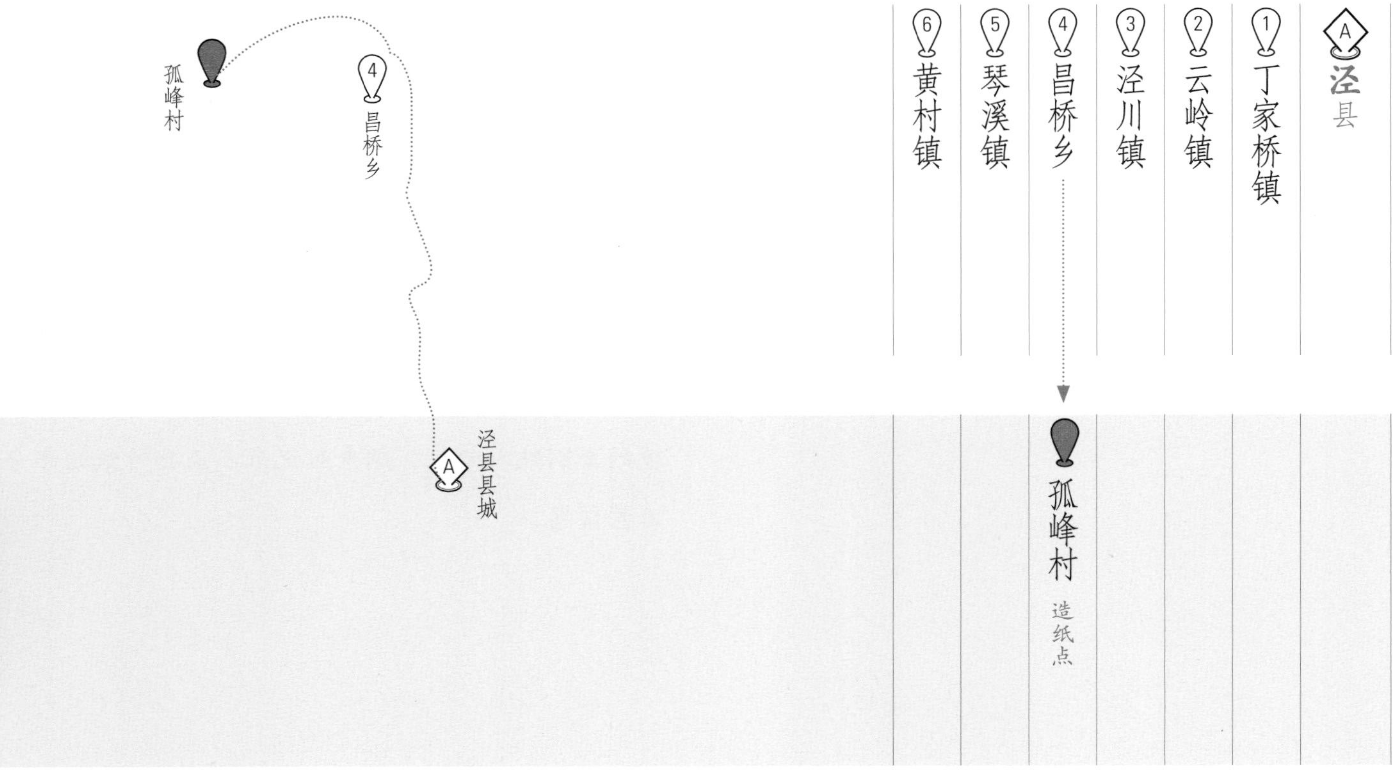

孤峰村 造纸点

孤峰村
昌桥乡
泾县县城

位置分布

市府、州府
县城
乡镇
村落
造纸点
历史造纸点
山
国家级自然保护区
S221 省道
G21 国道
昆河线 铁路
G 56 高速公路
线路

南陵县
泾县
青阳县
S322
G205
合福高铁

10 km
5 km
0
N

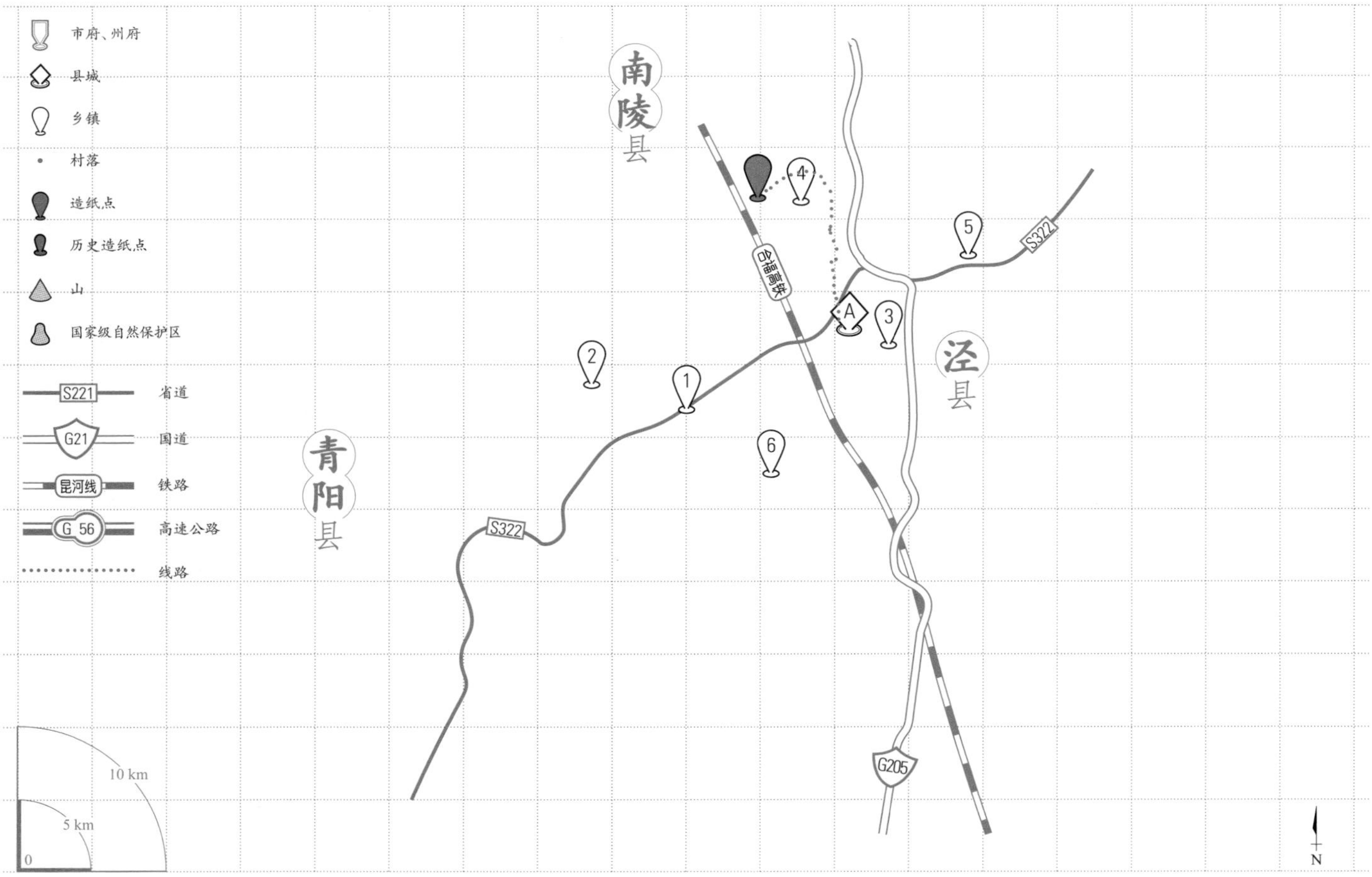

⊙1

节粗大、条杆匀称、纹理细腻，是制作团箕、竹筛等精细竹具的上佳材料，也成为泾县盛产竹纸的原料优势。调查期间，因竹纸活态生产完全中断已10余年，同时竹制品市场也大幅萎缩，原竹的砍伐量下降，蓄积量上升。2015年，全年砍伐量降至500万根以下，蓄积量升至7 939.5万根，蓄积量达到历史最高水平。





## 三 泾县竹纸的历史与传承情况

## 3 History and Inheritance of Bamboo Paper in Jingxian County

泾县产竹纸由来已久。1996年版《泾县志》[1]记载：明清之际泾县除宣纸外，土纸生产有6个品种：表芯纸（又称三六表）、干古纸、骨皮纸、黄纸、高廉纸、署纸，以表芯纸居多。清道光年间潘村、茶冲、棠叶村、巧峰一带普遍生产骨皮纸、黄纸、高廉纸等；孤峰、田坊、盘坑、梅家冲、清水塘、董家、李家塌、周家冲、殷家坑、水山坊等地主要生产表芯纸。这些纸棚槽户大多有商号。

据调查中了解的信息，泾县当地人又称竹纸为“草纸”“表芯纸”“大表纸”“媒纸”，因文献记载过略，现已无从追述其起源和最早制竹纸的历史。考察泾县竹纸的生产历史，可知泾县竹纸应属于皖南竹纸生产体系。

[1] 泾县地方志编纂委员会.泾县志［M］.北京：方志出版社，1996：215.

⊙1 泾县的山川林海 Mountains and forests in Jingxian County

最早描述泾县竹纸工艺的文献是《支那制纸业》（1878年，栖原陈政）、《清国制纸取调巡回日记》（1883年），上面记载了泾县清水塘制作竹纸的工艺信息，记载的原因可能是泾县盛产宣纸。日本的调查者在暗访宣纸技艺的同时也顺便将泾县竹纸工艺一并调查了，所以这些记载应由国外暗地收集经济技术情报而起。清水塘位于泾县昌桥镇孤峰村，其区域位于农耕社会时期九华古道泾县段的必经之路。清水塘制作竹纸信息除被日本人记录外，晚清政府对皖南造纸业也有关注，如在《科学杂志》（1908年）中载有《皖南制纸情形略》，介绍了泾县清（青）水塘制作竹纸的简单工艺信息。

《泾县志》载：民国二十三年(1934年)，全县表芯纸槽发展到386帘，从业人员1 930余人，年产表芯纸69 500担，占全县土纸产量的22%。民国二十六年（1937年）底，全县有纸棚81个，从业人员1 600余人。抗日战争开始后，时局动乱，土纸产量逐年下降。民国三十二年（1943年）五月，官办泾县中心工厂造纸设上郎坑（今古坝村），开始生产文化用纸，月产90令（每令3刀，每刀100张）。后改为泾县县营民生工厂，生产毛边纸、黄纸，后因每况愈下而停产。民国三十七年（1948年）全县土纸槽仅剩56帘，从业人员336人。

1949年中华人民共和国成立后，政府扶持槽户恢复表芯纸生产。1954年，全县有土纸槽180帘，从业人员1 100余人，年产纸700余吨。1957年，孤峰、田坊、盘坑、梅家冲、清水塘、董家、李家塌、周家冲、殷家坑、水山坊等地因区域较为集中，毛竹产量大，生产竹纸槽户多，成立了泾县建华造纸生产合作社，泾县的泉丰、潘石、茂林也建成了造纸厂，先后建成投产，受到政策促进等影响，以竹纸为主的土纸产量迅速上升。1959年全县有206帘槽，年产土纸1400 t，后因市场机制，卫生纸、文化用纸供应增加，土纸市场受到冲击，销量大减。全县土纸生产厂家先后缩小生产规模，年产量骤减。20世纪60年代末及整个70年代，每年泾县土纸产量均徘徊在200～300 t范围。80年

⊙1

⊙1 清代嘉庆版《泾县志》书影
*The Annals of Jingxian County* in Jiaqing Reign of the Qing Dynasty

⊙2

⊙3

代以后，泾县土纸产量更趋萎缩，部分区域虽有个体或农户生产，田坊、孤峰两地规模稍大的土纸生产厂家则被1972年建成的建华造纸厂统管，但全县总产量仍呈大幅下降趋势，1987年仅产176 t。

泾县在1949年后的30余年里，竹纸的生产范围非常广泛，按照调查组2014年调查期间泾县的9镇2乡行政区划，泾县昌桥乡、蔡村镇、汀溪乡、榔桥镇、泾川镇、琴溪镇、黄村镇、茂林镇等区域内均有竹纸生产，其中孤峰、田坊的竹纸最为有名，历史上规模最大，消亡的时间也最迟。在泾县20世纪50年代汇集的《泾县土纸样本》中，可以清晰地看到泾县所产竹纸的基本信息和纸的品质。

⊙2 九华古道泾县小岭段
Jiuhua ancient road in Xiaoling section of Jingxian County

⊙3 《泾县土纸样本》
*Samples of Handmade Paper in Jingxian County*

# 四 泾县竹纸生产的原料、工艺流程与技术分析

# 4 Raw Materials, Papermaking Techniques and Technical Analysis of Bamboo Paper in Jingxian County

## (一) 泾县竹纸生产的原料

### 1. 主料

泾县竹纸生产所需的主要原料是当地山上所产的嫩毛竹。砍伐当年生的嫩竹，砍伐时间为每年“小满”节气前后三天。据当地老工人介绍，此前的竹子由于没成长发育好，竹纤维会太嫩；“小满”过后太久，嫩竹纤维老化僵硬。太嫩或过老都难以做出较好品质的竹纸。“小满”前后三天的嫩竹最符合泾县优质竹纸的生产要求。

### 2. 辅料

（1）纸药。泾县竹纸制作时使用的纸药有榔叶（音名）、香叶子（音名）、杨桃藤（猕猴桃藤）。使用方式与使用季节如下：

榔叶：为当地的一种落叶乔木，当地人称榔叶。泾县竹纸使用的榔叶长在春、夏、秋三季，这三个季节在山上可以采集到榔叶。进入冬季后，榔叶落尽，只能使用香叶子或杨桃藤替代。制纸人采集榔叶下山后，放在锅里煮沸，浸泡12小时即可。使用时，用一细篾编成的箩筐放在浸泡榔叶的缸内，榔叶汁液从箩筐缝隙中渗出，完成过滤，取其过滤后的汁液再加入清水调至适用浓度即可使用。

香叶子：香叶子为常绿乔木，冬季上山采集，将采集下山的香叶子按照10∶1的比例掺入石灰膏（石灰膏即浸泡竹麻后沉淀在麻塘底部的石灰浆），加水焖煮三天三夜，得到的黏液叫蒸膏（音名），将蒸膏放入缸中存放。使用时，用细篾箩筐放在存放蒸膏的缸内，蒸膏汁液从箩筐缝隙中渗出，完成过滤。取过滤后的汁液再加入清水调至适用浓度即可使用。

⊙1

⊙1 访谈老纸工 Interviewing an old papermaker

杨桃藤：竹纸生产区域漫山都是竹子，少有生长杨桃藤，椰叶、香叶子都接济不上时才使用杨桃藤救急。砍下1～2年生的杨桃藤，用木槌将其捶开裂，用清水浸泡12小时左右，用挽钩在浸泡池内进行拉拽，使其汁液与水混合，拉拽到一定的程度，使浸泡的清水达到一定浓度时，将含汁液的水过滤后便可使用。

（2）水源、石灰。泾县竹纸对水源和石灰没有特别要求。对于水源，只要干净且流量相对充沛即可；对于石灰，只要没有熟化或消化的石灰即可。

⊙2

⊙3

## （二）泾县竹纸生产的工艺流程

因泾县竹纸现已无活态存在，调查组先后4次对竹纸生产工艺进行了实地调查。2014年3月19日访谈了昌桥乡孤峰行政村华侨自然村盘坑的造纸人郎金全；同年6月中旬对其进行回访后，又对泾川镇清水塘、殷家坑、古坝进行了实地考察与访谈；2016年5月12日，到竹纸老艺人王永兴家进行了补充调查；2016年5月13日，到黄村镇的九峰村进行了实地考察。经过对照梳理后，得到泾县竹纸生产的主要工艺流程为：

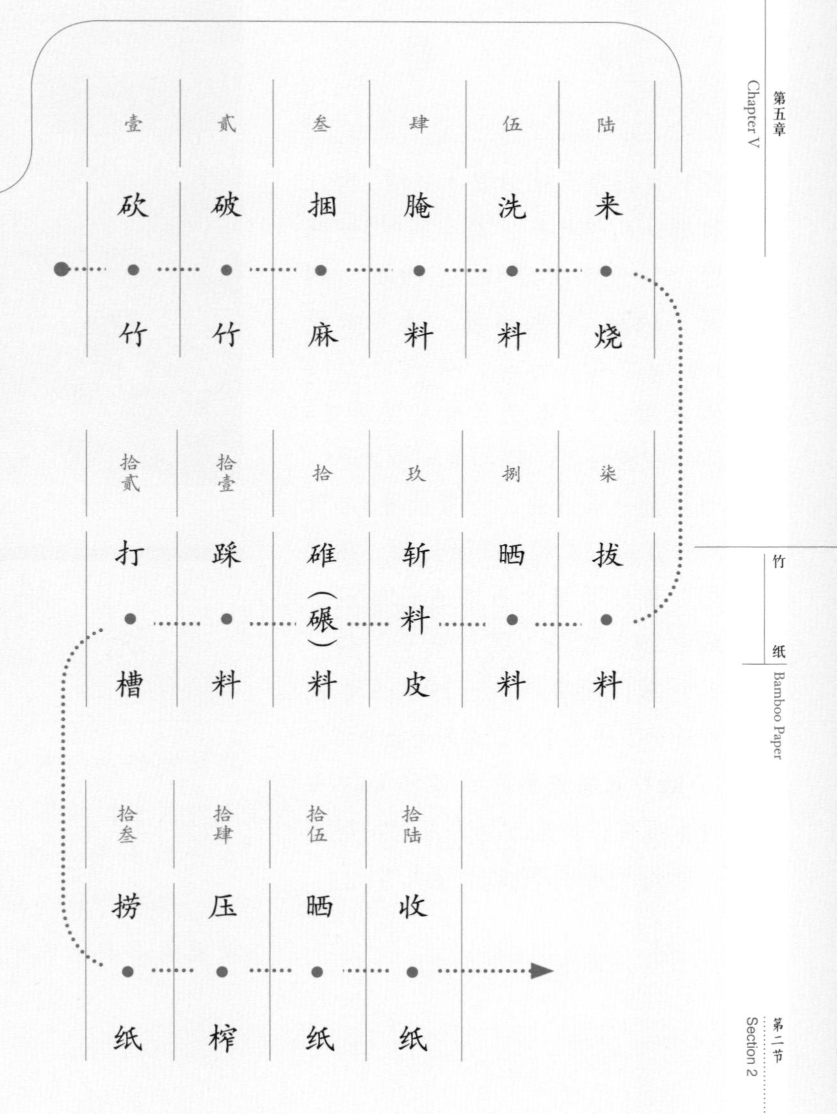


⊙2 槌杨桃藤 Beating kiwi fruit cane

⊙3 九峰村 Jiufeng Village

壹
## 砍　　竹
1

“小满”前后三天选嫩竹直接砍伐。因赶季节的原因，砍竹时，用砍刀将嫩毛竹直接砍倒，放在山上，可选择时间运下山或直接在山上破开后运下山。

贰
## 破　　竹
2

砍伐后集中在一起，然后用刀斩成5尺长一段，破成宽约3 cm的片状，一人一天可破1 500 kg左右的嫩竹。

叁
## 捆　　麻
3

破好的竹片称为竹麻，按照50 kg左右1捆用竹丝绳捆好，运回生产区进行下一步加工。

肆
## 腌　　料
4　⊙1

将竹片松散开来放置于麻塘[1]内，麻塘分几种规格，最小的叫5厢麻塘，其他依次为8厢、10厢、14厢、18厢。5厢麻塘一次可放置1万kg左右的竹麻，最大的为18厢麻塘，5尺长的竹麻可以放置3排。在腌料时，均匀铺一层竹子，均匀撒一层石灰，一般放置5层竹麻为宜。石灰和竹子放好后，用水充分淹没竹麻和石灰进行腌沤。腌沤20天后用工具“钎子”插一下竹麻，而后再腌沤10天左右就可进行下道工序了。据郎金全介绍，1万kg竹麻的腌制需要1 250 kg左右的生石灰；据王永兴介绍，一般按照15 kg生石灰/50 kg竹麻的配比。

⊙1

伍
## 洗　　料
5

将腌制好的竹麻用锄头、钉耙从池内捞出，边捞边在麻塘里就石灰水将竹片上附着的石灰清洗掉。所有竹麻全部洗完后，将麻塘里的石灰水放干并清理干净。

陆
## 来　　烧
6

麻塘清理干净后，用毛竹或松树打基脚，将清洗后的竹麻放置于毛竹或松树上，依次堆紧堆好后，在竹麻上面覆盖一层稻草，用竹片、石头将稻草压住，防止稻草被风吹乱、吹散后漏气。放置一个星期后，用手测试麻塘里的温度，感觉到竹麻发热（当地俗语称

⊙1
完全废弃的麻塘
Abandoned material pond

[1] 人工开挖的土坑，主要用于腌沤竹料。

"发烧""来烧")时，再往池里加水，直到水淹没最上层的竹麻为止。浸泡一个星期左右，将水放掉，等竹麻再次"来烧"，竹麻上"蒙"(当地俗语，类似"发霉")时，再往池里加水，直到取料加工时竹麻一直在水中浸泡。

⊙2

柒

## 拔料

7

由于竹麻有青皮和内芯(竹黄，也称幼料)之分，在腌沤过程中，内芯比青皮成熟得要快。通过人工分拣的方式，将青皮与内芯分开，内芯是可以直接捞纸的竹料；青皮需要进一步加工，也称为料皮。分拣幼料与料皮这道工序称为拔料。

捌

## 晒料

8

将通过拔料选出来的料皮放在太阳下晾晒，直至表皮显干即可。半干后的料皮直接搓成竹绳，其坚韧程度可比牛皮绳，在泾县竹纸制作体系里被用于扳榨和捆扎东西。这种绳子的缺陷就是不能沾水，一旦沾水，使用寿命就会大大降低。

玖

## 斩料皮

9

晒干后的料皮用刀砍短到3 cm左右，再进行下一步加工。

拾

## 碓(碾)料

10 ⊙2

将砍短的料皮通过舂碓的方式打碎，将其纤维束分解。竹纸制作体系里，在开始使用石碾后，大部分地区用碾料取代了碓料。将砍断的料皮用石碾碾压，碾压时间最少需要15小时。据介绍，在水源充裕的地方用水碾来碾压，在水源不足的地方一般以牛力带动石碾来碾压。

拾壹

## 踩料

11

料皮碾好后将料皮和幼料放在一起用脚踩踏，混合的比例按照拔料之前的配比还原。一次要踩一担多，够一人一天捞纸用，每次最少需要踩3小时，由踩料工掌握火候。踩料使用的工具叫板厢(又称踩槽)，规格为2 m×1.5 m×0.3 m(长×宽×高)。(打浆：泾县竹纸于1985年前后开始引进打浆机后，该工序被替代。)

拾贰

## 打槽

12

将踩好后的料放在槽里浸泡一夜，捞纸前先用通扒(音pa)按量拨拉到身边，用打槽棍反复划拨搅拌，将料充分打融后，加入纸药，然后进行捞纸。身边的纸浆被捞稀了，再用通扒按量拨料，如此往复。这道工序与捞纸交错进行。

⊙2
遗弃的石碾
Abandoned stone roller

## 拾叁 捞纸

### 13 ⊙3～⊙5

泾县竹纸采用单人捞纸法。先由外往里捞，称为挖水；再由右往左捞，水往左边倒出来，称为晃水，晃水的目的是使纸均匀。每捞一张纸均将其覆盖在纸槽边的木枕上，形成纸坨。每人每天一般捞两块纸坨。中午休息时，用废纸帘盖上，上放木板稍做挤压，去掉部分水分，下午捞纸时直接放在废纸帘上。一天下来，形成两块坨。每天捞纸工下班后，等纸槽里的水澄清，将澄清的水通过槽底部一小孔放出，放至肉眼可见料浆部位，再用滤网接住槽孔过滤残余料浆，把水沥干后再放进纸槽，混入新鲜料浆并加清水，于次日进行捞纸。

⊙3 ⊙4 ⊙5

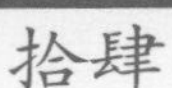

## 拾肆 压榨

### 14

泾县竹纸制作体系里将捞出来的湿纸堆叫纸坨，放在木枕上。先用一压杆，用手缓慢下压，后加两个小枕，换牛皮缆绳（后改用麻绳，也有用竹绳的）与木杠制成的绞杆，压到两面平整，纸坨基本挤不出水即可。为测试纸坨的干湿度，可用手按纸帖边，如按不进去就说明已干。然后按照先后顺序退榨，将纸坨直接背到纸焙屋里晒。

⊙6

## 拾伍 晒纸

### 15 ⊙6

泾县竹纸制作体系里称晒纸屋为“焙纸笼”，屋内有焙墙，两面都可以晒，一人负责一面墙。将湿纸从纸坨上分张揭下后，用纸刷将湿纸贴在焙墙上，从上到下晒4张。

## 拾陆 收纸

### 16

将干燥后的纸从纸焙上揭下，整齐地放于纸架上，按36张/刀分好，前后两刀错位间隔5 mm左右。全部晒完后，将每刀纸沿着长边对叠便为成品，然后按照144刀/担打捆后销售。

⊙3/5 捞纸演示 Showing papermaking procedures

⊙6 孤峰村的『焙纸笼』遗存 Abandoned paper drying cabin in Gufeng Village

（三）

## 泾县竹纸的性能分析

对采样自孤峰村的泾县竹纸的性能分析主要包括厚度、定量、紧度、抗张力、抗张强度、色度、纤维长度和纤维宽度等。按相应要求，每一指标都需重复测量若干次求平均值，通常情况下需要测量10次。测试小组对采集的泾县竹纸进行测试分析后，得到的其相关性能参数见表5.2。表中列出了各参数的最大值、最小值及测量10次后所得到的平均值。

表5.2　泾县竹纸相关性能参数
Table 5.2　Performance parameters of bamboo paper in Jingxian County

| 指标 | | 单位 | 最大值 | 最小值 | 平均值 |
| --- | --- | --- | --- | --- | --- |
| 厚度 | | mm | 0.208 | 0.075 | 0.141 |
| 定量 | | g/m² | — | — | 31.9 |
| 紧度 | | g/cm³ | — | — | 0.226 |
| 抗张力 | 纵向 | N | 5.8 | 0.7 | 3.3 |
| | 横向 | N | 5.5 | 1.1 | 2.6 |
| 抗张强度 | | kN/m | — | — | 0.197 |
| 色度 | | % | 24.4 | 23.1 | 24.0 |
| 纤维 | 长度 | mm | 1.81 | 0.08 | 0.57 |
| | 宽度 | μm | 54.7 | 2.6 | 12.7 |

由表5.2中的数据可知，所测泾县竹纸最厚约是最薄的2.77倍，经计算，其相对标准偏差为0.043。所测泾县竹纸的平均定量为31.9 g/m²。通过10次数据测量，测得泾县竹纸紧度为0.226 g/cm³，抗张强度为0.197 kN/m。

所测泾县竹纸平均色度为24.0%。色度最大值是最小值的1.06倍，相对标准偏差为0.391，差别较小。

所测泾县竹纸在10倍、20倍物镜下观测的纤维形态分别见图★1、图★2。所测泾县竹纸纤维长度：最长1.81 mm，最短0.08 mm，平均长度为0.57 mm；纤维宽度：最宽54.7 μm，最窄2.6 μm，平均宽度为12.7 μm。

★1

★2

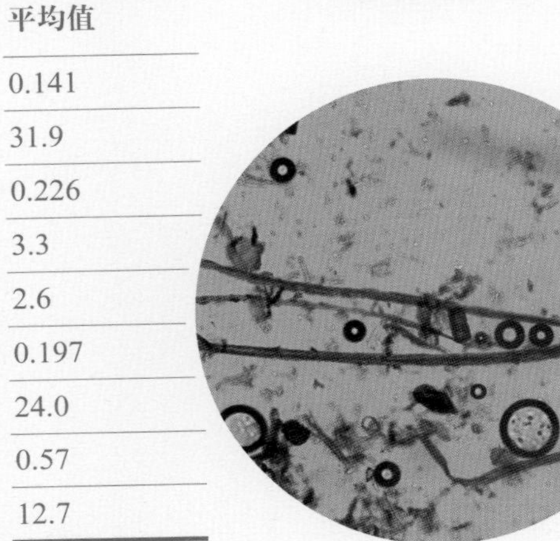

★1
泾县竹纸纤维形态图（10×）
Fibers of bamboo paper in Jingxian County (10× objective)

★2
泾县竹纸纤维形态图（20×）
Fibers of bamboo paper in Jingxian County (20× objective)

# 五
# 泾县竹纸的用途与销售情况

## 5
## Use and Sales Status of Bamboo Paper in Jingxian County







根据历史文献检索和对老艺人的访谈信息，传统时期泾县竹纸的用途是很广的，被当地人广泛用于祭祀、记事、生活等文化活动中，曾于20世纪50年代，与宣纸一道被编入《泾县土纸样本》，其主要文化功能用途归纳如下。

### （一）泾县竹纸的历史用途

#### 1. 祭祀

嘉庆十一年（1806年）《泾县志》载："清明：插柳于门，人簪一嫩柳，谓'辟邪'。具牲醪扫墓，以竹悬纸钱而插焉。或取青艾为饼，存禁烟寒食之意。……岁除，连日祀先，或送亲茶毕，又以食物相馈。除日换桃符春贴，各响爆竹，燔纸辟瘟，男女别岁。" 凡当地人在每年三月三、清明、七月半、冬至、除夕祭祖或行业祭祖时，烧"三六表"纸是一种重要的表现形式。除此之外，遇上丧事，烧"三六表"纸也是重要的送丧活动，其中丧者的女儿奉烧更多，需要烧化九斤四两的"三六表"纸。

#### 2. 记事

用于记事等文化活动的泾县竹纸称为表芯纸，黄中带白，纸面较为光滑，发墨洇彩慢。此纸重要的是价格低廉，广泛用作记事载体。泾县旧时学生练字、书契文书、文字凭证等大多选用廉价易得的本地产竹纸。

#### 3. 生活用纸

泾县竹纸除了用于卫生纸外，在传统包装中也选用此纸，如包中药、食品（如糖）、烟丝等。抽烟、引火、照明也用竹纸，将竹纸裁成小条，搓成纸杆，称之为"媒纸"或"媒子"，用以引火不易灭，节约火柴。

### （二）泾县竹纸的历史销售情况

据调查中王永兴介绍，1990年左右，泾县建华造纸厂的竹纸售价为95～100元/担，每刀

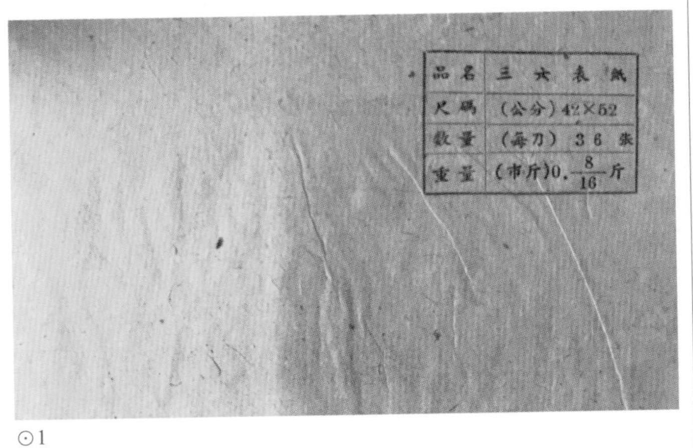


⊙1

约为0.65元，均销往各地供销社、土产公司。私人自制的竹纸因质量不稳定，销售渠道相对少，价格也偏低。当时的建华造纸厂分地块形成车间，分别有孤峰、盘坑、木坑等车间。

## 六 泾县竹纸的保护现状与发展思考

## 6 Current Status of Protection and Thoughts on Development of Bamboo Paper in Jingxian County

泾县竹纸除具有一般竹纸的历史、文化、经济价值外，其工艺更具有一定的特色。例如，料皮、幼料分开，经加工后重新组合；纸槽每天清槽。这两个特点在全国竹纸制作中较为少见，显示出泾县竹纸工艺的细腻，而这两点与泾县传承的宣纸技艺有相似之处。如今，泾县竹纸的活态已完全消失，仅存活在人们的记忆里。根据这一现状，结合多数刚刚消失不久的非物质文化遗产种类的保护措施，我们提出以下建议：

泾县竹纸手工技艺曾是一个时期内一个地方的文化记忆，当地政府在经济条件允许的情况下，可以采用“政府补贴、民众参与”的办法，建设一个小规模的公益性设施，保存当时的生产图片、资料和实物等。如果有条件，可以再现一些工艺，阶段性地在公职人员、中小学生中进行地方非物质文化遗产教育，使区域文化为更多的当地人所熟知。

在政府的支持与倡导下，将相关文化信息纳入当地民俗博物馆中，作为一项分支内容展出，这也是丰富中国手工造纸之乡产业信息内涵的一个有意义的选择。

⊙1 《泾县土纸样本》中关于『三六表』纸的信息 Information of "Sanliubiao paper" in *Samples of Handmade Paper in Jingxian County*

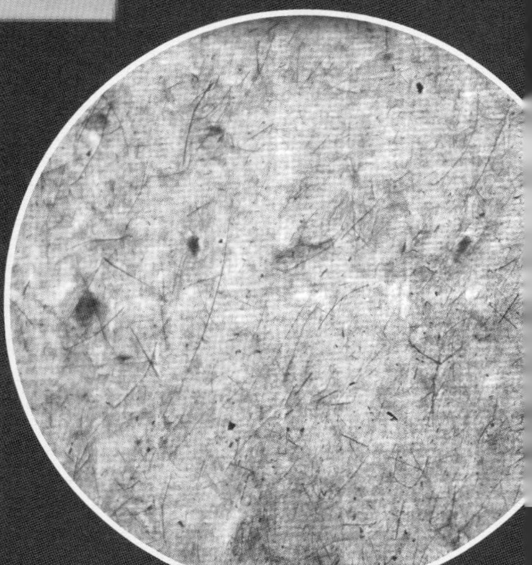

泾县孤峰村

# 竹纸

Bamboo Paper in Gufeng Village of Jingxian County

孤峰村竹纸透光摄影图

A photo of bamboo paper in Gufeng Village seen through the light

# 第三节

# 金寨县燕子河镇

安徽省
Anhui Province

六安市
Lu'an City

金寨县
Jinzhai County

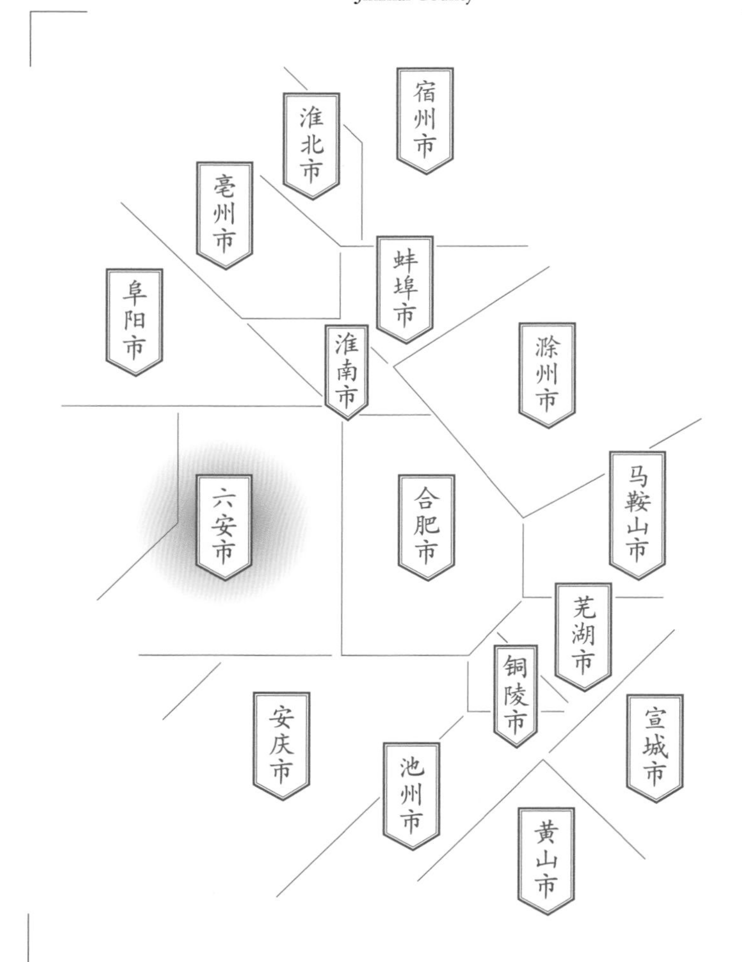


**调查对象**

燕子河镇
龙马村/燕溪村
竹纸

## Section 3
## Yanzihe Town in Jinzhai County

Subject

Bamboo Paper
in Longma Village / Yanxi Village
of Yanzihe Town

# 一
# 金寨县燕子河镇竹纸的基础信息与分布

## 1
## Basic Information and Distribution of Bamboo Paper in Yanzihe Town of Jinzhai County

金寨县为六安市辖县，地处安徽省西部的大别山腹心地域，地理坐标为：东经115° 22′ ~116° 11′，北纬31° 06′ ~31° 48′。全县总面积3 814 km²，县域面积居安徽省之首。历史上为红四方面军的主要诞生地，是红色旅游的代表性地区。

大别山山脉由西南向东北贯穿金寨县域全境，境内群山起伏，河流纵横，海拔千米以上的山峰有116座；平均海拔800 m以上的中山区，主要分布在南部及西部，面积近2 000 km²，占全县总面积的51.6%，坡度多在30° ~50° 范围，属于典型度很高的中东部深山区。县内盛产西洋参、茯苓、灵芝、天麻等贵重药材，为全国22个药材基地县之一。全县森林覆盖率为72.75%，林业用地面积2 940 km²，其中毛竹面积133 km²，年可产毛竹500万根，历史上为竹纸生产提供了丰富的资源。

2017年4月30日，调查组前往金寨县燕子河镇张畈村进行了竹纸业态现场调查，所获得的基础信息如下：张畈村委会原为乡建制，拆乡并镇后并入燕子河镇。造纸作坊遗址在张畈村的龙马村民组，区域大致地理坐标为：东经115° 49′ 36″，北纬31° 17′ 11″（海拔约 630 m）。

通过对纸工许修树、许修林、许修安、王文训等人进行现场访谈了解到，龙马村民组的竹纸生产设施毁于2008年的一场大水，此后因为手工竹纸本身的市场销售也已不好，龙马村的造纸户们失去了投资重建造纸设备的动力，竹纸生产业态就此中断。但龙马村里还残留了部分未被水毁的设备、厂房，年老的造纸人留守在家，年轻的村民们几乎都外出打工去了。

⊙1

⊙1
调查组成员与造纸人许修树（左二）、许修林（左三）合影
Researchers with papermaker Xu Xiushu (second from the left) and Xu Xiulin (third from the left)

4月30日晚上，调查组在向镇长和镇文化站站长进行补充调查后，委托燕子河镇文化站的负责人帮助了解有关竹纸的进一步信息。2017年6月下旬，镇文化站站长又寻访到了燕子河镇燕溪村的造纸户陈文德、孙家芝夫妇，他们提供的信息如下：

路线图
金寨县城
↓
龙马村/燕溪村
Road map from Jinzhai County centre to Longma / Yanxi Village

# 金寨县龙马村燕溪村位置示意图

# Location map of Longma / Yanxi Village in Jinzhai County

考察时间
2017年4月

Investigation Date
Apr. 2017

地域名称

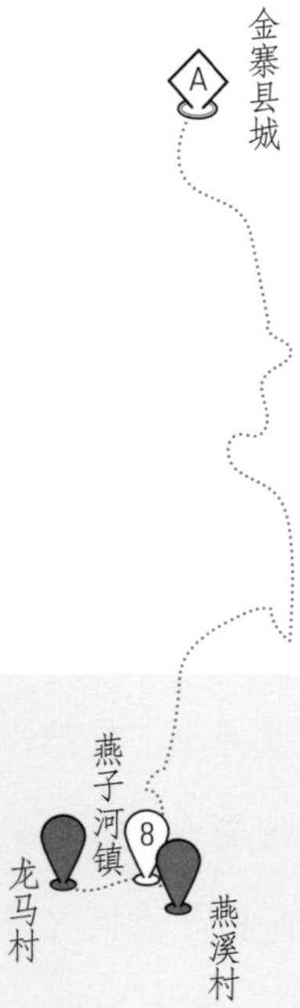

A
金寨县城
燕子河镇
8
龙马村
燕溪村


A 金寨县
1 白塔畈镇
2 梅山镇
3 全军乡
4 铁冲乡
5 汤家汇镇
6 双河镇
7 桃岭乡
8 燕子河镇

造纸点名称

龙马村
燕溪村
造纸点

位置分布

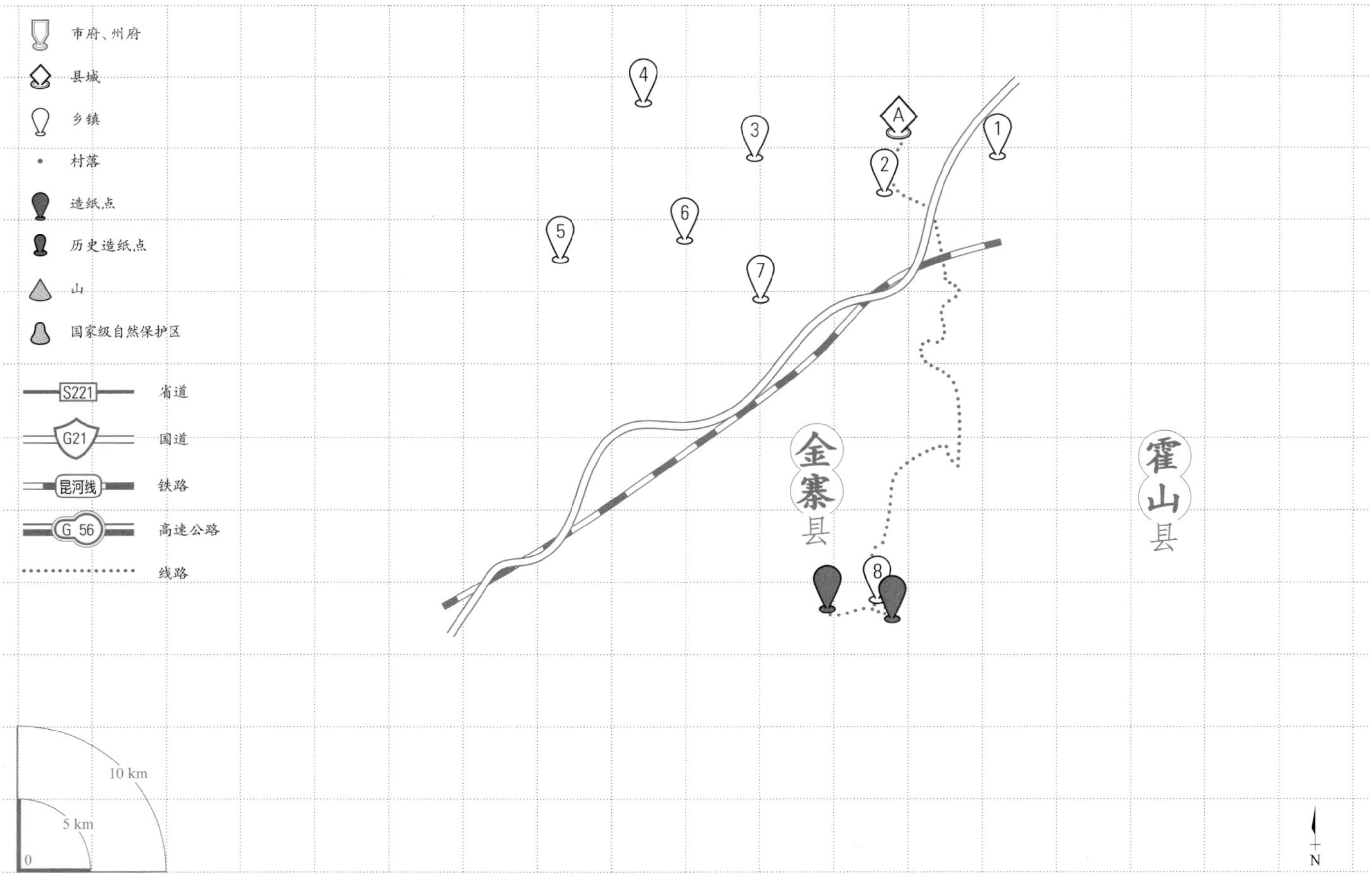

市府、州府
县城
乡镇
村落
造纸点
历史造纸点
山
国家级自然保护区
S221
省道
G21
国道
昆河线
铁路
G 56
高速公路
线路
A
1
2
3
4
5
6
7
8
金寨县
霍山县
10 km
5 km
0
N


Yanzihe Town in Jinzhai County

⊙1

燕溪村手工竹纸的制作已有250年左右的历史，陈文德祖上数代均以制作“山纸”（即竹纸，当地也称“火纸”，焚烧祭奠逝者以寄托哀思和怀念的竹纸）为主业，他自己从小就跟父亲学造纸，迄今已有50余年。直到5年前（2012~2013年）因为小型机械造的“火纸”以价廉的优势挤占了市场，而且二人年纪大了，才停止了火纸生产。6月28日，镇文化站站长经过和陈文德洽商，从老夫妇手中获得了原本两人准备去世后给自己使用的旧日自造的部分竹纸，寄给调查组作为分析时的采样纸。



## 二 金寨县燕子河镇竹纸的历史与传承情况

## 2 History and Inheritance of Bamboo Paper in Yanzihe Town of Jinzhai County

金寨县本地称竹纸为“草纸”“黄表纸”“火纸”，原制纸业态分布于金寨县的古碑镇、南溪镇、燕子河镇、张畈乡、桃岭乡、花石乡等乡镇，除陈文德夫妇提供的历史信息外，最早源于何时已无从查考。这其中的原因之一是金寨县建制时间较晚，民国二十一年（1932年），中华民国政府将分属于安徽省的六安、霍山、霍邱与河南省的固始、商城的5县划出55个保，设置了立煌县。民国三十六年（1947年），正式改名为金寨县，后分为金寨

⊙1 龙马村口的路 Road to Longma Village

（金西）、金东两县及金北办事处三个县级建制区域。1952年8月，安徽省人民政府将金寨县归属于六安专区，至今隶属未变。因历史上金寨县的名称出现较晚，官方文书少有集中记载区域内的重大事件，不仅竹纸制作信息难以查考，连该区域内其他重大历史信息查考也较为困难。其他地方即便有一些历史信息，也由于地名等各种信息不对称，历史溯源难度很大。1992年，上海人民出版社出版了金寨县地方志编纂委员会编制的《金寨县志》，在长达119万字的新版县志中，只能找到该区域内有过手工造纸的信息，但没有具体的地点和事项记载。调查中，陪同调查燕子河镇的村镇干部回忆，在他们很小的时候，金寨县除了造竹纸外，还曾造过用于糊窗户的“影皮纸”。

现场走访的许修林曾是薛河制纸作坊老板，调查时67岁。据许修林介绍，薛河制纸作坊在中华人民共和国成立前就有，新中国成立后因这种纸是迷信用纸而不再生产。后来许修林看见邻地在生产这种纸，便从1976年在生产香木的原址上开始制纸。许修林本人于1975年任张畈林场场长，1980年调任石灰厂厂长，1982年调任砖瓦厂厂长，1984年调任纸厂会计，次年又调任林场会计。1991年3月，出资7 000元收购薛河制纸作坊，雇用的人员主要是其兄许修树和许修树的儿子许永琪（音），许永琪技术比较全面，捞纸、榨纸、揭纸都会。该厂一直生产到2008年，毁于一场大水后停产。

据许修树介绍，龙马村里一共有十几户姓许的，从湖北迁来时只有老太爷一个人，到爷爷这一辈是6个人，到父亲这一辈是10个人。造纸技术是到薛家河以后开始的，老太爷买了别人的纸厂开始做纸。

许修树，1943年出生，他这一辈是“修”字辈。父亲是“怀”字辈，叫许怀珊。爷爷和父亲都是学造纸的，许修树20多岁开始跟父亲学造纸。父亲跟着爷爷学，爷爷跟着太爷爷学，一辈一辈传下来。

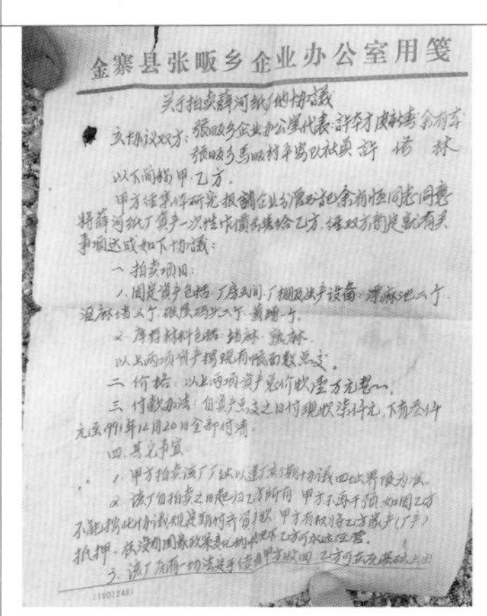
金寨县张畈乡企业办公室用笺

⊙1

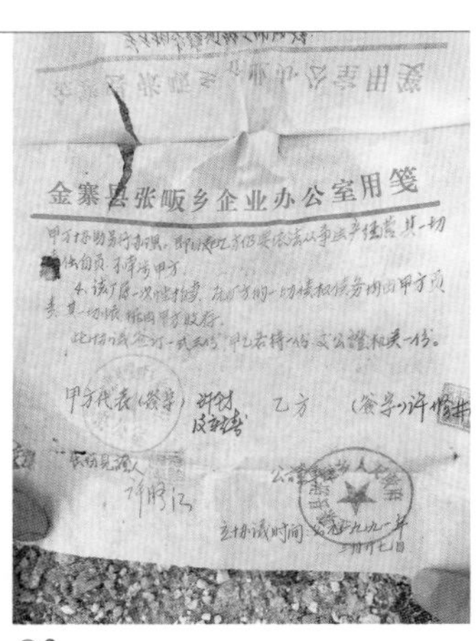
金寨县张畈乡企业办公室用笺

⊙2

⊙1/2 薛河制纸作坊转让协议
Transfer agreement of Xuehe Papermaking Mill

# 三
# 金寨县竹纸生产的原料、工艺流程与技术分析

3
Raw Materials, Papermaking Techniques and Technical Analysis of Bamboo Paper in Jinzhai County

## （一）
## 金寨县竹纸生产的原料

### 1. 主料

金寨县竹纸生产所需的主要原料是当地山上所产的毛竹、元竹。据许修林等人介绍，制作普通竹纸用老竹，制作黄表纸用嫩竹。无论制作什么纸，均不区分毛竹、元竹，将收购来的毛竹、元竹混合在一起制作。

⊙3

⊙3
厂房现场遗留的竹麻
Bamboo materials left in the factory site

### 2. 辅料

（1）纸药。金寨县竹纸制作需要用纸药，所用的纸药主要是杨桃藤（猕猴桃藤）。砍下1~2年生的杨桃藤在阴湿地存放，每天按量取用，一旦用水浸泡后就必须用完，其效用到次日就挥发殆尽。天暖用量稍多，天冷则用量少。将杨桃藤切成小段，放在木桶中，轻轻用木槌捣烂，加清水浸泡，再用手揉搓即成。也有用木槿条制成纸药的，将新发木槿条切成小段后，先要用榔头捶破，然后放在桶中水浸揉搓。此外，当地还用

“头痛花”的叶子制备纸药[1]。在调查中得知，当地也有用菜油做纸药的。

（2）水源、石灰。金寨县竹纸的纸棚大都建在有较大落差的溪水边，从上游高水位处沿着溪边另开一渠引水至纸棚，为沤料、漂洗等提供自流用水并为水碓提供动力。实测龙马村的水源pH为6~7。纸棚内的主要设备有纸槽和纸榨，水碓的动力机械部分放在棚外，碓臼则放在棚内，以便雨天也可照样生产，同时有效地利用棚内空间。当地所用水碓在运转时，水在水轮的顶部冲，而一般水碓水都在底部流。水碓的运转与制动是通过大渠上安置的闸阀的开关来控制的。龙马村所用的溪流名叫薛家河，也叫靴落河，传说乾隆皇帝四下江南的时候，经过这里掉了一只靴子而得名。

金寨县竹纸对石灰没有特别要求，以生石灰为主。



## （二）金寨县竹纸生产的工艺流程

据许修林等人的口述，将金寨县竹纸生产的工艺流程整理如下：

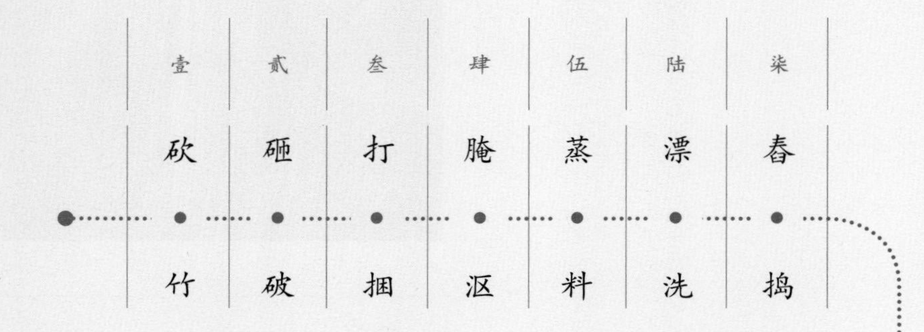


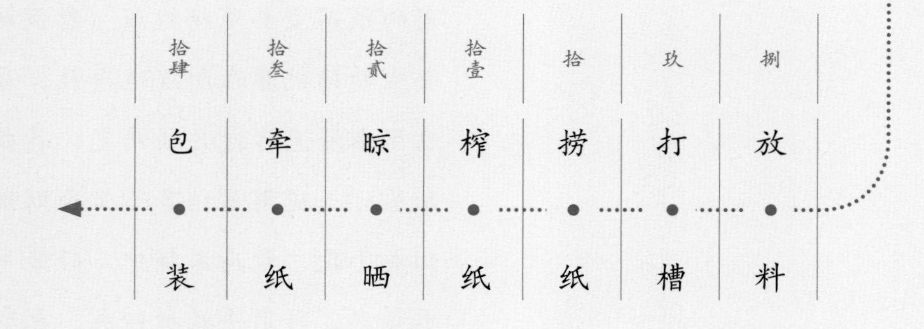




[1] 张秉伦，方晓阳，樊嘉禄.中国传统工艺大全：造纸与印刷[M].郑州：大象出版社，2005：115.

## 壹 砍竹

### 1

当地盛产毛竹、元竹，砍竹基本不分季节，只有在清明节前后不能砍，以免影响竹子生长。

## 贰 砸破

### 2 ⊙1

将砍下的竹子截成1.5 m左右，用大锤将竹子砸破，或直接用刀剖成约2 cm宽的竹片。这种被砸破的竹子称为竹麻。

## 叁 打捆

### 3

将竹麻按20～25 kg一捆捆扎好，便于腌沤。

## 肆 腌沤

### 4

将整捆竹麻放至麻凼[2]中，一层竹麻一层石灰，均匀堆放在麻凼中，堆好后在上面覆盖好稻草，草上压上木头或石块，向麻凼里放水，将竹麻全部淹入水中，至少腌沤半年。每个月都要用钢钎敲打一下竹麻，以便石灰更多地与竹麻混合。麻凼一般设在纸棚附近，便于操作。一麻凼可腌沤2 000～2 200捆竹麻，需用1.5～2 t石灰。

⊙1

⊙2

## 伍 蒸料

### 5 ⊙2

在楻甑[3]中，先用木或竹做好假底，将腌沤好的竹麻从麻凼里起出，挑装入楻甑的假底上，装满一甑后，用稻草或塑料薄膜封好甑口。在楻甑炉口烧火，蒸煮竹料。燃料用干木柴，连续蒸七八昼夜，停火后再焖一昼夜。每蒸一锅料要烧掉5 t干柴。蒸过的竹麻虽然外形变化不大，但已无“筋骨”，用手一揉，即成齑粉。

## 陆 漂洗

### 6 ⊙3

从甑中取出竹料，放入麻凼，用水淹过竹麻，放河里流动的清水入池，池水保持平满，多余的水从下边出水口处流出。如此浸泡1个月或更长时间，将竹料所带的石灰和污水完全漂洗掉。

⊙3

⊙1 许修安在演示砸破竹子
Xu Xiuan showing how to break the bamboo

⊙2 楻甑炉遗存
Abandoned Huangzeng stove

⊙3 洗料池旧址
Former site of cleaning pool

[2] 当地土语：腌沤竹麻的坑。

[3] 当地土语：指巨型蒸锅。

柒
## 舂　　捣
7　⊙4

将洗净的竹料放在水碓的石臼中舂捣，舂捣时专有一人坐在旁边，视火候将舂好的料取出。一般一臼料需要舂捣4～5小时。据许修林介绍，薛河制纸作坊后期使用石碾碾料，以水力带动石碾。

⊙4

捌
## 放　　料
8

按一天的产量将舂捣好的竹料放入纸槽，一天的产量在当地称一案纸。一般一案纸用10～12捆竹麻，一捆竹麻可产10～12刀纸。按量放好料后，便放入溪水，以水低于纸槽沿为宜，大约需水2.5 $m^3$。

玖
## 打　　槽
9　⊙5

⊙5

用打槽棍将身边的竹麻料通过搅拌方式打融，放入纸药后，稍稍搅拌一下，即可捞纸。待纸槽中竹麻料稀薄后，再次将沉在槽底的竹麻料掏到身边来打融，加入纸药。如此往复，直至纸槽中的竹麻料全部捞完。

拾
## 捞　　纸
10　⊙6

将帘子放在帘架上，两端分别用小竹片压住帘子，双手持帘长两端，由外向身边下水，纸浆布满整帘后，从右往左倒水，将帘架放在槽边，拿掉竹片，取下帘子，反扣在纸槽旁边的纸板上，从身边向外将纸帘揭起，把湿纸留在板上。如此反复，堆放在一起的湿纸在当地称为纸垛。金寨县竹纸有一帘抄1张纸的，有一帘抄2张纸的，也有一帘抄6～7张纸的。

⊙6

拾壹
## 榨　　纸
11

一案纸捞完后，将纸垛盖上旧帘子，再在旧帘子上盖上平整的木板。在木板上方压上大木杠，木杠一头穿在纸板一边固定的方框内，一头穿上连有滚子的麻绳，扳动滚子即可完成榨纸。

⊙4 已废弃的舂捣工具 Abandoned tools for pounding the bamboo materials

⊙5 废弃的纸槽与槽房 Abandoned papermaking trough and workshop

⊙6 许修树家中的旧纸帘 Old papermaking screen in Xu Xiushu's house

## 拾贰 晾晒 12

将压榨后的纸垛等分为四份，放在阳光下通风处晾晒至七八成干，再取回分张。

## 拾叁 牵纸 13

取两张与纸张大小相同的软牛皮或厚布，分别放在纸垛的上下两面，护住纸垛；用棍棒用力在纸垛上敲打，然后双手捏住纸垛的一端在平台上折纸垛，右手按住纸垛上面中心偏右位置，左手从纸垛左端向右上方推压，如此反复，使纸张错位，从而达到分离的目的。要分离一张纸，只需平摊在平整的河滩上使其自然晒干。

## 拾肆 包装 14

将晾晒干的纸按30张/刀或36张/刀，两边沿宽度方向往中间相对折叠成宽度的一半厚。按50刀/捆用竹篾捆好出售。

### (三) 金寨县竹纸生产的工具设备

## 壹 麻凼 1

用于腌沤的设备，在靠近溪水的地方采取平地开挖方式完成，既便于引水入凼，也便于排水。实测麻凼长8 m，宽7 m，深2 m，中间有8根桩子，将麻凼分成2个区。

⊙7

⊙8

⊙7/8 麻凼遗存 Former site of Madang (pond for fermenting the materials)

## 贰 楻 甑 2

用于蒸煮的设备，上方用石头垒成，下方有固定大铁锅，锅口与甑底相接，锅上纵横交错地铺上木杠（假底，当地也称锅箄），锅箄的作用是承担竹料的压力，并防止竹料落入锅中。实测楻甑的直径为6 m，甑沿宽约1.5 m，甑高2.6 m。甑下方有烧火炉口，炉口宽1.4 m，高1 m。

⊙9

⊙10

## 叁 纸 帘 3

用于捞纸的工具，一般从湖北英山购得，实测许修树家纸帘尺寸为：长92 cm，宽18.5 cm。每1 cm有13根帘丝。

## 肆 帘 架 4

用于捞纸的工具，实测许修树家帘架尺寸为：长108 cm，宽27 cm。

⊙11

## 伍 打钱模子 5

木质，自制，中有手柄。实测许修树家手柄长17 cm，模长20 cm，模宽19 cm。模子为外圆内方结构，外圆直径为1.4 cm，内方边长为0.5 cm。

⊙12

⊙13

⊙9 旧楻甑 Former Huangzeng

⊙10 许修树家的旧纸帘 Old papermaking screen in Xu Xiushu's house

⊙11 许修树家的旧捞纸帘架 Old screen frame for supporting the papermaking in Xu Xiushu's house

⊙12/13 打钱模子 Mold for making joss paper

## （四）金寨县竹纸的性能分析

金寨县竹纸的技术分析主要包括厚度、定量、紧度、抗张力、抗张强度、色度、纤维长度和纤维宽度等。按相应要求，每一指标都需重复测量若干次求平均值，通常情况下需要测量10次。调查组对采集的金寨县竹纸进行测试分析，得到的相关性能参数见表5.3。表中列出了各参数的最大值、最小值及测量10次后所得到的平均值。

表5.3　金寨县竹纸相关性能参数
Table 5.3　Performance parameters of bamboo paper in Jinzhai County

| 指标 | | 单位 | 最大值 | 最小值 | 平均值 |
|---|---|---|---|---|---|
| 厚度 | | mm | 0.260 | 0.205 | 0.227 |
| 定量 | | g/m² | — | — | 70.8 |
| 紧度 | | g/cm³ | — | — | 0.311 |
| 抗张力 | 纵向 | N | 20.1 | 12.8 | 18.0 |
| | 横向 | N | 17.0 | 3.7 | 11.8 |
| 抗张强度 | | kN/m | — | — | 0.993 |
| 色度 | | % | 12.1 | 11.8 | 11.9 |
| 纤维 | 长度 | mm | 1.69 | 0.36 | 0.79 |
| | 宽度 | μm | 47.1 | 1.8 | 15.8 |

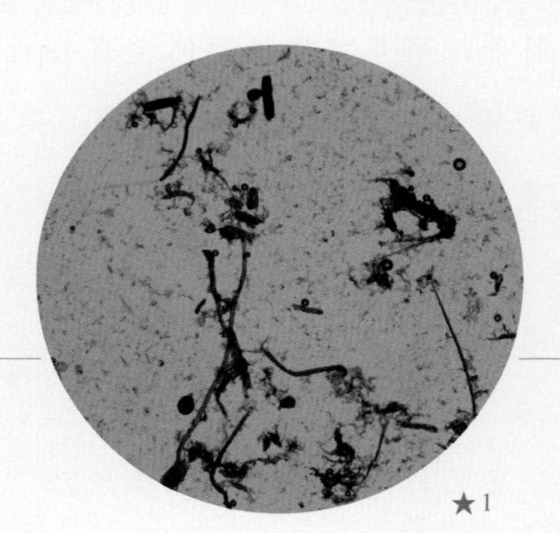

★1

★1 金寨县竹纸纤维形态图（10×）
Fibers of bamboo paper in Jinzhai County (10× objective)

★2

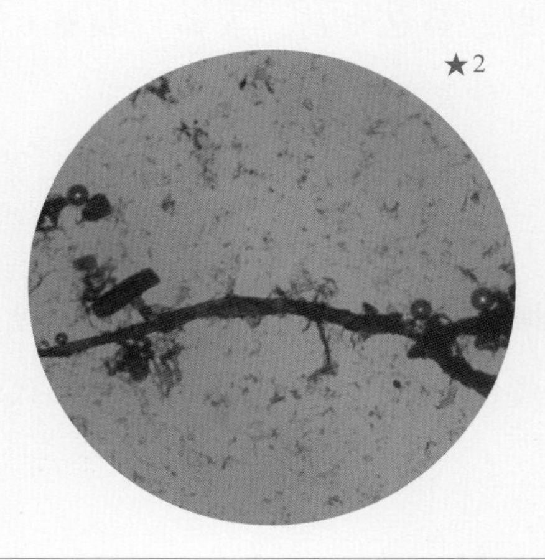

★2 金寨县竹纸纤维形态图（20×）
Fibers of bamboo paper in Jinzhai County (20× objective)

由表5.3中的数据可知，所测金寨县竹纸最厚约是最薄的1.27倍，经计算，其相对标准偏差为0.017，差别较小。所测金寨县竹纸的平均定量为70.8 g/m²。通过10次数据测量，测得金寨县竹纸紧度为0.311 g/cm³。金寨县竹纸抗张强度为0.993 kN/m。

所测金寨县竹纸平均色度为11.9%。色度最大值是最小值的1.03倍，相对标准偏差为0.145，差别较小。

所测金寨县竹纸在10倍、20倍物镜下观测的纤维形态分别见图★1、图★2。所测金寨县竹纸纤维长度：最长1.69 mm，最短0.36 mm，平均长度为0.79 mm；纤维宽度：最宽47.1 μm，最窄1.8 μm，平均宽度为15.8 μm。

# 五
# 金寨县竹纸的用途与销售情况

# 5
# Use and Sales Status of Bamboo Paper in Jinzhai County

⊙1

## （一）金寨县竹纸的用途

金寨县竹纸被当地人用于祭祀、记事、生活等文化活动，主要用途如下：

### 1. 祭祀

在现场调查中得知，当地人在大年三十吃年夜饭前、正月十五午饭前、清明节上坟、七月十五晚饭前均要祭祖，烧“草纸”是一种重要的表现形式。除此之外，遇上丧事，烧“草纸”也是重要的送丧活动之一。在金寨县农村婚俗中，新郎在娶亲前两三天，由父亲带着到祖坟上祭祖，也要烧纸。金寨县所烧的草纸需要用“打钱模子”将纸打成外圆内方的“纸钱”，且称这种纸钱才是真正的纸钱，否则“焚化的纸钱到阴间没用”。

### 2. 记事

用于记事等文化活动的金寨县竹纸称为黄表纸，黄中带白，纸面较为光滑，发墨洇彩慢，此纸重要的是价格低廉，民间被广泛用作记事载体。旧时学生练字、书契文书、文字凭证等大多选用廉价易得的竹纸。

### 3. 生活

金寨县竹纸除了用作卫生纸外，也用于传统包装中，如包中药、食品（如糖）、烟丝等。抽烟、引火、照明也用竹纸。

## （二）金寨县竹纸的销售历史

据许修林回忆，在2007年的时候，当地竹纸的售价为0.3元/刀，均销往当地代销点。

⊙1
造纸人家中的祭坛供桌
The altar table in a papermaker's house

# 六
# 金寨县竹纸的相关民俗与文化事项

## 6
## Folk Customs and Culture of Bamboo Paper in Jinzhai County

竹纸在金寨县所有盛产毛竹的地方均有生产，曾给当时金寨人的生活提供了便利，丰富了金寨县的民俗。据许修林、许修树介绍，制作竹纸是金寨县农民的一种副业，也听长辈们传诵曾有“祭蔡伦”等行业民俗，但这种民俗

⊙2
用破开的毛竹从山上接泉水的『自来水』
Water tube, using broken bamboo to transport mountain spring water

⊙2

消失已久，调查组没能了解到此行业祭祖的表现形式及内容。

金寨县竹纸在当地运用很广，在祭祀、记事、生活等方面均普遍使用。结合调查，也发现当地在农历七月十五有“不串门、不待客”的习俗。在当日，凡有来串门的乡邻、亲戚，定惹主人家不高兴，认为有 “吊孝”的嫌疑。除此之外，未获得其他民俗禁忌的口述信息。





## 七
## 金寨县竹纸的保护现状与发展思考

## 7
## Current Status of Protection and Thoughts on Development of Bamboo Paper in Jinzhai County

金寨县竹纸业态已消亡，部分地区虽有遗迹、废弃的设备和器具，却难以恢复曾经的业态，主要原因有三：一是现在机械竹纸兴起，使手工竹纸很难在后端市场上占据一席之地；二是这种竹纸毕竟不是当今文化生活的必需品；三是随着老纸工相继离世，很难恢复金寨县竹纸的技艺。据许修树介绍，其子许永琪尽管技术较为全面，但因在当地没有造纸氛围，只有远赴浙江打工。许永琪今年40多岁，膝下只有一女，当时在上大学，回归造纸几乎没有可能。许修林的两个儿子均外出求学后在外地工作，每月都寄生活费给他。从此可以看出，金寨县竹纸恢复的可能性暂时微乎其微。

2012年施行的《关于加强非物质文化遗产生产性保护的指导意见》中对“非物质文化遗产生

产性保护”做了如下定义：“非物质文化遗产生产性保护是指在具有生产性质的实践过程中，以保持非物质文化遗产的真实性、整体性和传承性为核心，以有效传承非物质文化遗产技艺为前提，借助生产、流通、销售等手段，将非物质文化遗产及其资源转化为文化产品的保护方式。”这种非活态的项目难以纳入非物质文化遗产代表作名录。但这一项目毕竟在金寨一地持续多年，具有独特、丰富的文化内涵，现虽已消亡，但制作器具或竹纸实物可以进入地方民俗文化展陈机构作为手工特色保存。

金寨县是革命老区，红色资源较为丰富，但乡土记忆也是必不可少的文化资源。因此，建议当地政府：如在条件允许的情况下，可以采用“政府补贴、社会参与”的办法，建设一个小规模的公益性设施，收集和保存当时的生产图片、资料、实物等，甚至可以再现一些工艺，阶段性地在公职人员、中小学生中进行地方爱国主义教育，推动该项目的相关文化知识普及，使这种具有区域文化特色的历史为更多人所熟知。

⊙1

⊙2

⊙1 山边废弃的造纸遗址 Abandoned papermaking site by the mountain

⊙2 旧日用的石头碾盘 Former stone roller

# 金寨县燕子河镇竹纸

Bamboo Paper in Yanzihe Town of Jinzhai County



龙马村竹纸透光摄影图

A photo of bamboo paper in Longma Village seen through the light

# 后 记





虽然《中国手工纸文库·安徽卷》的撰写工作是在云南、贵州、广西田野调查工作基本完成之后的2014年开始的，但因为《中国手工纸文库》的主持单位中国科学技术大学手工纸研究所在安徽省，中国手工造纸最具代表性的聚集地泾县与云南相比仿佛就在身边，因此不包括团队成员更早的人类学工艺社区调查及文化艺术之旅的积累，对宣纸之都——泾县的田野调查工作实际上从2008年起即已断断续续地在进行着。这部分工作已经带有明确的为《安徽卷》打基础的规划，也采集了若干纸样，形成了若干记录。

因为泾县宣纸、书画纸与加工纸业态的密集，《安徽卷》内容的丰富度几乎超过中国其他所有省份，因此分为上、中、下三卷。统稿的工作前后共进行了5轮，这确实源于素材采集与初步完成稿件的多样化和复杂性，以及不可避免地对若干内容需要多轮补充及优化，需要文献与调查现场反复印证。由于田野调查和文献研究基本上是以多位成员不同组合的方式参与的，而且多数章节前后多次的补充修订也不是由一人从头至尾完成的，因而即便工作展开之前制定了田野调查标准、撰稿标准，并提供了示范样稿，全卷的信息采集方式和初稿的表述风格依然存在诸多不统一、不规范之处。

# Epilogue

Field investigation of *Library of Chinese Handmade Paper: Anhui* started officially in the year of 2014, after the research team had finished their explorations in Yunnan, Guizhou Provinces and Guangxi Zhuang Autonomous Region. The grantee of the project, Handmade Paper Research Institute of University of Science and Technology of China locates in Anhui Province. So compared to Yunnan Province, Jingxian County which boasts the representative papermaking gathering place, is closer and more convenient for our investigation. Actually, our fieldworks can be traced back as early as 2008, and have lasted ever since intermittently in Jingxian County, the capital of Xuan paper, not to mention our even earlier anthropological survey of techniques, cultural and art investigation of the area. We had purposefully accumulated data and sample paper as the basis of this volume.

*Library of Chinese Handmade Paper*: *Anhui* is further divided into three sub-volumes for it includes extensive data on Xuan paper in Jingxian County, calligraphy and painting paper, and processed paper, which is unparalleled for all other provinces in China. The researchers have put into five rounds of sedulous efforts to modify the manuscript, and revisit the papermaking sites for more information and verification due to the diverse and complex materials. Field investigation and literature studies

初稿合成后，统稿与补充调查工作由汤书昆、黄飞松、朱赟、朱正海主持。从2016年3月开始，共进行了5轮统稿，到2019年8月才最终形成了现在的定稿。虽然我们感觉安徽手工造纸调查与研究还有不少需要进一步完善之处，但《安徽卷》的工作从2009年7月汤书昆、王祥、陈彪、黄飞松、周先稠几次组队的预调查开始，已历经10年，从2014年9月正式启动调查至定稿也有5年整。其间，纸样测试、英文翻译、编辑与设计等工作团队成员尽心尽力，使《安徽卷》的品质一天天得到改善，变得更有阅读价值和表达魅力。转眼已到2019年底，各界同仁和团队成员对《安徽卷》均有很高的出版期待，若干未尽事宜只能期待今后有修订缘分时再来完善了。

《安徽卷》书稿的完成和完善有赖于团队成员全心全意的投入与持续不懈的努力，在即将出版付印之际，除了向所有参与成员表达衷心的感谢外，特在后记中对各位同仁的工作做如实的记述。



of each section and chapter are accomplished by the cooperative efforts of multiple researchers, and even the modification was undertaken by many. Therefore, investigation rules, writing norms and format set beforehand may still fail to make amends for the possible deviation in our way of information collection and the writing style of the first manuscript.

Modification and supplementary investigation were headed by Tang Shukun, Huang Feisong, Zhu Yun and Zhu Zhenghai. Ten years have passed since Tang Shukun, Wang Xiang, Chen Biao, Huang Feisong and Zhou Xianchou started the preliminary investigation in July 2009; and another 5 years have passed since the formal investigation started in September 2014. Since March 2016, five rounds of modification contributed to the final version in August 2019. Of course, we admit that the volume should never claim perfection, yet finally, through meticulous works in sample testing, translation, editing and designing , the book actually has been increasingly polished day by day. We can be positive that the book, with fluent writing and intriguing pictures, is worth reading, and ready for publication with best wishes from the academia and our researchers, though we still harbor expectation for further and deeper exploration and modification.

This volume acknowledges the consistent efforts and wholehearted contribution of the following researchers:

一、 田野调查、文稿撰写与修订

| 第一章 安徽省手工造纸概述 | |
|---|---|
| 撰稿 | 初稿主执笔：黄飞松、朱赟、汤书昆<br>修订与补充完稿：汤书昆、陈敬宇、黄飞松<br>参与撰稿：郑久良、孙舰、尹航、何瑗 |

| 第二章 宣纸 | |
|---|---|
| 第一节 | 中国宣纸股份有限公司（地点：泾县榔桥镇乌溪村） |
| 田野调查 | 汤书昆、黄飞松、朱赟、陈彪、郑久良、程曦、许骏、王怡青、何瑗 |
| 撰稿 | 初稿主执笔：黄飞松<br>修订补稿：汤书昆、黄飞松、朱赟<br>参与撰稿：刘伟、王圣融、王怡青 |
| 第二节 | 泾县汪六吉宣纸有限公司（地点：泾县泾川镇茶冲村） |
| 田野调查 | 朱正海、汤书昆、黄飞松、朱赟、郑久良、罗文伯、程曦、何瑗、王圣融、王怡青、沈佳斐 |
| 撰稿 | 初稿主执笔：朱赟<br>修订补稿：黄飞松、汤书昆<br>参与撰稿：王圣融 |
| 第三节 | 安徽恒星宣纸有限公司（地点：泾县丁家桥镇后山村） |
| 田野调查 | 汤书昆、朱正海、黄飞松、朱赟、郑久良、罗文伯、汪梅、程曦、许骏、何瑗、王圣融、王怡青、沈佳斐 |
| 撰稿 | 初稿主执笔：朱赟<br>修订补稿：黄飞松、汤书昆<br>参与撰稿：郑久良、王圣融 |
| 第四节 | 泾县桃记宣纸有限公司（地点：泾县汀溪乡上漕村） |
| 田野调查 | 汤书昆、黄飞松、朱正海、朱大为、朱赟、郑久良、程曦、许骏、刘伟、何瑗、沈佳斐 |
| 撰稿 | 初稿主执笔：刘伟<br>修订补稿：黄飞松、汤书昆<br>参与撰稿：朱赟、沈佳斐 |
| 第五节 | 泾县汪同和宣纸厂（地点：泾县泾川镇古坝村） |
| 田野调查 | 黄飞松、朱正海、朱大为、何瑗、朱赟、郑久良、程曦、许骏、刘伟、王圣融、王怡青、沈佳斐 |
| 撰稿 | 初稿主执笔：程曦<br>修订补稿：黄飞松、汤书昆<br>参与撰稿：朱赟、王圣融 |
| 第六节 | 泾县双鹿宣纸有限公司（地点：泾县泾川镇城西工业集中区） |
| 田野调查 | 汤书昆、黄飞松、朱赟、郑久良、罗文伯、刘伟、何瑗、王圣融、王怡青、沈佳斐 |
| 撰稿 | 初稿主执笔：罗文伯<br>修订补稿：黄飞松、汤书昆<br>参与撰稿：郑久良、王圣融 |
| 第七节 | 泾县金星宣纸有限公司（地点：泾县丁家桥镇工业园区） |
| 田野调查 | 黄飞松、朱正海、汤书昆、郑久良、朱赟、程曦、刘伟、何瑗、王圣融、王怡青、沈佳斐 |
| 撰稿 | 初稿主执笔：郑久良<br>修订补稿：黄飞松、朱正海、朱赟<br>参与撰稿：王圣融、朱赟 |
| 第八节 | 泾县红叶宣纸有限公司（地点：泾县丁家桥镇枫坑村） |
| 田野调查 | 汤书昆、何瑗、朱赟、刘伟、王圣融、沈佳斐 |
| 撰稿 | 初稿主执笔：刘伟、黄飞松<br>修订补稿：汤书昆、黄飞松、朱赟<br>参与撰稿：王圣融、王怡青 |
| 第九节 | 安徽曹氏宣纸有限公司（地点：泾县丁家桥镇枫坑村） |
| 田野调查 | 汤书昆、黄飞松、朱赟、何瑗、郑久良、许骏、刘伟、程曦、王圣融、王怡青、沈佳斐 |
| 撰稿 | 初稿主执笔：许骏、黄飞松、汤书昆<br>修订补稿：汤书昆、黄飞松<br>参与撰稿：王怡青 |

| | |
|---|---|
| 第十节 | 泾县千年古宣宣纸有限公司（地点：泾县丁家桥镇小岭村） |
| 田野调查 | 汤书昆、朱赟、刘伟、何瑗、王圣融、王怡青、沈佳斐 |
| 撰稿 | 初稿主执笔：朱赟、黄飞松<br>修订补稿：黄飞松、汤书昆<br>参与撰稿：王怡青 |
| 第十一节 | 泾县小岭景辉纸业有限公司（地点：泾县丁家桥镇小岭村） |
| 田野调查 | 汤书昆、朱大为、朱赟、郑久良、何瑗、许骏、刘伟、王圣融、王怡青、沈佳斐 |
| 撰稿 | 初稿主执笔：朱赟、汤书昆<br>修订补稿：汤书昆、黄飞松<br>参与撰稿：刘伟、许骏 |
| 第十二节 | 泾县三星纸业有限公司（地点：泾县丁家桥镇李园村） |
| 田野调查 | 汤书昆、黄飞松、朱正海、朱赟、郑久良、许骏、刘伟、何瑗、王圣融、王怡青、沈佳斐 |
| 撰稿 | 初稿主执笔：刘伟、黄飞松<br>修订补稿：汤书昆、黄飞松<br>参与撰稿：朱赟 |
| 第十三节 | 安徽常春纸业有限公司（地点：泾县丁家桥镇工业园区） |
| 田野调查 | 汤书昆、黄飞松、朱赟、郑久良、罗文伯、汪梅、何瑗、王圣融、王怡青、沈佳斐 |
| 撰稿 | 初稿主执笔：朱赟<br>修订补稿：黄飞松、汤书昆<br>参与撰稿：郑久良、王圣融 |
| 第十四节 | 泾县玉泉宣纸纸业有限公司（地点：泾县丁家桥镇李园村） |
| 田野调查 | 汤书昆、黄飞松、朱大为、朱赟、刘伟、郑久良、何瑗、王圣融、王怡青、沈佳斐 |
| 撰稿 | 初稿主执笔：朱赟<br>修订补稿：黄飞松、汤书昆<br>参与撰稿：王圣融 |
| 第十五节 | 泾县吉星宣纸有限公司（地点：泾县泾川镇上坊村） |
| 田野调查 | 朱正海、黄飞松、汤书昆、朱赟、郑久良、程曦、许骏、刘伟、何瑗、沈佳斐 |
| 撰稿 | 初稿主执笔：程曦<br>修订补稿：汤书昆、黄飞松<br>参与撰稿：刘伟、朱赟 |
| 第十六节 | 泾县金宣堂宣纸厂（地点：泾县榔桥镇大庄村） |
| 田野调查 | 汤书昆、朱赟、郑久良、罗文伯、汪梅、何瑗、钟一鸣、王圣融、王怡青 |
| 撰稿 | 初稿主执笔：程曦<br>修订补稿：黄飞松、汤书昆<br>参与撰稿：何瑗、钟一鸣、朱赟 |
| 第十七节 | 泾县小岭金溪宣纸厂（地点：泾县丁家桥镇小岭村金坑村民组） |
| 田野调查 | 黄飞松、汤书昆、郑久良、何瑗、王圣融、朱赟、王怡青、沈佳斐 |
| 撰稿 | 初稿主执笔：刘伟<br>修订补稿：黄飞松、汤书昆<br>参与撰稿：朱赟、王圣融 |
| 第十八节 | 黄山白天鹅宣纸文化苑有限公司（地点：黄山市黄山区新明乡、耿城镇） |
| 田野调查 | 汤书昆、朱赟、郑久良、许骏 |
| 撰稿 | 初稿主执笔：朱赟、汤书昆<br>修订补稿：汤书昆 |

## 第三章　书画纸

| | |
|---|---|
| 第一节 | 泾县载元堂工艺厂（地点：泾县泾川镇城西工业集中区） |
| 田野调查 | 朱正海、朱赟、郑久良、罗文伯、程曦、何瑗、沈佳斐 |
| 撰稿 | 初稿主执笔：郑久良、程曦<br>修订补稿：黄飞松、汤书昆<br>参与撰稿：王圣融、朱赟 |

| | |
|---|---|
| 第二节 | 泾县小岭强坑宣纸厂（地点：泾县丁家桥镇小岭村） |
| 田野调查 | 刘伟、何瑗、钟一鸣、王圣融、郭延龙、沈佳斐 |
| 撰稿 | 初稿主执笔：刘伟<br>修订补稿：黄飞松、汤书昆<br>参与撰稿：何瑗、钟一鸣、郭延龙 |
| 第三节 | 泾县雄鹿纸厂（地点：泾县丁家桥镇李园村） |
| 田野调查 | 汤书昆、黄飞松、朱赟、郑久良、何瑗、王圣融、沈佳斐 |
| 撰稿 | 初稿主执笔：郑久良<br>修订补稿：黄飞松、汤书昆<br>参与撰稿：刘伟、王圣融 |
| 第四节 | 泾县紫光宣纸书画社（地点：泾县丁家桥镇后山村） |
| 田野调查 | 朱正海、黄飞松、朱赟、郑久良、刘伟、沈佳斐 |
| 撰稿 | 初稿主执笔：郑久良<br>修订补稿：黄飞松、汤书昆、朱赟 |
| 第五节 | 泾县小岭西山宣纸工艺厂（地点：泾县丁家桥镇小岭村） |
| 田野调查 | 汤书昆、黄飞松、朱赟、郑久良、程曦、许骏、刘伟、沈佳斐 |
| 撰稿 | 初稿主执笔：朱赟<br>修订补稿：黄飞松、汤书昆<br>参与撰稿：何瑗 |
| 第六节 | 安徽澄文堂宣纸艺术品有限公司（地点：泾县黄村镇九峰村） |
| 田野调查 | 黄飞松、朱赟、王圣融、王怡青、沈佳斐 |
| 撰稿 | 初稿主执笔：黄飞松<br>修订补稿：汤书昆、黄飞松<br>参与撰稿：王怡青 |

**第四章　皮纸**

| | |
|---|---|
| 第一节 | 泾县守金皮纸厂（地点：泾县泾川镇园林村） |
| 田野调查 | 朱正海、黄飞松、朱赟、郑久良、罗文伯、何瑗、王圣融、郭延龙、沈佳斐 |
| 撰稿 | 初稿主执笔：郑久良<br>修订补稿：黄飞松、汤书昆、朱赟 |
| 第二节 | 泾县小岭驰星纸厂（地点：泾县丁家桥镇小岭村） |
| 田野调查 | 汤书昆、黄飞松、朱正海、朱赟、郑久良、罗文伯、王圣融、沈佳斐 |
| 撰稿 | 初稿主执笔：黄飞松、朱赟<br>修订补稿：汤书昆<br>参与撰稿：罗文伯 |
| 第三节 | 潜山县星杰桑皮纸厂（地点：安庆市潜山县官庄镇坛畈村） |
| 田野调查 | 汤书昆、朱赟、郑久良、刘伟、程曦 |
| 撰稿 | 初稿主执笔：汤书昆、郑久良<br>修订补稿：汤书昆 |
| 第四节 | 岳西县金丝纸业有限公司（地点：安庆市岳西县毛尖山乡板舍村） |
| 田野调查 | 汤书昆、朱赟、汪淳、郑久良、刘伟、王圣融、尹航 |
| 撰稿 | 初稿主执笔：王圣融、汤书昆<br>修订补稿：汤书昆<br>参与撰稿：朱赟、刘伟 |
| 第五节 | 歙县深渡镇棉溪村（地点：黄山市歙县深渡镇棉溪村） |
| 田野调查 | 汤书昆、陈琪、刘靖、朱赟、王秀伟、朱岱、沈佳斐、孙燕、叶婷婷 |
| 撰稿 | 初稿主执笔：汤书昆、朱赟<br>修订补稿：汤书昆、陈琪 |
| 第六节 | 黄山市三昕纸业有限公司（地点：黄山市休宁县海阳镇晓角村） |
| 田野调查 | 汤书昆、刘靖、陈政、李宪奇、朱赟、王秀伟、郑久良、陈琪、朱岱、沈佳斐 |
| 撰稿 | 初稿主执笔：汤书昆、郑久良<br>修订补稿：汤书昆 |

| 第七节 | 歙县六合村（地点：黄山市歙县杞梓里镇六合村） |
|---|---|
| 田野调查 | 陈琪、汤书昆、陈政、孙燕、叶婷婷、沈佳斐 |
| 撰稿 | 初稿主执笔：汤书昆、孙燕<br>修订补稿：汤书昆、陈琪 |

第五章　竹纸

| 第一节 | 歙县青峰村（地点：黄山市歙县青峰村） |
|---|---|
| 田野调查 | 陈琪、汤书昆、贡斌、李宪奇、沈佳斐 |
| 撰稿 | 初稿主执笔：汤书昆、王圣融<br>修订补稿：汤书昆、陈琪<br>参与撰稿：陈琪 |
| 第二节 | 泾县孤峰村（地点：泾县昌桥乡孤峰村、泾川镇古坝村、黄村镇九峰村） |
| 田野调查 | 陈彪、黄飞松、朱正海、沈佳斐 |
| 撰稿 | 初稿主执笔：黄飞松、陈彪<br>修订补稿：黄飞松、汤书昆 |
| 第三节 | 金寨县燕子河镇（地点：金寨县燕子河镇龙马村 / 燕溪村） |
| 田野调查 | 汤书昆、黄飞松、张静明、蓝强 |
| 撰稿 | 初稿主执笔：汤书昆、黄飞松<br>修订补稿：汤书昆、黄飞松 |

第六章　加工纸

| 第一节 | 安徽省掇英轩书画用品有限公司（地点：巢湖市黄麓镇） |
|---|---|
| 田野调查 | 汤书昆、陈彪、李宪奇、朱赟、郑久良、刘伟、王圣融、程曦、叶珍珍、沈佳斐 |
| 撰稿 | 初稿主执笔：汤书昆、钟一鸣<br>修订补稿：汤书昆、李宪奇<br>参与撰稿：叶珍珍 |
| 第二节 | 泾县艺英轩宣纸工艺品厂（地点：泾县琴溪镇赤滩街道） |
| 田野调查 | 汤书昆、朱赟、刘伟、郑久良、程曦、许骏、王圣融、郭延龙、王怡青、沈佳斐 |
| 撰稿 | 初稿主执笔：朱赟<br>修订补稿：汤书昆 |
| 第三节 | 泾县艺宣阁宣纸工艺品有限公司（地点：泾县泾川镇城西工业集中区） |
| 田野调查 | 黄飞松、朱大为、朱赟、郑久良、程曦、许骏、刘伟、王圣融、沈佳斐 |
| 撰稿 | 初稿主执笔：黄飞松、朱赟<br>修订补稿：黄飞松、汤书昆<br>参与撰稿：王圣融 |
| 第四节 | 泾县宣艺斋宣纸工艺厂（地点：泾县泾川镇城西工业集中区） |
| 田野调查 | 黄飞松、郑久良、程曦、朱赟、王圣融、郭延龙、沈佳斐 |
| 撰稿 | 初稿主执笔：程曦<br>修订补稿：汤书昆、黄飞松<br>参与撰稿：郑久良、刘伟、王圣融 |
| 第五节 | 泾县贡玉堂宣纸工艺厂（地点：泾县黄村镇紫阳村） |
| 田野调查 | 朱正海、朱大为、朱赟、郑久良、程曦、许骏、刘伟、王圣融、郭延龙、沈佳斐 |
| 撰稿 | 初稿主执笔：刘伟<br>修订补稿：汤书昆、黄飞松<br>参与撰稿：郭延龙 |
| 第六节 | 泾县博古堂宣纸工艺厂（地点：泾县丁家桥镇小岭村） |
| 田野调查 | 汤书昆、朱正海、朱赟、郑久良、王圣融、郭延龙、沈佳斐 |
| 撰稿 | 初稿主执笔：朱赟<br>修订补稿：汤书昆、黄飞松 |
| 第七节 | 泾县汇宣堂宣纸工艺厂（地点：泾县泾川镇曹家村） |
| 田野调查 | 汤书昆、朱正海、朱赟、郑久良、刘伟、王圣融、郭延龙、沈佳斐 |
| 撰稿 | 初稿主执笔：郑久良<br>修订补稿：汤书昆、黄飞松<br>参与撰稿：王圣融、朱赟 |

| 第八节 | 泾县风和堂宣纸加工厂（地点：泾县泾川镇五星村） |
| --- | --- |
| 田野调查 | 汤书昆、黄飞松、沈佳斐 |
| 撰稿 | 初稿主执笔：沈佳斐、汤书昆、黄飞松<br>修订补稿：汤书昆、黄飞松 |

第七章　工具

| 第一节 | 泾县明堂纸帘工艺厂（地点：泾县丁家桥镇） |
| --- | --- |
| 田野调查 | 朱赟、刘伟、王圣融、郭延龙、沈佳斐 |
| 撰稿 | 初稿主执笔：刘伟<br>修订补稿：黄飞松、汤书昆<br>参与撰稿：王圣融 |
| 第二节 | 泾县全勇纸帘工艺厂（地点：泾县丁家桥镇工业园区） |
| 田野调查 | 黄飞松、朱赟、刘伟、王圣融、郭延龙、沈佳斐 |
| 撰稿 | 初稿主执笔：刘伟<br>修订补稿：汤书昆、黄飞松<br>参与撰稿：王圣融 |
| 第三节 | 泾县后山大剪刀作坊（地点：泾县丁家桥镇后山村） |
| 田野调查 | 朱正海、朱赟、黄飞松、刘伟、王圣融、廖莹文、沈佳斐 |
| 撰稿 | 初稿主执笔：黄飞松<br>修订补稿：黄飞松、汤书昆<br>参与撰稿：朱赟 |

二、 技术与辅助工作

| 手工纸分布示意图绘制 | 主持：郭延龙、陈龑<br>参与绘制：郭延龙、朱赟、何瑗、姚的卢、叶珍珍 |
| --- | --- |
| 实物纸样测试分析 | 主持：朱赟、陈龑<br>测试：朱赟、陈龑、何瑗、郑久良、程曦、汪宣伯、钟一鸣、叶婷婷、郭延龙、王圣融、王怡青、黄立新、赵梦君、王裕玲、宋福星 |
| 实物纸样拍摄 | 黄晓飞 |
| 实物纸样整理 | 汤书昆、朱赟、倪盈盈、郑斌、付成云、蔡婷婷、刘伟、何瑗、王圣融、叶珍珍、陈龑、王怡青 |
| 实物纸样透光纤维图制作 | 朱赟、陈龑、何瑗、刘伟、王怡青、廖莹文 |

三、 总序、编撰说明、附录与后记部分

总序

| 撰稿 | 汤书昆 |
| --- | --- |

编撰说明

| 撰稿 | 汤书昆、朱赟 |
| --- | --- |

附录

| 术语整理编制 | 朱赟、陈登航、付成云、秦庆 |
| --- | --- |
| 图目整理编制 | 朱赟、王怡青、王圣融、叶珍珍、廖莹文 |
| 表目整理编制 | 朱赟、王怡青、王圣融、叶珍珍、廖莹文 |

后记

| 撰稿 | 汤书昆 |
| --- | --- |

四、 统稿与翻译

| 统稿主持 | 汤书昆 |
| --- | --- |
| 统稿规划 | 朱赟、朱正海 |
| 翻译主持 | 方媛媛 |
| 其他参与翻译人员 | 刘丽、汪晓婧、高倩、高丁祎、胡昕、刘惠敏 |

1. Field investigation, writing and modification

Chapter I　Introduction to Handmade Paper in Anhui Province

| | |
|---|---|
| Writers | First manuscript written by: Huang Feisong, Zhu Yun, Tang Shukun<br>Modified by: Tang Shukun, Chen Jingyu, Huang Feisong<br>Zheng Jiuliang, Sun Jian, Yin Hang, He Ai have also contributed to the writing |

Chapter II　Xuan Paper

| | |
|---|---|
| Section 1 | China Xuan Paper Co., Ltd. (Location: Wuxi Village in Langqiao Town of Jingxian County) |
| Field investigators | Tang Shukun, Huang Feisong, Zhu Yun, Chen Biao, Zheng Jiuliang, Cheng Xi, Xu Jun, Wang Yiqing, He Ai |
| Writers | First manuscript written by: Huang Feisong<br>Modified by: Tang Shukun, Huang Feisong, Zhu Yun<br>Liu Wei, Wang Shengrong, Wang Yiqing have also contributed to the writing |
| Section 2 | Wangliuji Xuan Paper Co., Ltd. in Jingxian County (Location: Chachong Village in Jingchuan Town of Jingxian County) |
| Field investigators | Zhu Zhenghai, Tang Shukun, Huang Feisong, Zhu Yun, Zheng Jiuliang, Luo Wenbo, Cheng Xi, He Ai, Wang Shengrong, Wang Yiqing, Shen Jiafei |
| Writers | First manuscript written by: Zhu Yun<br>Modified by: Huang Feisong, Tang Shukun<br>Wang Shengrong has also contributed to the writing |
| Section 3 | Anhui Hengxing Xuan Paper Co., Ltd. (Location: Houshan Village in Dingjiaqiao Town of Jingxian County) |
| Field investigators | Tang Shukun, Zhu Zhenghai, Huang Feisong, Zhu Yun, Zheng Jiuliang, Luo Wenbo, Wang Mei, Cheng Xi, Xu Jun, He Ai, Wang Shengrong, Wang Yiqing, Shen Jiafei |
| Writers | First manuscript written by: Zhu Yun<br>Modified by: Huang Feisong, Tang Shukun<br>Zheng Jiuliang, Wang Shengrong have also contributed to the writing |
| Section 4 | Taoji Xuan Paper Co., Ltd. in Jingxian County (Location: Shangcao Village in Tingxi Town of Jingxian County) |
| Field investigators | Tang Shukun, Huang Feisong, Zhu Zhenghai, Zhu Dawei, Zhu Yun, Zheng Jiuliang, Cheng Xi, Xu Jun, Liu Wei, He Ai, Shen Jiafei |
| Writers | First manuscript written by: Liu Wei<br>Modified by: Huang Feisong, Tang Shukun<br>Zhu Yun, Shen Jiafei have also contributed to the writing |
| Section 5 | Wangtonghe Xuan Paper Factory in Jingxian County (Location: Guba Village in Jingchuan Town of Jingxian County) |
| Field investigators | Huang Feisong, Zhu Zhenghai, Zhu Dawei, He Ai, Zhu Yun, Zheng Jiuliang, Cheng Xi, Xu Jun, Liu Wei, Wang Shengrong, Wang Yiqing, Shen Jiafei |
| Writers | First manuscript written by: Cheng Xi<br>Modified by: Huang Feisong, Tang Shukun<br>Zhu Yun, Wang Shengrong have also contributed to the writing |
| Section 6 | Shuanglu Xuan Paper Co., Ltd. in Jingxian County (Location: Chengxi Industrial Park in Jingchuan Town of Jingxian County) |
| Field investigators | Tang Shukun, Huang Feisong, Zhu Yun, Zheng Jiuliang, Luo Wenbo, Liu Wei, He Ai, Wang Shengrong, Wang Yiqing, Shen Jiafei |
| Writers | First manuscript written by: Luo Wenbo<br>Modified by: Huang Feisong, Tang Shukun<br>Zheng Jiuliang, Wang Shengrong have also contributed to the writing |
| Section 7 | Jinxing Xuan Paper Co., Ltd. in Jingxian County (Location: Industrial Zone in Dingjiaqiao Town of Jingxian County) |
| Field investigators | Huang Feisong, Zhu Zhenghai, Tang Shukun, Zheng Jiuliang, Zhu Yun, Cheng Xi, Liu Wei, He Ai, Wang Shengrong, Wang Yiqing, Shen Jiafei |
| Writers | First manuscript written by: Zheng Jiuliang<br>Modified by: Huang Feisong, Zhu Zhenghai, Zhu Yun<br>Wang Shengrong, Zhu Yun have also contributed to the writing |
| Section 8 | Hongye Xuan Paper Co., Ltd. in Jingxian County (Location: Fengkeng Village in Dingjiaqiao Town of Jingxian County) |
| Field investigators | Tang Shukun, He Ai, Zhu Yun, Liu Wei, Wang Shengrong, Shen Jiafei |
| Writers | First manuscript written by: Liu Wei, Huang Feisong<br>Modified by: Tang Shukun, Huang Feisong, Zhu Yun<br>Wang Shengrong, Wang Yiqing have also contributed to the writing |
| Section 9 | Anhui Caoshi Xuan Paper Co., Ltd. (Location: Fengkeng Village in Dingjiaqiao Town of Jingxian County) |
| Field investigators | Tang Shukun, Huang Feisong, Zhu Yun, He Ai, Zheng Jiuliang, Xu Juan, Liu Wei, Cheng Xi, Wang Shengrong, Wang Yiqing, Shen Jiafei |
| Writers | First manuscript written by: Xu Jun, Huang Feisong, Tang Shukun<br>Modified by: Tang Shukun, Huang Feisong<br>Wang Yiqing has also contributed to the writing |
| Section 10 | Millennium Xuan Paper Co., Ltd. in Jingxian County (Location: Xiaoling Village in Dingjiaqiao Town of Jingxian County) |
| Field investigators | Tang Shukun, Zhu Yun, Liu Wei, He Ai, Wang Shengrong, Wang Yiqing, Shen Jiafei |
| Writers | First manuscript written by: Zhu Yun, Huang Feisong<br>Modified by: Huang Feisong, Tang Shukun<br>Wang Yiqing has also contributed to the writing |

| | |
|---|---|
| Section 11 | Xiaoling Jinghui Paper Co., Ltd. in Jingxian County (Location: Xiaoling Village in Dingjiaqiao Town of Jingxian County) |
| Field investigators | Tang Shukun, Zhu Dawei, Zhu Yun, Zheng Jiuliang, He Ai, Xu Jun, Liu Wei, Wang Shengrong, Wang Yiqing, Shen Jiafei |
| Writers | First manuscript written by: Zhu Yun, Tang Shukun<br>Modified by: Tang Shukun, Huang Feisong<br>Liu Wei, Xu Jun have also contributed to the writing |
| Section 12 | Sanxing Paper Co., Ltd. in Jingxian County (Location: Liyuan Village in Dingjiaqiao Town of Jingxian County) |
| Field investigators | Tang Shukun, Huang Feisong, Zhu Zhenghai, Zhu Yun, Zheng Jiuliang, Xu Jun, Liu Wei, He Ai, Wang Shengrong, Wang Yiqing, Shen Jiafei |
| Writers | First manuscript written by: Liu Wei, Huang Feisong<br>Modified by: Tang Shukun, Huang Feisong<br>Zhu Yun has also contributed to the writing |
| Section 13 | Anhui Changchun Paper Co., Ltd. (Location: Industrial Zone in Dingjiaqiao Town of Jingxian County) |
| Field investigators | Tang Shukun, Huang Feisong, Zhu Yun, Zheng Jiuliang, Luo Wenbo, Wang Mei, He Ai, Wang Shengrong, Wang Yiqing, Shen Jiafei |
| Writers | First manuscript written by: Zhu Yun<br>Modified by: Huang Feisong, Tang Shukun<br>Zheng Jiulaing, Wang Shengrong have also contributed to the writing |
| Section 14 | Yuquan Xuan Paper Co., Ltd. in Jingxian County (Location: Liyuan Village in Dingjiaqiao Town of Jingxian County) |
| Field investigators | Tang Shukun, Huang Feisong, Zhu Dawei, Zhu Yun, Liu Wei, Zheng Jiuliang, He Ai, Wang Shengrong, Wang Yiqing, Shen Jiafei |
| Writers | First manuscript written by: Zhu Yun<br>Modified by: Huang Feisong, Tang Shukun<br>Wang Shengrong has also contributed to the writing |
| Section 15 | Jixing Xuan Paper Co., Ltd. in Jingxian County (Location: Shangfang Village in Jingchuan Town of Jingxian County) |
| Field investigators | Zhu Zhenghai, Huang Feisong, Tang Shukun, Zhu Yun, Zheng Jiuliang, Cheng Xi, Xu Jun, Liu Wei, He Ai, Shen Jiafei |
| Writers | First manuscript written by: Cheng Xi<br>Modified by: Tang Shukun, Huang Feisong<br>Liu Wei, Zhu Yun have also contributed to the writing |
| Section 16 | Jinxuantang Xuan Paper Factory in Jingxian County (Location: Dazhuang Village in Langqiao Town of Jingxian County) |
| Field investigators | Tang Shukun, Zhu Yun, Zheng Jiuliang, Luo Wenbo, Wang Mei, He Ai, Zhong Yiming, Wang Shengrong, Wang Yiqing |
| Writers | First manuscript written by: Cheng Xi<br>Modified by: Huang Feisong, Tang Shukun<br>He Ai, Zhong Yiming, Zhu Yun have also contributed to the writing |
| Section 17 | Xiaoling Jinxi Xuan Paper Factory in Jingxian County (Location: Jinkeng Villages' Group of Xiaoling Village in Dingjiaqiao Town of Jingxian County) |
| Field investigators | Huang Feisong, Tang Shukun, Zheng Jiuliang, He Ai, Wang Shengrong, Zhu Yun, Wang Yiqing, Shen Jiafei |
| Writers | First manuscript written by: Liu Wei<br>Modified by: Huang Feisong, Tang Shukun<br>Zhu Yun, Wang Shengrong have also contributed to the writing |
| Section 18 | Huangshan Baitian'e Xuan Paper Cultural Garden Co., Ltd. (Location: Xinming Town and Gengcheng Town of Huangshan District in Huangshan City) |
| Field investigators | Tang Shukun, Zhu Yun, Zheng Jiuliang, Xu Jun |
| Writers | First manuscript written by: Zhu Yun, Tang Shukun<br>Modified by: Tang Shukun |

Chapter III Calligraphy and Painting Paper

| | |
|---|---|
| Section 1 | Zaiyuantang Xuan Paper Factory in Jingxian County (Location: Chengxi Industrial Zone in Jingchuan Town of Jingxian County) |
| Field investigators | Zhu Zhenghai, Zhu Yun, Zheng Jiuliang, Luo Wenbo, Cheng Xi, He Ai, Shen Jiafei |
| Writers | First manuscript written by: Zheng Jiuliang, Cheng Xi<br>Modified by: Huang Feisong, Tang Shukun<br>Wang Shengrong, Zhu Yun have also contributed to the writing |
| Section 2 | Xiaoling Qiangkeng Xuan Paper Factory in Jingxian County (Location: Xiaoling Village in Dingjiaqiao Town of Jingxian County) |
| Field investigators | Liu Wei, He Ai, Zhong Yiming, Wang Shengrong, Guo Yanlong, Shen Jiafei |
| Writers | First manuscript written by: Liu Wei<br>Modified by: Huang Feisong, Tang Shukun<br>He Ai, Zhong Yiming, Guo Yanlong have also contributed to the writing |
| Section 3 | Xionglu Xuan Paper Factory in Jingxian County (Location: Liyuan Village in Dingjiaqiao Town of Jingxian County) |
| Field investigators | Tang Shukun, Huang Feisong, Zhu Yun, Zheng Jiuliang, He Ai, Wang Shengrong, Shen Jiafei |
| Writers | First manuscript written by: Zheng Jiuliang<br>Modified by: Huang Feisong, Tang Shukun<br>Liu Wei, Wang Shengrong have also contributed to the writing |

| | |
|---|---|
| Section 4 | Ziguang Xuan Paper Factory in Jingxian County (Location: Houshan Village in Dingjiaqiao Town of Jingxian County) |
| Field investigators | Zhu Zhenghai, Huang Feisong, Zhu Yun, Zheng Jiuliang, Liu Wei, Shen Jiafei |
| Writers | First manuscript written by: Zheng Jiuliang<br>Modified by: Huang Feisong, Tang Shukun, Zhu Yun |
| Section 5 | Xiaoling Xishan Xuan Paper Factory in Jingxian County (Location: Xiaoling Village in Dingjiaqiao Town of Jingxian County) |
| Field investigators | Tang Shukun, Huang Feisong, Zhu Yun, Zheng Jiuliang, Cheng Xi, Xu Jun, Liu Wei, Shen Jiafei |
| Writers | First manuscript written by: Zhu Yun<br>Modified by: Huang Feisong, Tang Shukun<br>He Ai has also contributed to the writing |
| Section 6 | Chengwentang Xuan Paper Co., Ltd. in Anhui Province (Location: Jiufeng Village in Huangcun Town of Jingxian County) |
| Field investigators | Huang Feisong, Zhu Yun, Wang Shengrong, Wang Yiqing, Shen Jiafei |
| Writers | First manuscript written by: Huang Feisong<br>Modified by: Tang Shukun, Huang Feisong<br>Wang Yiqing has also contributed to the writing |

Chapter IV Bast Paper

| | |
|---|---|
| Section 1 | Shoujin Bast Paper Factory in Jingxian County (Location: Yuanlin Village in Jingchuan Town of Jingxian County) |
| Field investigators | Zhu Zhenghai, Huang Feisong, Zhu Yun, Zheng Jiuliang, Luo Wenbo, He Ai, Wang Shengrong, Guo Yanlong, Shen Jiafei |
| Writers | First manuscript written by: Zheng Jiuliang<br>Modified by: Huang Feisong, Tang Shukun, Zhu Yun |
| Section 2 | Xiaoling Chixing Paper Factory in Jingxian County (Location: Xiaoling Village in Dingjiaqiao Town of Jingxian County) |
| Field investigators | Tang Shunkun, Huang Feisong, Zhu Zhenghai, Zhu Yun, Zheng Jiuliang, Luo Wenbo, Wang Shengrong, Shen Jiafei |
| Writers | First manuscript written by: Huang Feisong, Zhu Yun<br>Modified by: Tang Shukun<br>Luo Wenbo has also contributed to the writing |
| Section 3 | Xingjie Mulberry Bark Paper Factory in Qianshan County (Location: Tanfan Village in Guanzhuang Town of Qianshan County in Anqing City) |
| Field investigators | Tang Shukun, Zhu Yun, Zheng Jiuliang, Liu Wei, Cheng Xi |
| Writers | First manuscript written by: Tang Shukun, Zheng Jiuliang<br>Modified by: Tang Shukun |
| Section 4 | Jinsi Paper Co., Ltd. in Yuexi County (Location: Banshe Village in Maojianshan Town of Yuexi County in Anqing City) |
| Field investigators | Tang Shukun, Zhu Yun, Wang Chun, Zheng Jiuliang, Liu Wei, Wang Shengrong, Yin Hang |
| Writers | First manuscript written by: Wang Shengrong, Tang Shukun<br>Modified by: Tang Shukun<br>Zhu Yun, Liu Wei have also contributed to the writing |
| Section 5 | Mianxi Village in Shendu Town of Shexian County (Location: Mianxi Village in Shendu Town of Shexian County in Huangshan City) |
| Field investigators | Tang Shukun, Chen Qi, Liu Jing, Zhu Yun, Wang Xiuwei, Zhu Dai, Shen Jiafei, Sun Yan, Ye Tingting |
| Writers | First manuscript written by: Tang Shukun, Zhu Yun<br>Modified by: Tang Shukun, Chen Qi |
| Section 6 | Sanxin Paper Co., Ltd. in Huangshan City (Location: Xiaojiao Village in Haiyang Town of Xiuning County in Huangshan City) |
| Field investigators | Tang Shukun, Liu Jing, Chen Zheng, Li Xianqi, Zhu Yun, Wang Xiuwei, Zheng Jiuliang, Chen Qi, Zhu Dai, Shen Jiafei |
| Writers | First manuscript written by: Tang Shukun, Zheng Jiuliang<br>Modified by: Tang Shukun |
| Section 7 | Liuhe Village in Shexian County (Location: Liuhe Village in Qizili Town of Shexian County in Huangshan City) |
| Field investigators | Chen Qi, Tang Shukun, Chen Zheng, Sun Yan, Ye Tingting, Shen Jiafei |
| Writers | First manuscript written by: Tang Shukun, Sun Yan<br>Modified by: Tang Shukun, Chen Qi |

Chapter V Bamboo Paper

| | |
|---|---|
| Section 1 | Qingfeng Village in Shexian County (Location: Qingfeng Village in Shexian County of Huangshan City) |
| Field investigators | Chen Qi, Tang Shukun, Gong Bin, Li Xianqi, Shen Jiafei |
| Writers | First manuscript written by: Tang Shukun, Wang Shengrong<br>Modified by: Tang Shukun, Chen Qi<br>Chen Qi has also contributed to the writing |
| Section 2 | Gufeng Village in Jingxian County (Location: Gufeng Village in Changqiao Town, Guba Village in Jingchuan Town, Jiufeng Village in Huangcun Town, Jingxian County) |
| Field investigators | Chen Biao, Huang Feisong, Zhu Zhenghai, Shen Jiafei |
| Writers | First manuscript written by: Huang Feisong, Chen Biao<br>Modified by: Huang Feisong, Tang Shukun |

| Section 3 | Yanzihe Town in Jinzhai County (Location: Longma / Yanxi Village in Yanzihe Town of Jinzhai County) |
| --- | --- |
| Field investigators | Tang Shukun, Huang Feisong, Zhang Jingming, Lan Qiang |
| Writers | First manuscript written by: Tang Shukun, Huang Feisong<br>Modified by: Tang Shukun, Huang Feisong |



## Chapter VI Processed Paper

| Section 1 | Duoyingxuan Calligraphy and Painting Supplies Co., Ltd. in Anhui Province (Location: Huanglu Town of Chaohu City) |
| --- | --- |
| Field investigators | Tang Shukun, Chen Biao, Li Xianqi, Zhu Yun, Zheng Jiuliang, Liu Wei, Wang Shengrong, Cheng Xi, Ye Zhenzhen, Shen Jiafei |
| Writers | First manuscript written by: Tang Shukun, Zhong Yiming<br>Modified by: Tang Shukun, Li Xianqi<br>Ye Zhenzhen has also contributed to the writing |
| Section 2 | Yiyingxuan Xuan Paper Craft Factory in Jingxian County (Location: Chitan Street in Qinxi Town of Jingxian County) |
| Field investigators | Tang Shukun, Zhu Yun, Liu Wei, Zheng Jiuliang, Cheng Xi, Xu Jun, Wang Shengrong, Guo Yanlong, Wang Yiqing, Shen Jiafei |
| Writers | First manuscript written by: Zhu Yun<br>Modified by: Tang Shukun |
| Section 3 | Yixuange Xuan Paper Craft Co., Ltd. in Jingxian County (Location: Chengxi Industrial Zone in Jingchuan Town of Jingxian County) |
| Field investigators | Huang Feisong, Zhu Dawei, Zhu Yun, Zheng Jiuliang, Cheng Xi, Xu Jun, Liu Wei, Wang Shengrong, Shen Jiafei |
| Writers | First manuscript written by: Huang Feisong, Zhu Yun<br>Modified by: Huang Feisong, Tang Shukun<br>Wang Shengrong has also contributed to the writing |
| Section 4 | Xuanyizhai Xuan Paper Craft Factory in Jingxian County (Location: Chengxi Industrial Zone in Jingchuan Town of Jingxian County) |
| Field investigators | Huang Feisong, Zheng Jiuliang, Cheng Xi, Zhu Yun, Wang Shengrong, Guo Yanlong, Shen Jiafei |
| Writers | First manuscript written by: Cheng Xi<br>Modified by: Tang Shukun, Huang Feisong<br>Zheng Jiuliang, Liu Wei, Wang Shengrong have also contributed to the writing |
| Section 5 | Gongyutang Xuan Paper Craft Factory in Jingxian County (Location: Ziyang Village in Huangcun Town of Jingxian County) |
| Field investigators | Zhu Zhenghai, Zhu Dawei, Zhu Yun, Zheng Liangjiu, Cheng Xi, Xu Jun, Liu Wei, Wang Shengrong, Guo Yanlong, Shen Jiafei |
| Writers | First manuscript written by: Liu Wei<br>Modified by: Tang Shukun, Huang Feisong<br>Guo Yanlong has also contributed to the writing |
| Section 6 | Bogutang Xuan Paper Craft Factory in Jingxian County (Location: Xiaoling Village in Dingjiaqiao Town of Jingxian County) |
| Field investigators | Tang Shukun, Zhu Zhenghai, Zhu Yun, Zheng Jiuliang, Wang Shengrong, Guo Yanlong, Shen Jiafei |
| Writers | First manuscript written by: Zhu Yun<br>Modified by: Tang Shukun, Huang Feisong |
| Section 7 | Huixuantang Xuan Paper Craft Factory in Jingxian County (Location: Caojia Village in Jingchuan Town of Jingxian County) |
| Field investigators | Tang Shukun, Zhu Zhenghai, Zhu Yun, Zheng Jiuliang, Liu Wei, Wang Shengrong, Guo Yanlong, Shen Jiafei |
| Writers | First manuscript written by: Zheng Jiuliang<br>Modified by: Tang Shukun, Huang Feisong<br>Wang Shengrong, Zhu Yun have also contributed to the writing |
| Section 8 | Fenghetang Xuan Paper Craft Factory in Jingxian County (Locaion: Wuxing Village in Jingchuan Town of Jingxian County) |
| Field investigators | Tang Shukun, Huang Feisong, Shen Jiafei |
| Writers | First manuscript written by: Shen Jiafei, Tang Shukun, Huang Feisong<br>Modified by: Tang Shukun, Huang Feisong |

## Chapter VII Tools

| Section 1 | Mingtang Papermaking Screen Craft Factory in Jingxian County (Location: Dingjiaqiao Town in Jingxian County) |
| --- | --- |
| Field investigators | Zhu Yun, Liu Wei, Wang Shengrong, Guo Yanlong, Shen Jiafei |
| Writers | First manuscript written by: Liu Wei<br>Modified by: Huang Feisong, Tang Shukun<br>Wang Shengrong has also contributed to the writing |
| Section 2 | Quanyong Papermaking Screen Craft Factory in Jingxian County (Location: Industrial Zone in Dingjiaqiao Town of Jingxian County) |
| Field investigators | Huang Feisong, Zhu Yun, Liu Wei, Wang Shengrong, Guo Yanlong, Shen Jiafei |
| Writers | First manuscript written by: Liu Wei<br>Modified by: Tang Shukun, Huang Feisong<br>Wang Shengrong has also contributed to the writing |

| Section 3 | Houshan Shears Workshop in Jingxian County (Location: Houshan Village in Dingjiaqiao Town of Jingxian County) |
|---|---|
| Field investigators | Zhu Zhenghai, Zhu Yun, Huang Feisong, Liu Wei, Wang Shengrong, Liao Yingwen, Shen Jiafei |
| Writers | First manuscript written by: Huang Feisong<br>Modified by: Huang Feisong, Tang Shukun<br>Zhu Yun has also contributed to the writing |



## 2. Technical Analysis and Other Related Works

| | |
|---|---|
| Handmade paper distribution maps | Headed by: Guo Yanlong, Chen Yan<br>Drawn by: Guo Yanlong, Zhu Yun, He Ai, Yao Dilu, Ye Zhenzhen |
| Sample paper test | Headed by: Zhu Yun, Chen Yan<br>Members: Zhu Yun, Chen Yan, He Ai, Zheng Jiuliang, Cheng Xi, Liu Wei, Wang Xuanbo, Zhong Yiming, Ye Tingting, Guo Yanlong, Wang Shengrong, Wang Yiqing, Huang Lixin, Zhao Mengjun, Wang Yuling, Song Fuxing |
| Paper sample pictures | Photographed by: Huang Xiaofei |
| Paper sample | Sorted by: Tang Shukun, Zhu Yun, Ni Yingying, Zheng Bin, Fu Chengyun, Cai Tingting, Liu Wei, He Ai, Wang Shengrong, Ye Zhenzhen, Chen Yan, Wang Yiqing |
| Paper pictures showing the paper fiber | Produced by: Zhu Yun, Chen Yan, He Ai, Liu Wei, Wang Yiqing, Liao Yingwen |

## 3. Preface, Introduction to the Writing Norms, Appendices and Epilogue

| | |
|---|---|
| **Preface** | |
| Writer | Tang Shukun |
| **Introduction to the Writing Norms** | |
| Writers | Tang Shukun, Zhu Yun |
| **Appendices** | |
| Terminology | Zhu Yun, Chen Denghang, Fu Chengyun, Qin Qing |
| List of figures | Zhu Yun, Wang Yiqing, Wang Shengrong, Ye Zhenzhen, Liao Yingwen |
| List of tables | Zhu Yun, Wang Yiqing, Wang Shengrong, Ye Zhenzhen, Liao Yingwen |
| **Epilogue** | |
| Writer | Tang Shukun |

## 4. Modification and Translation

| | |
|---|---|
| Director of modification and verification | Tang Shukun |
| Modification planner | Zhu Yun, Zhu Zhenghai |
| Chief translator and director of Translation | Fang Yuanyuan |
| Other translators | Liu Li, Wang Xiaojing, Gao Qian, Gao Dingyi, Hu Xin, Liu Huimin |

在历时3年半的多轮修订、增补与统稿工作中，汤书昆、黄飞松、朱赟、朱正海、方媛媛、陈敬宇、郭延龙等作为主持人或重要内容模块的负责人，对文稿内容、图片与示意图的修订增补，代表性纸样的测试分析，英文翻译，文献注释考订，表述格式的规范化，数据与表述的准确性核实等方面做了大量扎实而辛苦的工作。而责任编辑团队、北京敬人工作室设计团队、北京雅昌艺术印刷有限公司印制团队精益求精、力求完美的反复打磨，都是《安徽卷》书稿从最初的田野记录式提炼整理，到以今天的面貌和质量展现不容忽视的工作。

在《安徽卷》的田野调查过程中，先后得到中国宣纸股份有限公司胡文军先生、黄山市地方志办公室陈政先生、岳西县文化馆汪淳先生、金寨县政府王玉华先生等多位手工造纸传统技艺和非物质文化遗产研究与保护专家的帮助，在《中国手工纸文库·安徽卷》正式出版之际，我谨代表田野调查和文稿撰写团队，向记名与未曾记名的支持者表达真诚的谢意！



汤书昆

2019年12月于中国科学技术大学

Tang Shukun, Huang Feisong, Zhu Yun, Zhu Zhenghai, Fang Yuanyuan, Chen Jingyu, Guo Yanlong et al., who were in charge of the writing, modification and other related works, all contributed their efforts to the completion of this book in the past three and half years. Their meticulous efforts in writing, drawing or photographing, mapping, technical analysis, translating, format modifying, noting and proofreading should be recognized and eulogized in the achievement of the high-quality work. The editors of the book, Beijing Jingren Book Design Studio, Bejing Artron Art Printing Co., Ltd. have been dedicated to the polishing and publication of the book, whose efforts enable a field investigation-based research to be presented in a stylish and quality way.

Many experts from the field of handmade paper production and intangible cultural heritage research and protection have helped in our field investigations: Hu Wenjun from China Xuan Paper Co., Ltd., Chen Zheng from the Office of Chronicles in Huangshan City, Wang Chun from Cultural Center in Yuexi County, Wang Yuhua from local government in Jinzhai County, et al. On the verge of publication, sincere gratitude should go to all those who have supported and recognized our efforts!

Tang Shukun
University of Science and Technology of China
December 2019